From Data to Dollars

Getting Started with Data Analytics and AI in Startups

Piotr Sidoruk

Foreword by Matt Dancho

Apress®

From Data to Dollars: Getting Started with Data Analytics and AI in Startups

Piotr Sidoruk
New York, NY

ISBN-13 (pbk): 979-8-8688-1897-4 ISBN-13 (electronic): 979-8-8688-1898-1
https://doi.org/10.1007/979-8-8688-1898-1

Managing Director, Apress Media LLC: Welmoed Spahr
Acquisitions Editor: Shaul Elson
Development Editor: Laura Berendson
Coordinating Editor: Gryffin Winkler

Cover image designed by Freepik (www.freepik.com)

Distributed to the book trade worldwide by Springer Science+Business Media New York, 1 New York Plaza, New York, NY 10004. Phone 1-800-SPRINGER, fax (201) 348-4505, e-mail orders-ny@springer-sbm.com, or visit www.springeronline.com. Apress Media, LLC is a Delaware LLC and the sole member (owner) is Springer Science + Business Media Finance Inc (SSBM Finance Inc). SSBM Finance Inc is a **Delaware** corporation.

For information on translations, please e-mail booktranslations@springernature.com; for reprint, paperback, or audio rights, please e-mail bookpermissions@springernature.com.

Apress titles may be purchased in bulk for academic, corporate, or promotional use. eBook versions and licenses are also available for most titles. For more information, reference our Print and eBook Bulk Sales web page at http://www.apress.com/bulk-sales.

If disposing of this product, please recycle the paper

To the startups that failed. Thank you for the lessons.

Table of Contents

About the Author

Piotr Sidoruk is a startup data expert with a unique experience of being the first data hire at several startups across both the United States and Europe. He has built data infrastructure from scratch, helping startups scale and secure around $20M in funding and revenue through strategic recommendations and data storytelling that convinces investors to fund their growth (e.g., he helped Playbook build data storytelling to raise $9.3M).

With a degree in Quantitative Methods in Economics and Information Systems and a background in Psychology, Piotr brings a multidisciplinary approach to data-driven problem-solving. He has worked across a range of industries, including fitness, retail, FMCG, financial services, pharma, iGaming, crypto, and real estate, collaborating with major corporations such as IBM, Bain, and Roche. He also has extensive experience in the startup world, having contributed to ventures such as Playbook, a New York-based fitness and creator economy platform; Rentier, a European real estate valuation platform; and MyMenu, an AI-powered meal planning engine that was purchased and implemented by a large German international discount retail chain.

About the Technical Reviewers

 **Bartosz Konieczny** is a freelance data engineer who's been writing code since 2010. He's the author of *Data Engineering Design Patterns* (O'Reilly) and has spent much of his career working with major public cloud platforms and open source tools like Apache Spark, Kafka, Airflow, and Delta Lake to solve real-world data challenges—ranging from ingestion and cleansing to sessionization, ordered processing, and large-scale migrations.

Beyond client work, Bartosz is active in the data engineering community. He shares what he's learned through blog posts on *waitingforcode*, delivers training (both online and in person), and speaks at conferences such as the Data+AI Summit, Spark+AI Summit, and the Big Data Technology Warsaw Summit—all with the goal of helping fellow engineers navigate the evolving data landscape.

 **Srik Gorthy** is a data scientist with more than 11 years of experience at companies including Google, TikTok, and AMD, as well as multiple startups. His work spans trust and safety, fraud prevention, content moderation, and large-scale experimentation, where he has built machine learning solutions and analytics frameworks that combine technical rigor with business impact. Beyond his industry contributions, he is a mentor, conference speaker, and technical reviewer of books in data science and AI, committed to advancing the professional community and shaping responsible, impactful applications of technology.

Acknowledgments

This book would not have been possible without the vibrant community of thinkers, writers, and practitioners who have shared their knowledge so openly. It is the result of years spent learning from an incredible online community. If this book provides you with a map, it is only because I was guided by the work of countless others who charted the territory first.

I owe a huge thank-you to everyone whose articles, books, podcasts, social media posts, and YouTube videos are referenced throughout these pages.

Special mention must go to the work of Andrew Chen, for his essays on growth, product analytics, and the creator economy; Sean Ellis, for his pioneering work on growth hacking; and Alistair Croll and Benjamin Yoskovitz, whose work on *Lean Analytics* was foundational. I also drew significant inspiration from the startup philosophies of Eric Ries, Steve Blank, and Paul Graham; the insights from Sequoia Capital; and the strategic frameworks of Alexander Osterwalder and Yves Pigneur.

For their invaluable insights into startup metrics and turning insight into impact, I am indebted to John Doerr, Lenny Rachitsky, David Sacks, Speedinvest, and the wealth of material from Y Combinator. On the topics of startup valuation, the work of Aswath Damodaran, Dave Berkus, Bill Payne, and Leonis Capital (specifically on AI startup valuation) was instrumental.

I must also acknowledge the technical leaders and entrepreneurs who make the lives of data professionals easier and offer unique perspectives on the future of data in startups, including Maxime Beauchemin (creator of Airflow and Superset), Tristan Handy (co-founder of dbt Labs), Gian Segato (founding data scientist and engineer at Replit), and Thomas Dohmke (former CEO of GitHub). My understanding of data science for business has been shaped by authors like Foster Provost and Tom Fawcett, as well as the team at a16z for their essential articles on modern data infrastructure, AI, and startups.

Finally, this book is inspired by the numerous startup founders who lead by example with data-driven decisions, such as Brian Chesky of Airbnb and Evan Spiegel of Snapchat.

There are many more—authors of technical books, creators of data strategy frameworks, and those documenting case studies in blogs and magazines—who deserve thanks. This book is my contribution to the conversation they all started. My hope is that it honors their work and encourages you, the reader, to seek out their original content.

Foreword

You don't need cutting-edge AI to grow—most companies need better tracking and better decisions.

That might sound almost too simple in a world obsessed with "AI transformation" and shiny tech stacks. But if you've spent any real time inside a normal business—the kind of business that actually pays the bills in this economy—you know what I'm talking about. Most companies aren't high-tech automation engines. They're held together by smart people, a basic ERP system, spreadsheets, and a lot of tribal knowledge. Data exists, but it's incomplete, inconsistent, or not captured at all. And when that's the case, the business ends up making decisions on gut feel, the loudest voice, or whatever fire is burning hottest that week.

That's the world I came up in.

I don't have a degree in data science. My background is mechanical engineering. What I did have was an analytical mind, curiosity, and the willingness to get my hands dirty. I started where a lot of people start: Excel. Then I learned how to do better analysis. Then I learned some programming. Then I learned how to store information in a database so we could actually track what mattered. Over time, I earned my stripes by using data to help the last full-time company I worked for grow from about $3M a year in revenue to about $15M.

And here's the part most people miss: The company didn't grow because we built some futuristic AI system. It grew because we started tracking the right things, we created basic feedback loops, and we used that information to focus on the projects that materially moved the business. In other words, data became a tool for making better decisions, not a report card or a vanity dashboard.

I also made plenty of mistakes along the way. That's part of why I appreciate this book so much.

One of my biggest detours was spending time learning the wrong things at the wrong time. I went deep into deep learning—spent about six months on it—only to realize it wasn't what the business needed. The bottleneck wasn't fancy models. The bottleneck was that we weren't collecting enough useful data, the data we had wasn't structured for decision-making, and we didn't have a clear system for turning information into action.

Eventually, I figured it out. But I learned it the hard way: by building, breaking, rebuilding, and wasting time chasing what sounded impressive instead of what produced results.

That's why I'm excited about *From Data to Dollars*. It is the road map I never had.

Piotr doesn't write from the perspective of a textbook or a research lab. He writes from the reality of startups and growth-stage companies—where time is limited, teams are small, data is messy, priorities change weekly, and the pressure to show progress is real. This book doesn't just tell you that "data is important." It shows you how to build a practical system that connects data work to business outcomes.

What I like most is the sequence. First, it grounds you in strategy: what data is for, how to think about value, and how to avoid drowning in random requests. Then it gets brutally practical about metrics—how to focus on what actually matters, how to avoid vanity metrics, and how to build clarity around growth. From there, it moves into the infrastructure and execution side: the minimum viable data stack, the mindset of building what you need now (not what you might need someday), and how to operationalize data work so it reliably produces decisions, not just reports.

And yes—there's AI here. But it's positioned the right way: as a force multiplier on top of fundamentals, not a replacement for them. That's exactly how it should be. Because AI won't save a company that can't measure its own operations. If you don't have tracking, definitions, and feedback loops, you don't have the fuel. You just have noise.

If you're reading this as a founder or operator, here's what I want you to hear: You don't need to become a data scientist to benefit from this book. You need to become someone who respects measurement, builds simple systems, and uses data to make decisions consistently. This book will help you do that.

If you're reading this as the first "data person" in a startup—analyst, scientist, engineer, or the accidental data person because you're the one who knows spreadsheets—this book will save you from a lot of pain. It will help you stop trying to boil the ocean. It will help you pick the right problems. It will help you build trust in your work. And it will help you translate analysis into impact.

And if you're like me—someone who didn't start in data but learned it because the business needed it—this book is going to feel familiar in the best way. It will put words and structure around things you may already be doing, and it will give you a clearer path forward on the things you haven't solved yet.

One more point I'll make, because I think it matters: you might think a lot has changed in the past ten years. In some ways, it has. We have more tools than ever, and AI has expanded what's possible. But the core problem hasn't changed at all. Many companies still don't track the right data. They still live in unstructured messes. They still lack basic definitions and feedback loops. And because of that, they still can't use data consistently to make better decisions. That's why the ideas in this book are not just relevant—they're urgent.

If I had this road map back when I was building my own "data-to-dollars" journey, I would have moved faster, made fewer detours, and created impact sooner. I believe it can do the same for you.

Read it with a pencil. Pick a few ideas and implement them immediately. Build the habit of tracking what matters. Make the decisions that the data is pointing to— especially when it challenges your assumptions. And you'll find that "data science" isn't some mysterious discipline reserved for high-tech companies. It's simply the process of turning information into better decisions—and better decisions into dollars.

—*Matt Dancho | Founder, CEO | Business Science*

Introduction

I thought I knew data. With a degree in Quantitative Methods and Information Systems and professional experience in the corporate world, I felt well equipped for any data-related challenge. But when I stepped into the startup environment, I quickly realized how much I had to learn. The established theories and best practices I had studied seemed to crumble in the face of the unique pressures and limitations of a new venture.

If you are a data professional moving into the startup world, you're in for a shock. The structured, predictable environment of a large corporation is gone, replaced by a landscape of constant change and ambiguity. Big teams of people specialized in their narrow fields are gone. Instead, you have very small teams. Or no teams at all.

It is a different reality, one that Steve Blank's Customer Development framework so aptly describes: A startup is not a miniature version of a large company. It is a temporary organization searching for a scalable business model. And that changes everything when it comes to data too.[1]

My mind raced with questions: How do you prioritize an overwhelming number of tasks in this environment? How do you know which way to go? How do you overcome the daily obstacles of incomplete data (or no data at all), limited time and budget, and a dynamically changing business model? How can you minimize time spent on your work while maximizing its business value? How do you present data to investors, and how can you even begin to estimate what your startup is worth—especially an AI-based one?

I couldn't find a single book that would guide me through these challenges. So, after years of collecting notes from articles, reports, and conversations with experienced practitioners, I decided to write it myself. This book is the result of that journey, a comprehensive guide to getting started with data analytics and AI in startups.

---

[1] Steve Blank, "Driving Corporate Innovation: Design Thinking vs. Customer Development," SteveBlank.com, July 30, 2014, https://steveblank.com/2014/07/30/driving-corporate-innovation-design-thinking-customer-development/.

The Content of the Book

This book is structured into four parts to guide you from foundational concepts to advanced applications, mirroring the journey of building a data-driven startup from the ground up.

Part 1: The Foundations
We begin with the essentials, establishing the strategic mindset and technical groundwork necessary for success. This part covers how to align your work with core business goals, build a lean data strategy, and navigate the technical landscape from day one to avoid common and costly mistakes.

- **Chapter 1: The Role of Data in Startups** explains why data matters for startups, drawing on lessons learned from the journeys of iconic companies such as Robinhood, Spotify, Airbnb, Uber, Duolingo, Stripe, Canva, and ElevenLabs.

- **Chapter 2: Building a Data-Driven Strategy** provides a practical guide to creating a proactive data road map, drawing on proven frameworks like the Data Value Chain, the Data Flywheel, and the Lean Analytics Cycle to foster a truly data-informed culture.

- **Chapter 3: Data Infrastructure Basics** demystifies the modern data stack by walking through its five foundational elements—Data Sources, Ingestion, Storage, Transformation, and Output—and guiding you through the critical "build vs. buy" decisions and the principles of designing a lean infrastructure.

Part 2: Metrics That Matter
Next, we move from strategy to measurement, focusing on the metrics that truly define a startup's health and potential. This section teaches you how to look beyond surface-level numbers to find the deep, actionable insights that drive growth by combining quantitative analysis with a rich understanding of user behavior.

- **Chapter 4: The Founder's Metrics Toolkit** teaches you how to select the right metrics, distinguishing between actionable KPIs and misleading vanity metrics. It covers how to define a North Star Metric and applies popular frameworks like AARRR (Acquisition, Activation, Retention, Referral, Revenue).

- **Chapter 5: Advanced Growth Analysis** moves beyond foundational metrics to explore the "why" behind your data, covering three core techniques: cohort analysis to assess business health, Power User Curves to identify your most valuable customers, and methods for measuring network effects.

- **Chapter 6: From Insight to Impact** teaches you how to translate data into strategic action by always asking "So what?". It provides frameworks for aligning your team with OKRs and for building a data-driven narrative that speaks directly to what investors need to see at every stage of your startup's growth.

Part 3: Tools and Skills

Here, we take a deeper dive into the modern data professional's toolkit. This part covers the practical, hands-on skills and technologies required to execute your strategy, from essential programming languages to the platforms that will empower your entire organization to make data-informed decisions.

- **Chapter 7: Programming Skills for Data Professionals** defines the role of the modern, adaptable data generalist. It moves beyond just code to cover the essential skills of ruthless prioritization, effective data storytelling for high-stakes scenarios like preparing investor data rooms, and how to leverage AI as a powerful assistant rather than a replacement.

- **Chapter 8: Data Engineering and Orchestration** introduces a pragmatic "startup data engineering mindset" that prioritizes progress over perfection. It details how to use industry-standard tools like dbt and Airflow to build reliable, automated data pipelines and ensure data quality without creating bureaucracy.

- **Chapter 9: Business Intelligence Platforms** provides a durable framework for navigating the crowded BI market. It teaches you how to cut through the noise, ask the right questions, and choose the right platform for your startup's specific stage to avoid costly "BI debt" and empower your entire team to act on data.

- **Chapter 10: Product Analytics and Event Tracking** shifts from business metrics to the user journey. It teaches you how to choose the right product analytics platform, combine quantitative tools like session replays with qualitative methods like *The Mom Test* to understand the "why" behind every click, and build a robust event tracking plan to capture it all.

Part 4: Advanced Topics and the Future

Finally, we'll move from foundational skills to advanced applications, tackling the most forward-thinking challenges and opportunities in the startup ecosystem. This section will guide you through building a disciplined experimentation practice, valuing your company, applying sophisticated data science techniques, and preparing for the AI-driven future. You will learn how to translate data capabilities into measurable financial value and position yourself at the forefront of the industry's evolution.

- **Chapter 11: Experimentation and A/B Testing** frames experimentation as a strategic tool to accelerate learning and avoid the *zombie startup* trap. It provides a disciplined framework for designing high-impact tests, knowing when not to test, and building a trustworthy practice that values learning over *always winning*.

- **Chapter 12: Startup Valuation Methods** provides a step-by-step guide to valuation, covering the journey from qualitative, pre-revenue methods (like the Berkus Method) to quantitative, post-revenue models (like DCF), and tackles the unique factors driving the valuations of modern AI startups.

- **Chapter 13: Advanced Techniques in Data Science and AI** explains why there is no universal data science and AI playbook for startups, teaching you instead how to prioritize projects with frameworks like RICE, avoid the common failures that doom most AI initiatives, and apply essential machine learning techniques to solve real business problems.

- **Chapter 14: The Future of Data in Startups** explores the new startup era where AI acts as a force multiplier and the most enduring companies are those that build proprietary data flywheels. It looks ahead to the most promising frontiers where data and AI are being applied to solve the complex challenges of the physical world.

Who Should Read This Book and What You Will Learn

While this book is written primarily for data professionals—analysts, scientists, and engineers—who are transitioning into the unique environment of a startup, it is by no means exclusive to them. I wrote it to be a universal guide for anyone involved in building a startup who wants to harness the power of data.

This book is also for

- **Startup Founders** who want to understand how data can be used to build and grow their company, what skills their first data hire should possess, and what kind of challenges that person will face

- **Startup Professionals** such as managers, UX designers, and programmers who want to become more data literate, collaborate more effectively with data teams, and use insights to make better decisions in their own roles

Given this broad audience, it is natural that your journey through this book will be personal. Depending on your background, some topics may seem more challenging or less immediately relevant than others. My goal is that every reader, regardless of their specific role, will find sections that speak directly to their needs and challenges.

To serve this diverse audience, I have made a conscious decision to use language that is accessible to everyone, regardless of their technical background. You will not find coding exercises in this book. This was a deliberate choice to ensure the content remains digestible and focused on strategy, process, and practical application rather than implementation details.

Given the wide scope of topics covered in these pages, we simply don't have the space to delve deeply into the code behind each concept. Instead, this book provides the map and the compass. For readers who wish to explore specific territories in greater technical detail, many chapters include "Additional Resources" sections that point to the books, courses, and articles that I have found most valuable for continued learning.

For the Data Professional

This book provides a road map for data professionals to become indispensable leaders in a startup environment. You will learn to

- **Adopt a Startup Mindset:** Transition from a technical specialist to a strategic generalist, focusing on business impact and speed over academic perfection.

- **Build the Foundation:** Create a lean data strategy from the ground up, making smart "build vs. buy" decisions and selecting the right tools for scalable data infrastructure.

- **Drive Growth and Retention:** Identify essential startup metrics, master experimentation methods to accelerate learning, and use techniques like cohort analysis to uncover the real drivers of user retention.

- **Apply Advanced Techniques:** Leverage data science and AI to solve critical business problems, and learn to build proprietary data flywheels that create a defensible competitive moat.

- **Translate Work into Value:** Master startup valuation methods to connect your technical work to financial impact and tell compelling data stories that influence stakeholders.

- **Future-Proof Your Career:** Cultivate a mindset of agency, using AI as a force multiplier to stay ahead and lead in the evolving data landscape.

For the Founder and Non-Data Professional

This book demystifies data, empowering you to lead a data-driven startup with confidence. You will learn to

- **Build Your Data Capability:** Understand the role of data in a startup, know what to expect from your first data hire, and learn how to partner with them effectively to drive results.

- **Make Smart Tech Investments:** Demystify the modern data stack—from data pipelines to analytics tools—so you can make informed decisions about your company's technology.

- **Measure What Matters:** Grasp the key metrics and experimentation techniques used to measure business health, user retention, and product engagement, enabling you to ask the right questions and interpret the answers.

- **Master Strategic Levers:** Harness AI as a force multiplier for lean teams and understand the key drivers of startup valuation to build a compelling narrative for investors.

- **Foster a Data-Driven Culture:** Build an organization where data initiatives are tied to strategic goals and insights are consistently translated into tangible action.

A Note on the Comparisons in This Book

It is important to frame the tool comparisons presented in this book (such as tool comparisons in Chapter 3). The data technology landscape is in a constant state of flux, and the analysis here represents a subjective snapshot based on my own criteria and a startup-centric perspective at the time of writing. Every startup has unique requirements, and what is a critical feature for one may be a minor detail for another. Therefore, please treat these comparisons as a framework for your own thinking rather than a definitive guide.

In an effort to provide a clear overview, I have made generalizations—for example, rating maintenance effort as *low*, *medium*, or *high*—which inevitably simplify complex realities. You may find that some criteria used are not important for your use case or that other critical ones are missing. Furthermore, while some open source tools are described as *free*, this refers only to licensing fees; the true cost of ownership must include the very real expenses of hosting, infrastructure, and engineering maintenance. Ultimately, these tables are intended to provide a simplified map to help you navigate the landscape, not to serve as a substitute for your own deep-dive research tailored to your specific context.

The Role of AI and the "Human-in-the-Loop"

Artificial intelligence and automation are profoundly reshaping the startup landscape. We hear stories of "one-person startups" and see how AI enables small teams to achieve more with fewer resources. Does this mean that data professionals are becoming obsolete?

Far from it. The message of this book is simple: you are more essential than ever. AI doesn't replace the need for skilled professionals; it redefines their role. AI is rapidly becoming integrated into virtually every job function, and professionals who resist adopting these tools will find themselves at a competitive disadvantage. Those who embrace AI to amplify their efficiency and capabilities will outpace those who don't. The new reality requires you to become a holistic specialist, capable of orchestrating numerous automated tools and acting as the critical "human-in-the-loop."

This concept is a central theme of the book, one we will dive into deeply in Chapter 14. There, I argue that as AI-driven automation becomes more powerful, the need for skilled human oversight grows right along with it. The key is to become an expert "human-in-the-loop" who can ensure that these complex systems remain accountable, secure, and reliable.

The process of writing this book served as a perfect illustration of this principle. While AI tools were invaluable for improving language and readability—a significant help for a non-native English speaker like me—the core work remained fundamentally human. It was my responsibility to select, interpret, and cite hundreds of external materials; to verify every piece of information for accuracy, guarding against the "hallucinations" that even the best AI tools can produce; and ultimately, to provide the unique opinions, personal anecdotes, and startup-specific insights that form the heart of this book.

This book is the culmination of a complex and time-consuming human journey. It required years of working with companies of all sizes, learning from real people, and gathering firsthand experience. It is built on lessons learned not just from things that worked well but also from bad decisions and failures. AI can be a powerful assistant, but it cannot replicate the creativity, judgment, and lived experience that are essential for this kind of work. The "human-in-the-loop" remains the indispensable author, just as you will be the indispensable professional in your startup.

Building a startup is challenging, but it is also uniquely gratifying—it is where the future is built. My hope is that this book will be your trusted companion on that journey, giving you the ideas, confidence, and tools to not just navigate the chaos but to harness it.

Let's get started.

PART I

The Foundations

CHAPTER 1

The Role of Data in Startups

Errors using inadequate data are much less than those using no data at all.

—Charles Babbage[1]

Charles Babbage, the 19th-century pioneer often dubbed the "father of the computer," operated in a world of steam engines and mechanical looms—far removed from today's digital startups. Yet his timeless insight on data's value cuts straight to the heart of modern entrepreneurship. Startups live or die by their ability to pivot, innovate, and scale. But without data, even the most visionary teams risk becoming architects of guesswork, building on assumptions rather than evidence.

This chapter explores how data shapes every facet of startup success: why it matters as a foundational pillar, how to understand customers beyond surface-level metrics, and the art of building a data-driven culture that prioritizes evidence over ego. We'll explore the importance of data storytelling when communicating with investors, trace the evolution of data-driven decision-making from mainframes to AI, and reveal how startups leverage data as a competitive advantage—turning insights into moats. Finally, we'll spotlight data-driven success stories from industry disruptors like Robinhood, Spotify, Airbnb, Uber, Duolingo, Stripe, Canva, and ElevenLabs, whose journeys illustrate data's transformative power.

---

[1] This quotation is widely attributed to Charles Babbage, but its origin has not been definitively traced to his writings. However, these words seem to reflect his views, such as those that can be read in his semi-autobiographical book, *Passages from the Life of a Philosopher*, published in 1864.

© Piotr Sidoruk 2026
P. Sidoruk, *From Data to Dollars*, https://doi.org/10.1007/979-8-8688-1898-1_1

It's almost poetic that Babbage, who conceptualized the first mechanical computer, had no inkling of the startup ecosystems that would one day rely on his legacy. His era's "startups" were workshops and industrial ventures, yet his words transcend time: Imperfect data, he argued, still outshines the void of ignorance. For today's founders, this truth is existential. Startups face compressed timelines, scarce resources, and markets that reward speed over deliberation. In this environment, decisions fueled by instinct alone aren't just risky—they're catastrophic. Data, even incomplete or messy, becomes the compass for navigating uncertainty.

But data's role isn't meant to be merely utilitarian. It should be the lifeblood of a startup's culture. Success hinges not on hoarding spreadsheets or dashboards but on fostering a mindset where curiosity and evidence eclipse hunches and hierarchy. A data-driven culture asks: What story does this metric tell? Where are the gaps in our knowledge? How can we test our boldest assumptions? In contrast, a culture not driven by data tends to rely on precedent or authority, asking instead: What has always been done? What does our gut feeling tell us? What does the highest-paid person in the room think? It's not about waiting for pristine datasets—it's about acting decisively with the signals you have, refining your approach as new insights emerge.

Data's influence extends beyond internal decisions. For many modern startups, it's the core of their product. Imagine YouTube without personalized recommendations, Uber without real-time traffic analysis, or social platforms like X (formerly Twitter) devoid of algorithmic curation. These companies didn't just use data—they weaponized it, embedding it into their offerings to create seamless, addictive experiences. Data isn't a backstage advisor here; it's the star performer.

This duality—data as both navigator and product—creates a competitive edge. Startups that bake data into their DNA do more than optimize operations; they unlock unique value for customers. A fitness app that adapts to user behavior, a fintech tool that predicts cash flow gaps, a logistics platform that learns from delivery patterns—these aren't features. They're moats.

Yet for many founders, the hurdle is starting. Early-stage teams often dismiss data as a problem for later, once they've "scaled." But Babbage's wisdom applies here too: Begin small. Track one critical metric. Run micro-experiments. Let each dataset, however modest, challenge your biases and illuminate next steps. Over time, these fragments coalesce into clarity—guiding smarter pivots, sharper products, and a culture where data isn't feared, but embraced.

The startups that endure aren't those with the most data, but those that treat it as oxygen: invisible, essential, and transformative. From day one, they ask not "Can we afford to prioritize data?" but "Can we afford not to?" In doing so, they honor a truth as old as Babbage's gears and as urgent as tomorrow's pitch deck: In the race to innovate, data isn't just power—it's survival.

Why Data Matters for Startups

If there's one thing experienced investors and entrepreneurs agree on, it's that data is not a magic wand—but it is a powerful tool when used correctly. As Andreessen Horowitz (a16z), one of Silicon Valley's most prominent venture capital firms, puts it:

> *Data is fundamental to many software companies' product strategies, and there are ways it can contribute to defensibility — but don't rely on it as a magic wand.*
>
> —a16z[2]

The idea that simply collecting more data will automatically create a competitive advantage is a myth. This approach often leads to the creation of a data swamp—a data lake so disorganized and poorly managed that it becomes nearly impossible to navigate or extract value from. Instead, startups must be intentional about how they use data to improve their products, refine their decision-making, and build a sustainable business.[3,4]

Startups that successfully leverage data often do so in ways that go beyond mere accumulation. In some cases, more data leads to significantly better products—think of an AI-powered cancer screening tool where increasing accuracy from 80% to 85% could be the difference between life and death. But a16z warns that data scale effects don't last forever, and relying solely on data as a moat is risky. Instead, startups should focus on how data helps them make better decisions, optimize their products, and support growth in a way that aligns with their broader strategy.

---

[2] Martin Casado and Peter Lauten, "The Empty Promise of Data Moats," Andreessen Horowitz (a16z), accessed January 31, 2025, `https://a16z.com/the-empty-promise-of-data-moats`.

[3] Martin Casado and Peter Lauten, "The Empty Promise of Data Moats," Andreessen Horowitz (a16z), accessed January 31, 2025, `https://a16z.com/the-empty-promise-of-data-moats`.

[4] "Data Swamp: Is It Sinking You In?," Atlan, November 21, 2023, `https://atlan.com/data-swamp-explained/`.

Data-driven strategies may be the key to avoiding missteps. An illustrative example comes from Wistia—a startup that accidentally disabled its pop-up feature, anticipating a decline in user engagement. Contrary to expectations, data revealed an increase in user retention during this period. This "happy accident" led the team to permanently remove the pop-ups, resulting in sustained user engagement improvements.[5]

These narratives demonstrate that, without data, startups risk navigating blindly, potentially missing opportunities or making detrimental decisions. Embracing data not only illuminates the path forward but also empowers startups to pivot effectively, seize opportunities, and foster sustainable growth.

The Power of Informed Decision-Making

In the early days of a startup, decisions are often driven by intuition. Before the explosion of data, even the most successful products were built primarily on gut feelings, experience, and unstructured insights. As Sequoia Capital, one of the most respected venture capital firms, puts it, "Intuition can be grounded in data but not necessarily in a structured form." While valuable, intuition has its limits—it is shaped by personal biases and incomplete information, which can lead to flawed decision-making and significant risks.[6]

This doesn't mean intuition has no place in product development. In fact, during the earliest stages, when no data exists, founders rely heavily on their instincts, domain knowledge, and design expertise to shape their vision. But as a startup scales, relying solely on intuition becomes increasingly risky. The evolution of data analytics has shown that informed decision-making—rooted in rigorous testing, real-world user behavior, and continuous iteration—leads to far better outcomes.

Take Facebook and Instagram as examples. Features like Facebook Live, Instagram's Feed ranking, and the now-ubiquitous "stories" format weren't just the result of creative brainstorming. They emerged from deep analytical insights, user behavior studies, and relentless experimentation. Sequoia notes that the failure of intuition is one of the key

---

[5] "Four Stories From the Weird World of Startup Analytics," Amplitude Blog, accessed January 31, 2025, https://amplitude.com/blog/four-stories-startup-analytics.

[6] "The Building Blocks of a Data-Informed Company," Sequoia Capital Articles, accessed January 31, 2025, https://articles.sequoiacap.com/the-building-blocks-of-a-data-informed-company.

reasons why a "test and learn" culture is essential. Experimentation—rather than gut feeling—helps startups validate their ideas, optimize products, and discover the small, compounding improvements that drive long-term success.

Furthermore, data empowers startups to act quickly. In a fast-moving market, reacting to changes in real time can be the difference between thriving and failing. By continuously analyzing user behavior, market trends, and customer feedback, startups can refine their strategies, allocate resources more effectively, and build products that truly resonate with their audience. The best startups understand that data isn't just about measuring success—it's about shaping it.

Understanding Your Customers

In the early stages, founders often operate on instinct, making assumptions about who their customers are and what they crave. Data cuts through these assumptions, offering clarity. Metrics like demographics, behavioral patterns, and conversion rates reveal who engages with a product, how they use it, and where they drop off. This quantitative foundation helps startups refine their ideal customer profile, tailor experiences, and allocate resources to the most effective marketing channels.

But data alone paints an incomplete picture. Andrew Chen, a partner at Andreessen Horowitz and author of *The Cold Start Problem*, cautions that consumer tech companies often over-rely on numbers like page views or click-through rates while overlooking the human motivations behind them. Chen, who has spent years studying how startups scale through network effects, argues that founders who aren't part of their own target audience face a unique risk: building products in a vacuum. Without firsthand empathy for their users, they may misinterpret data or miss subtle psychological drivers that shape decisions.[7,8]

This is where qualitative research bridges the gap. Startups should blend analytics with direct, open-ended conversations—even before a product exists. Early engagement with potential customers is not a luxury but a necessity. Simple tactics, such as recruiting target users through surveys with small incentives, can uncover

---

[7] Andrew Chen, *The Cold Start Problem: How to Start and Scale Network Effects* (Harper Business, 2021).

[8] Andrew Chen, "Talk to Your Target Customer in 4 Easy Steps," Andrew Chen's Blog, accessed January 31, 2025, https://andrewchen.com/talk-to-your-target-customer-in-4-easy-steps.

frustrations, habits, and unmet needs that spreadsheets might obscure. For example, a startup analyzing app usage data might notice high drop-off rates during onboarding. But only through interviews would they discover that users feel overwhelmed by too many steps or confused by jargon—insights that reshape the product's design and messaging.

Chen's framework emphasizes curiosity over validation. Rather than asking leading questions to confirm hypotheses, founders should explore how people live, work, and make choices. A gaming startup, for instance, might assume players prioritize cutting-edge graphics. Yet conversations could reveal that what truly hooks users is the ability to compete with friends or share achievements socially. These emotional and psychological nuances, invisible in quantitative dashboards, often determine whether a product resonates or falls flat.

The synergy between data and empathy is where innovation thrives. Quantitative metrics guide startups toward efficiency—optimizing ads, improving retention funnels, or identifying high-value user segments. Qualitative insights, on the other hand, spark creativity. They reveal why a feature delights, how a message connects, or when a pain point becomes unbearable. Together, they help founders avoid the trap of building features users don't care about or campaigns that miss the mark.

In the end, customer understanding is not a one-time exercise but a mindset. Startups that embrace both data and human stories don't just solve problems—they anticipate needs, foster loyalty, and create products that feel less like tools and more like indispensable parts of their users' lives. The most enduring companies aren't those with the most advanced algorithms, but those that listen deeply, question assumptions, and let their customers' voices shape every decision.

Driving Growth and Innovation with the Data Bridge

Data is the engine of reinvention. While many founders chase growth through isolated tactics or viral gimmicks, lasting success demands a deeper discipline. Growth, as Chen articulates, is not a patchwork of hacks but a carefully designed system. It requires understanding how each piece of the business—user acquisition, engagement, retention—interlocks to create compounding loops. A viral product, for instance, isn't just about novelty; it's built on mechanics that turn users into advocates, where every interaction nudges them to invite others. Data's role here is to map these invisible threads, ensuring each loop strengthens the next.[9]

Yet too often, startups fixate on surface-level metrics: downloads, page views, flashy spikes in traffic. These vanity numbers can obscure the truth. What matters is whether growth is sustainable. Churn rates, retention curves, and customer lifetime value reveal far more about a startup's health than install counts ever will. A food delivery app might celebrate a surge in sign-ups after a marketing blitz, but if users abandon the platform after one order, the growth is hollow. Data must serve strategy, not distraction. It should answer whether the product solves a real problem, whether users return because they love it, and whether the business model can endure.

This shift from tactical thinking to systemic rigor mirrors the evolution of growth teams themselves. The early days of "growth hacking"—a term coined by Sean Ellis to describe marketers who blurred the lines between product and analytics—have given way to a more collaborative ethos. Today, sustainable growth is a cross-functional endeavor. Engineers, designers, and product managers work in tandem, embedding growth into the product's architecture. A fitness app, for example, might weave social sharing into its core experience, transforming workouts into moments of connection that organically attract new users. The goal isn't to bolt on features but to design products that inherently encourage adoption and loyalty.[10]

Scaling, however, demands a delicate balance. Startups often begin with an almost obsessive focus on individual customers—listening to feedback, refining details, perfecting the minutiae. But as Chen observes, there comes a pivotal moment when founders must transition from craftsmanship to scale. Data becomes the bridge. It helps identify patterns: which behaviors predict long-term retention, which markets hold untapped potential, which operational inefficiencies quietly drain resources. A fintech startup might discover through usage data that customers who set savings goals within their first week are three times more likely to stay active. This insight doesn't just inform a tweak; it reshapes onboarding flows, marketing messages, and even product roadmaps.

---

[9] Andrew Chen, "Growth is a System, Not a Bag of Tricks," Andrew Chen's Blog, accessed January 31, 2025, https://andrewchen.com/growth-is-a-system-not-a-bag-of-tricks.

[10] Sean Ellis and Morgan Brown, *Hacking Growth: How Today's Fastest-Growing Companies Drive Breakout Success* (Crown Currency, 2017).

Innovation, too, thrives on this interplay of intuition and evidence. Data-driven product development isn't about guessing what customers want—it's about creating feedback loops that turn behavior into insight. When a productivity tool's users consistently ignore a "premium features" prompt, the lesson isn't to push harder. It's to ask why. Interviews might reveal that the pricing feels opaque, or the value isn't clear. Operational data, meanwhile, can expose bottlenecks invisible to the naked eye. A logistics startup analyzing delivery times might uncover that a single bottleneck in its supply chain adds days to shipments—a fixable problem that, once resolved, boosts customer satisfaction and margins.

In the end, data's greatest power lies in its ability to turn ambiguity into action. For investors, metrics like retention and unit economics signal whether a startup understands its own story. For competitors, a data-fluent company isn't just chasing trends—it's anticipating them, building products that evolve with their customers' lives. And for founders, growth is not a race to the top of a chart. It's the quiet work of aligning systems, listening deeply, and letting the numbers tell a story worth scaling.

Building a Data-Driven Culture

To fully harness the power of data, startups must foster a data-driven culture. This means encouraging experimentation, collaboration, and continuous learning. It is not merely about collecting data but about creating an environment where decisions are guided by evidence rather than intuition or tradition. A company that embraces data as a core pillar of its culture unlocks greater potential for innovation, efficiency, and sustained growth.

A data-driven culture starts with a mindset shift. Instead of relying on gut feelings or conventional wisdom, teams should be empowered to test their hypotheses, measure outcomes, and iterate based on insights derived from data. This iterative approach not only leads to better business decisions but also fosters adaptability, allowing companies to navigate uncertainty with greater confidence. When employees see the impact of data-informed choices, they become more engaged in the process of discovery and refinement, further embedding this culture into the fabric of the organization.

Data serves as a common language within a startup, enabling more effective communication and alignment. Grounding discussions in data eliminates ambiguity and personal biases, ensuring that teams are focused on shared objectives. When data becomes the foundation of strategic conversations, decision-making becomes more transparent and accountable. This common framework strengthens collaboration, as it provides clarity on what success looks like and how it can be measured objectively.

Sequoia highlights the importance of data in defining metrics and goals that align with a company's mission. The most successful data-informed cultures, they observe, exhibit a maniacal focus on impact, creating an organizational commitment to measurement and truth-seeking. This commitment extends beyond setting goals—it permeates the very structure of the company, influencing everything from incentive systems to team dynamics. Promotions, compensations, and rewards are tied to measurable contributions, reinforcing a culture where success is not just about effort but about tangible results.[11]

The pursuit of a data-driven culture is not simply about installing dashboards or hiring analysts. It is about instilling a belief that every decision, big or small, can be improved through rigorous analysis and continuous learning. By embedding this ethos into the organization, startups can position themselves to be more resilient, innovative, and, ultimately, more successful in achieving their mission.

Data Storytelling for Investors

Venture capitalists are in the business of betting on the future. They seek startups that demonstrate not just potential but evidence—real, data-backed proof that their vision is viable, scalable, and defensible. In *Secrets of Sand Hill Road*, Scott Kupor underscores that while passion and vision are essential, they are not sufficient. VCs prioritize companies that can translate their vision into measurable traction using data to de-risk their investment. The art of securing venture capital, therefore, lies in blending hard data with a compelling narrative.[12]

Product-market fit is the holy grail of startup success. Founders often feel they have achieved it based on anecdotal customer feedback, but VCs require more than qualitative conviction. They need quantitative evidence. Metrics such as customer retention, Net Promoter Score (NPS), conversion rates, and month-over-month revenue growth provide tangible proof that customers not only want the product but are actively

---

[11] "The Building Blocks of a Data-Informed Company," Sequoia Capital Articles, accessed January 31, 2025, `https://articles.sequoiacap.com/the-building-blocks-of-a-data-informed-company`.

[12] Scott Kupor, *Secrets of Sand Hill Road: Venture Capital and How to Get It* (Portfolio, 2019).

engaging with it. Cohort analysis can show how retention rates improve over time as product iterations align with user needs. Churn and engagement rates highlight customer stickiness, proving that users are not just trying but adopting the product. Revenue growth and CAC/LTV ratios demonstrate that customer acquisition costs are sustainable relative to customer lifetime value, indicating long-term profitability.

VCs don't invest in static businesses; they invest in companies poised for exponential growth. The burden is on the founder to prove not only that their startup is gaining traction but that it can scale efficiently. A consistently upward revenue growth trajectory, particularly if revenue is accelerating rather than linear, signals strong potential. Unit economics must show that customers contribute positively to the bottom line and that margins improve at scale. Operational leverage is key—if revenue grows but operational costs do not increase proportionally, profitability improves over time. Startups should craft a story that conveys how their current growth translates into future potential, making it clear that they are not just growing but scaling efficiently.

No VC wants to fund a startup that can be easily copied or outcompeted. A strong competitive advantage ensures that growth remains defensible. Data can serve as crucial proof of this. Network effects, proprietary data or AI models, and high customer loyalty metrics all contribute to a startup's moat. Founders should be prepared to demonstrate how user growth strengthens the platform, how accumulated data makes the product smarter or more efficient over time, and how strong engagement rates and customer referrals signal a defensible position. If a platform's recommendation engine has improved conversion rates by 40% over the past year due to a proprietary dataset, that tells a compelling story of differentiation.

VCs invest in founders who can paint a bold vision of the future while substantiating their claims with data. Numbers alone won't inspire confidence, but a well-crafted narrative without data feels empty. The most successful pitches are those where data and storytelling converge—where every statistic reinforces the founder's vision, making the startup's success feel inevitable. By leveraging data to validate product-market fit, scalability, and competitive advantage, startups can not only attract investors but also build a business with the foundations to endure. As Scott Kupor emphasizes, VCs want to see proof that a startup is more than an idea—it's a data-driven opportunity that's too compelling to ignore.

In essence, data empowers startups to move beyond guesswork and make strategic decisions based on evidence. It provides the insights needed to allocate resources efficiently, achieve product-market fit, and, ultimately, increase the chances of success.

For startups, data isn't just a tool—it's a competitive advantage. But how did we get here? To fully appreciate the role of data in today's startup ecosystem, it's worth exploring how data-driven decision-making has evolved over the decades, shaping the way businesses operate and innovate.

Evolution of Data-Driven Decision-Making

The story of data-driven decision-making (DDDM) is deeply intertwined with the history of computing and the relentless march of technological progress. From the bulky mainframes of the 1950s to the AI-powered platforms of today, the journey of DDDM reflects a broader shift in how businesses—especially startups—leverage data to survive and thrive. Let's take a closer look at how this evolution unfolded.[13]

The Past: From Math to Millions

The 1950s marked the birth of digital computing. Early computers like the UNIVAC and IBM's first commercial machines were primarily used for tasks like payroll and inventory management. Data analysis was rudimentary, with businesses relying on mainframes to process batches of data for operational efficiency.[14]

For small businesses, access to such technology was rare due to its prohibitive cost. Decision-making was predominantly intuition based, supported by basic statistics and financial ledgers. Data, in its modern sense, was still a distant concept.

The 1960s saw the emergence of Management Information Systems (MIS), which allowed businesses to generate reports from structured data stored in centralized databases. For the first time, organizations could use data systematically for planning and control, marking a significant step forward in decision-making.

While large corporations began to invest in MIS, small businesses and startups were still on the sidelines. The high cost of computing infrastructure kept advanced tools out of reach for most. However, the gradual adoption of more sophisticated office equipment, like early calculators and ledger systems, began to improve operational decision-making, even if only incrementally.

---

[13] Thomas Haigh and Paul E. Ceruzzi, *A New History of Modern Computing* (The MIT Press, 2021).
[14] David Hemmendinger and Paul A. Freiberger, "UNIVAC," in Encyclopædia Britannica, last modified September 24, 2025, https://www.britannica.com/technology/computer/UNIVAC.

The 1970s brought a game-changing innovation: relational databases. Pioneered by Edgar F. Codd, this new way of storing and querying data revolutionized how businesses interacted with information. Structured Query Language (SQL) made it easier to access and manipulate data, laying the groundwork for modern analytics.

Startups began to emerge as early adopters of these technologies, using data for niche applications. However, most small companies still relied on manual processes, as computing resources remained expensive and complex. The seeds of data-driven decision-making had been planted, but they were far from flourishing.

The 1980s marked the rise of Business Intelligence (BI) tools. Platforms like Oracle and early versions of SAS enabled organizations to extract actionable insights from their data. At the same time, personal computers became more accessible, democratizing data analysis for smaller firms.

Spreadsheets, particularly Lotus 1-2-3 and later Microsoft Excel, became indispensable tools for startups. Entrepreneurs used them to model business scenarios, track finances, and conduct basic forecasting. For the first time, small businesses could engage in data-driven decision-making without needing a mainframe in the basement.

The 1990s ushered in the era of data warehousing. Tools like Informatica and Teradata allowed businesses to consolidate data from multiple sources, enabling deeper and more comprehensive analysis. The rise of the internet also opened new doors for startups, as digital data became a valuable resource.[15]

Ecommerce startups, in particular, began to leverage web analytics to track customer behavior. Early versions of tools like Google Analytics (launched in 2005 but rooted in principles developed in the 1990s) laid the foundation for data-driven marketing strategies. Startups were no longer just using data—they were building their businesses around it.

The 2000s were defined by the explosion of "big data." Open source technologies like Hadoop and NoSQL databases made it possible to store and analyze massive datasets. Cloud computing further lowered the barrier to entry, allowing even the smallest startups to harness the power of data.

---

[15] Katie Hafner and Matthew Lyon, *Where Wizards Stay Up Late: The Origins of the Internet* (Simon & Schuster, 1998).

During this period, startups embraced tools like Mixpanel, Google Analytics, and Tableau to track metrics, optimize user experiences, and drive growth. The Lean Startup methodology, popularized by Eric Ries, emphasized iterative development based on data-driven insights, cementing data's role as a cornerstone of modern entrepreneurship.[16]

By the 2010s, data science and machine learning had emerged as key disciplines. Programming languages like Python and R empowered data scientists to build predictive models, while cloud platforms like AWS, Azure, and Google Cloud provided scalable infrastructure for analytics.

Startups used these advancements to outpace competitors. Companies like Uber and Airbnb became icons of data-driven decision-making, leveraging real-time data to optimize operations, improve customer experiences, and scale globally. Smaller firms adopted tools like Looker, Power BI, and Metabase to democratize data access within their teams, ensuring that insights were no longer confined to a handful of experts.[17]

The current decade is defined by AI-enhanced decision-making. Large language models (LLMs) like those delivered by OpenAI and DeepSeek, predictive analytics, and AI-driven automation are redefining how businesses operate. Startups and small companies now have unprecedented access to advanced tools, often powered by APIs, that enable natural language querying, customer segmentation, and personalized marketing at scale.

Platforms like Snowflake, Databricks, dbt, and modern AI-driven analytics tools allow even small teams to harness the power of data science. Startups increasingly rely on AI to identify trends, predict customer behaviors, and automate routine decision-making processes. The line between data and strategy has never been blurrier—or more powerful. As we stand on the cusp of a new era in data-driven innovation, the question becomes not just how startups will use these tools but how they will redefine entire industries and markets in the process.

---

[16] Eric Ries, *The Lean Startup: How Today's Entrepreneurs Use Continuous Innovation to Create Radically Successful Businesses* (Crown Currency, 2011).

[17] Alistair Croll and Benjamin Yoskovitz, *Lean Analytics: Use Data to Build a Better Startup Faster* (O'Reilly, 2013).

The Future: Where Startups Thrive

History is littered with predictions that never came to pass—flying cars, colonies on Mars, paperless offices. Futurists and experts alike have consistently underestimated the messy, nonlinear trajectory of progress. If there's one lesson to heed, it's this: the future resists simplicity. Yet here we are, compelled to imagine it anyway. What seems certain is that startups will navigate a landscape where data-driven decision-making is both a lifeline and a battleground. The road ahead will be defined not by flawless foresight but by adaptability. Tools will grow more powerful, yes, but competition will sharpen. For every door automation opens, a hundred others will slam shut.

The evolution of data-driven decision-making mirrors humanity's own dance with technology. In the 1950s, hulking mainframes processed punch cards to answer basic questions about efficiency or inventory. Today, algorithms parse petabytes of data to predict customer behavior, optimize supply chains, and even draft marketing copy. What began as a postmortem analysis of what happened has become a dynamic conversation about what could happen next. Startups, once hamstrung by limited budgets and manpower, now wield tools that rival those of Fortune 500 companies. The cloud democratizes supercomputing; open source libraries offer machine learning to anyone with curiosity and grit. But this isn't just a story of access—it's a shift in philosophy. Data is no longer a static record. It's a compass.

Three seismic shifts underpin this transformation. The first is the leap from reactive to proactive thinking. Early data systems were historians, chronicling successes and failures after the fact. Modern AI, however, acts as a strategist. It identifies patterns invisible to the human eye, forecasting market shifts or customer churn before they fully materialize. For startups, this predictive edge is existential. A five-person team can pivot a business model overnight when algorithms flag a coming trend—something a corporate goliath, bogged down by committees and legacy code, might miss entirely.

The second shift is automation's quiet revolution. Gone are the days of manual data entry and spreadsheet sorcery. Machine learning models now clean, analyze, and visualize data in real time, transforming raw numbers into narratives. This isn't about replacing humans but liberating them. Founders spend less time wrestling with pivot tables and more time testing bold ideas. A/B testing, once a weeks-long ordeal, unfolds in hours. Customer feedback loops tighten from quarters to days. Speed becomes a startup's oxygen.

The final shift is the rise of prescriptive analytics. It's no longer enough to know what happened or even why. The critical question is how—how to act, how to optimize, how to win. Advanced platforms now recommend actions, simulating outcomes like a chess engine plotting moves. For a fledgling company, this is transformative. Imagine a tool that doesn't just flag a drop in user engagement but suggests personalized interventions, calculates their ROI, and automates their rollout. Decision-making shifts from intuition-driven gambles to calculated experiments.

Yet for all its promise, this future is a double-edged sword. Democratized tools mean lower barriers to entry—but also market saturation. When every startup can harness AI, differentiation hinges on creativity and execution. The competitive edge moves from access to application. Startups must not only adopt technology but reinvent its use. A niche ecommerce platform might leverage computer vision for virtual try-ons, while a fintech startup deploys natural language processing to detect fraud in customer service chats. The winners will be those who treat data not as a crutch but as a collaborator.

There's also a darker undercurrent: the weight of responsibility. With great data comes great vulnerability—to breaches, biases, and ethical missteps. A startup's agility can become a liability if it outpaces its governance. Algorithms trained on flawed data perpetuate discrimination; automated decisions made without transparency erode trust. The companies that thrive will be those that bake ethics into their code, literally and figuratively, recognizing that innovation without integrity is a ticking clock.

The path forward is neither easy nor guaranteed. For every startup that leverages data to disrupt an industry, a dozen will drown in the noise. But therein lies the thrill. The future belongs to the nimble, the curious, and the resilient—those who see data not as a crystal ball but as a clay to be shaped. It will be messy, unpredictable, and alive with possibility. Just like every revolution worth remembering.

Data in a Startup's Early Stages

Every startup begins with an idea—a spark of inspiration promising to solve a problem or create something new. But not every idea is destined to succeed. Most startups fail not due to lack of creativity or effort, but because they build something nobody wants. This is where data becomes crucial, helping validate ideas and refine visions long before a single line of code is written.

Validating Your Idea: Does It Make Sense?

For every Airbnb or Dropbox that reshapes an industry, countless ventures collapse under the weight of a fatal flaw: solving problems that exist only in their creators' minds. The bridge between imagination and reality isn't luck or charisma; it's data. In the fragile early days of a startup, data acts as both compass and critic, grounding ambition in reality.

The Lean Startup methodology, Eric Ries' manifesto for evidence-based entrepreneurship, treats businesses as a series of hypotheses to be tested. Is the problem urgent enough that people will pay to fix it? Does your solution align with how they live or work? These aren't questions for debate—they're experiments waiting to be run. At its core, the methodology operates on a simple feedback loop: Build-Measure-Learn. The process begins with building a minimum viable product (MVP), which is the most basic version of the product that can still test a key assumption. This MVP is then put in front of customers to measure their behavior and gather feedback. The final step is to learn from this data using the insights to either continue on the current path (persevere) or make a fundamental change in strategy (pivot). This iterative cycle is designed to reduce waste and accelerate learning, ensuring that startups build something customers actually want.[18,19]

What predates and heavily influenced the Lean Startup movement is Steve Blank's Customer Development framework. Rather than an academic theory, the framework was forged from the direct experience of its creator. A serial entrepreneur and startup veteran, Blank worked on numerous tech projects in Silicon Valley between 1978 and 1999. After retiring, he began to codify the lessons from his 21-year career, resulting in this influential model. His framework insists that entrepreneurs "get out of the building" early. Its core principle is that startups are not miniature versions of large companies but are instead temporary organizations in search of a repeatable and scalable business model. Through surveys, interviews, or social media scraping, you gather the raw material of truth: the language customers use to describe their frustrations. A developer dreaming of a productivity app might assume users want AI-driven automation—only to discover, via Reddit threads or survey data, that their audience craves simplicity

---

[18] Eric Ries, *The Lean Startup: How Today's Entrepreneurs Use Continuous Innovation to Create Radically Successful Businesses* (Crown Currency, 2011).

[19] Eric Ries, "Good enough never is (or is it?)," Startup Lessons Learned, September 27, 2010, https://www.startuplessonslearned.com/2010/09/good-enough-never-is-or-is-it.html.

instead. The framework is structured around four iterative steps: Customer Discovery, Customer Validation, Customer Creation, and Company Building. By systematically testing hypotheses about customers and their problems before and during product development, the model aims to reduce the immense risks associated with building something nobody wants.[20,21,22,23,24]

Clayton Christensen's Jobs to Be Done theory sharpens this insight: What job would customers "hire" your product to do? Data reveals the answer in their behavior, not their polite assurances.[25]

Market demand, however, is a fickle beast. Tools like SEMrush (for search volume) or Statista (for industry trends) quantify interest, but numbers alone can deceive. Here, Alexander Osterwalder's Business Model Canvas forces clarity: Who exactly is your customer? What channels reach them? How does your revenue model align with their willingness to pay? When Dropbox's Drew Houston tested demand with a simple explainer video—no product, just a promise—the surge in sign-ups became his North Star. Data, in this case, wasn't abstract; it was 75,000 people raising their hands.[26,27]

Yet even the most compelling data demands skepticism. Every founder's vision is built on assumptions, and assumptions are liars. The Lean Startup's Build-Measure-Learn loop turns these risks into rituals. Launch a minimum viable product—a landing page, a manual service masquerading as tech, a Kickstarter campaign—and measure

---

[20] Steve Blank and Bob Dorf, *The Startup Owner's Manual: The Step-By-Step Guide for Building a Great Company* (Wiley, 2020).

[21] Steve Blank, *The Four Steps to the Epiphany: Successful Strategies for Products That Win* (K & S Ranch, 2013).

[22] Steve Blank, "What's A Startup? First Principles," SteveBlank.com, January 25, 2010, `https://steveblank.com/2010/01/25/whats-a-startup-first-principles/`.

[23] Steve Blank, "Driving Corporate Innovation: Design Thinking vs. Customer Development," SteveBlank.com, July 30, 2014, `https://steveblank.com/2014/07/30/driving-corporate-innovation-design-thinking-customer-development/`.

[24] Steve Blank, "About Steve," SteveBlank.com, accessed October 11, 2025, `https://steveblank.com/about/`.

[25] Clayton M. Christensen, Taddy Hall, Karen Dillon, and David S. Duncan, *Competing Against Luck: The Story of Innovation and Customer Choice* (Harper Collins, 2016).

[26] Eric Ries, "How DropBox Started As A Minimal Viable Product," TechCrunch, October 19, 2011, `https://techcrunch.com/2011/10/19/dropbox-minimal-viable-product/`.

[27] Alexander Osterwalder and Yves Pigneur, *Business Model Generation: A Handbook for Visionaries, Game Changers, and Challengers* (John Wiley and Sons, 2010).

what happens. When Airbnb's founders started with air mattresses in their loft, they weren't testing a platform; they were testing whether strangers would pay to sleep in a stranger's home. The bookings that trickled in weren't just revenue—they were validation etched in ink.

This is the quiet power of frameworks like Ash Maurya's Lean Canvas. By compressing your business model into a single page, it forces you to articulate your riskiest assumptions: What if our pricing alienates our core users? What if competitors replicate this in weeks? Data fills these blanks. Pre-orders, waitlist sign-ups, or A/B tests transform vague fears into percentages and probabilities.[28]

In the end, data is more than spreadsheets and surveys—it's the art of seeing your idea through the market's eyes. It's the difference between a founder who says "I believe" and one who can say "I know." As Y Combinator—one of the world's leading startup accelerators—reminds us with its motto, the goal isn't to build something great. It's to build something wanted. Data is how you learn to listen.

Choosing the Right Business Model

Startups must leverage data and analytics to determine the most effective business model by understanding the unique characteristics of their target audience. By gathering and analyzing demographic information and customer feedback, entrepreneurs can uncover which pricing strategies or revenue models best resonate with potential users. This approach echoes the principles in *Business Model Generation* by Osterwalder and Pigneur, where knowing your customer is central to crafting a compelling value proposition.[29]

Competitive analysis is equally essential. Studying how other companies in your sector structure their revenue streams and engage their customers provides vital context. For example, if data reveals that competitors successfully deploy subscription models, this may indicate a market preference for recurring payments—a concept also explored in *Zero to One* by Peter Thiel, where innovation often springs from understanding and outperforming existing paradigms.[30]

---

[28] Ash Maurya, *Running Lean: Iterate from Plan A to a Plan That Works* (O'Reilly, 2012).

[29] Alexander Osterwalder and Yves Pigneur, *Business Model Generation: A Handbook for Visionaries, Game Changers, and Challengers* (John Wiley and Sons, 2010).

[30] Peter Thiel, *Zero to One: Notes on Startups, or How to Build the Future* (Crown Currency, 2014).

My experience in the Creator Economy industry further underscores the importance of data-driven business model selection. I've observed that different creators thrive with varied approaches: some excel at selling long-term subscriptions, others at short-term commitments, while some find success with one-time purchases. Each creator essentially caters to a distinct user group with unique preferences and behaviors. This diversity highlights the critical role of experimentation and number crunching. Even for solo influencers or small businesses with limited growth prospects, understanding the data behind user engagement and revenue patterns is crucial. It allows these "one-man armies" to optimize their offerings, maximize their income, and build sustainable careers. The power of data analytics isn't reserved for tech giants or high-growth startups; it's an essential tool for anyone looking to carve out their niche in today's digital landscape.

Financial projections based on historical data further inform business model decisions. Startups can simulate various scenarios—adjusting pricing or customer acquisition costs—to forecast revenue potential. Such an evidence-based approach to modeling aligns with the iterative learning process detailed in *Measure What Matters* by John Doerr, ensuring that decisions are backed by concrete, quantifiable insights.[31]

Flexibility is a key advantage of a data-driven strategy. Initial business models serve as hypotheses to be tested, refined, or even completely pivoted based on continuous user behavior monitoring. This build–measure–learn cycle allows companies to adjust their models rapidly and effectively in response to real-world feedback.

Ultimately, using data and analytics in this way transforms business model selection into a dynamic, evidence-based process that not only identifies the optimal model but also continuously optimizes it for sustained growth and success.

Building a Data-Driven Foundation

The early stages of a startup are all about laying a strong foundation. By using data to validate your idea and choose the right business model, you're not just increasing your chances of success—you're also building a culture of evidence-based decision-making that will serve you well as your startup grows.

---

[31] John Doerr, *Measure What Matters: How Google, Bono, and the Gates Foundation Rock the World with OKRs* (Portfolio, 2018).

Remember, data isn't just a tool for answering questions; it's a way of thinking. It's about approaching every decision with curiosity, humility, and a commitment to learning. Whether you're exploring a new idea, testing a hypothesis, or refining your business model, data is your most reliable guide.

As your startup evolves, the role of data evolves with it. What begins as a tool for validation soon becomes a powerful engine for growth and innovation. This transformation is where many startups find their true competitive edge. Let's explore how data can elevate your startup from a promising idea to a market disruptor.

Data As a Competitive Advantage

Startups operate in a challenging environment, contending with limited resources, small teams, and fierce competition from established players. Yet one of the most powerful levers at their disposal is data. When used effectively, data can level the playing field, empowering startups to compete with and even surpass larger competitors. As outlined in *Competing on Analytics* by Davenport and Harris, a startup that rigorously uses data-driven insights can rapidly uncover market opportunities and optimize its strategies despite a tight budget.[32]

In today's fast-paced business landscape, agility is key. Startups inherently possess the flexibility to act quickly on data insights. Instead of cumbersome layers of bureaucracy, startup founders can immediately pivot based on real-time information. Frameworks described in *Data Science for Business* by Provost and Fawcett emphasize how data should drive decision-making. Startups enable rapid course corrections that large organizations might take weeks or months to execute.[33]

A startup's concentrated focus on a single product or niche is another distinct advantage. By dedicating resources to one core area, founders can collect and analyze data that directly impacts their primary objectives. For instance, a startup developing a fitness app can harness user data to continuously refine engagement and retention metrics, mirroring the focused, iterative approach discussed in data-centric methodologies.

---

[32] Thomas H. Davenport and Jeanne Harris, *Competing on Analytics: The New Science of Winning* (Harvard Business Review Press, 2017).

[33] Foster Provost and Tom Fawcett, *Data Science for Business: What You Need to Know About Data Mining and Data-Analytic Thinking* (O'Reilly, 2013).

Moreover, startups enjoy the freedom of starting from scratch without legacy systems. They can design modern, scalable data architectures using cloud-based solutions and open source tools. This fresh slate enables them to implement robust data infrastructures—a concept that *Measure What Matters* by John Doerr reinforces by showing how well-defined metrics and objectives can drive exponential growth even in resource-constrained environments.

Embedding a data-driven culture from day one is also critical. In a startup, every team member—from engineers to marketers—can align around data-backed insights, ensuring that each decision is informed by objective evidence. This cultural emphasis on measurable outcomes not only streamlines operations but also bolsters the startup's narrative to investors and customers alike.

Furthermore, startups often enjoy closer relationships with their customers. By blending qualitative customer feedback with quantitative user behavior data, they gain a deep understanding of their audience's needs. This dual approach is fundamental to the modern data science toolkit and enables startups to iterate quickly, ensuring that their product continuously evolves to meet market demands.

Finally, data is a compelling storytelling tool. Startups can use real-time analytics to showcase their growth trajectory through cohort analyses and other metrics that highlight customer success stories. This transparent, data-driven narrative builds trust with investors and clients, underscoring the startup's potential to scale.

By leveraging these advantages—and drawing on frameworks such as those found in *Competing on Analytics*, *Data Science for Business*, and *Measure What Matters*—startups can effectively use data not just as a tool for decision-making but as a strategic asset that creates sustainable competitive differentiation. While larger companies may have more resources, a startup's agility, focus, and innovative data practices can be the key to thriving in today's fast-paced, data-driven world.[34,35,36]

The power of data in startups isn't just theoretical—it's been proven time and again by companies that have disrupted industries and achieved remarkable growth. These success stories serve as both inspiration and practical guides for entrepreneurs looking

---

[34] Thomas H. Davenport and Jeanne Harris, *Competing on Analytics: The New Science of Winning* (Harvard Business Review Press, 2017).

[35] Foster Provost and Tom Fawcett, *Data Science for Business: What You Need to Know About Data Mining and Data-Analytic Thinking* (O'Reilly, 2013).

[36] John Doerr, *Measure What Matters: How Google, Bono, and the Gates Foundation Rock the World with OKRs* (Portfolio, 2018).

to harness data's potential. Let's explore some concrete examples of how startups have turned data into their secret weapon, transforming challenges into opportunities and ideas into industry-changing innovations.

Data-Driven Success in Startups

Data has become the cornerstone of modern business success, and startups are increasingly leveraging it to achieve remarkable outcomes. These companies prove that even with limited resources, a reality that extends beyond small teams and tight budgets to include constraints like scarce market data, a lack of brand recognition, and the intense pressure of a finite runway, a strategic approach to data can drive innovation, improve customer experiences, and create a sustainable competitive advantage. Here are some specific examples of startups that have harnessed the power of data to fuel their growth and success.

Robinhood: Unconventional Launch

Robinhood's early journey is a textbook case of how raw numbers and smart analytics can create a powerful narrative—even when no finished product exists. The founders launched a simple landing page that boldly promised "$0 commission stock trading, stop paying up to $10 for every trade."[37]

Instead of bombarding users with technical details or overwhelming product features, the page leveraged human psychology through the power of FOMO (fear of missing out). The strategy was straightforward yet ingenious. Users were asked only to enter their email address—no lengthy forms, no jargon, just a quick sign-up that placed them in a visible, real-time waitlist. On the thank-you page, each new registrant could see exactly where they stood in line, a gamified element that not only heightened anticipation but also spurred individuals to share their unique referral links.

---

[37] Josh Ledgard, "Setting up a Waitlist Like Robinhood - Complete Guide," KickoffLabs Blog, accessed January 27, 2025, https://kickofflabs.com/blog/how-to-setup-a-viral-waiting-list-launch-page-like-robinhood-with-kickofflabs/.

The message was: "Interested in priority access? Get early access by referring your friends. The more friends that join, the sooner you'll get access." This simple approach turned a basic landing page into a viral acquisition engine.[38]

One memorable moment illustrates the potency of this approach. Co-founder Vlad Tenev recalls waking up one Saturday morning to discover an unexpected surge on Google Analytics—about 600 concurrent visitors, the vast majority of whom had been driven to the site by a Hacker News post. That day, Robinhood's story ranked third on Hacker News, and in a single day, they amassed 10,000 sign-ups. Over the first week, the numbers soared to over 50,000, eventually reaching nearly one million waitlisted users before the app was even launched. This remarkable traction was achieved solely through the power of analytics and the simplicity of their landing page.[39]

The extraordinary waitlist numbers didn't just serve as a vanity metric; they became the cornerstone of Robinhood's pitch to investors. Despite not having a fully built app, the founders had concrete evidence that an immense and enthusiastic audience was ready to trade commission-free. When it came time to secure funding, Robinhood had already closed a seed round of about $3 million—a round led by prominent investors such as Index Ventures, Google Ventures, and Andreessen Horowitz. In an environment where regulatory requirements mandated a minimum level of capital on the balance sheet, having nearly one million eager sign-ups validated the team's vision and demonstrated genuine market demand.

The story goes further. Tenev described how, despite facing around 75 rejections from venture capitalists, the team's relentless determination eventually paid off. Investors were not only impressed by the raw numbers; they saw the potential for a disruptive model in an industry that had long relied on exorbitant fees and opaque practices. The fact that early backers were willing to invest in a company with no product yet underscored the transformative power of leveraging analytics to build an organic, enthusiastic user base. Even celebrity investors like Snoop Dogg later joined the roster, highlighting that this bold, data-driven approach had captured the imagination of a diverse array of supporters.

---

[38] Ryan Kaufman, "How Robinhood Turned a Landing Page into a 1M Person Waitlist," The Growth Playbook Substack, accessed February 3, 2025, `https://thegrowthplaybook.substack.com/p/how-robinhood-turned-a-landing-page`.

[39] Anna Mazarakis and Alyson Shontell , "The founders of Robinhood, a no-fee stock-trading app, were initially rejected by 75 venture capitalists — now their startup is worth $1.3 billion," Business Insider, accessed February 9, 2025, `https://www.businessinsider.com/robinhood-app-vlad-tenev-founder-free-stock-trading-valuation-2017-7?IR=T`.

This narrative is a lesson for any startup aiming to convince investors: you don't always need a fully developed product to prove your concept. With a clear, compelling value proposition and smart use of analytics, you can demonstrate market potential and secure the capital necessary to build a revolutionary service. Robinhood's journey—from a minimalist landing page to a multi-billion-dollar company—remains a powerful example of how great ideas, backed by real user data, can change the rules of an industry.

Spotify: Personalization at Scale

Spotify, now a global music streaming giant, began its journey as a modest Swedish startup that placed an unwavering focus on harnessing data. From its earliest days, the company understood that data analytics would be central to delivering a highly personalized user experience. By meticulously analyzing user listening habits, Spotify developed innovative features that set it apart from competitors and fostered deep customer loyalty.

A cornerstone of Spotify's strategy is its ability to tailor recommendations to each individual user. In 2015, the company introduced Discover Weekly—a feature that curates a personalized playlist of new music every Monday based on the user's unique listening history. Discover Weekly proved to be a huge success, with Spotify users streaming billions of hours from their personalized playlists. This remarkable engagement underscored the power of data-driven personalization in increasing user satisfaction and retention.[40]

Another testament to Spotify's innovative use of data is the Daily Mix feature, which artfully blends a user's favorite tracks with fresh recommendations. This adaptive playlist not only enhances listening time but also continuously evolves as user preferences change. Behind these sophisticated recommendation systems lies an extensive culture of experimentation. Spotify has built a robust framework for A/B testing, with hundreds

---

[40] Spotify Advertising Team, "Five years of discovery and engagement through Discover Weekly," Spotify Advertising, July 15, 2020, https://ads.spotify.com/en-US/news-and-insights/five-years-of-discovery-and-engagement-through-discover-weekly/.

of teams running tens of thousands of experiments each year. These experiments, which measure metrics such as click-through rates and listening duration, enable the company to refine its algorithms and improve overall user experience.[41]

Spotify's journey is not without its challenges. As the company pushes the boundaries of algorithmic personalization, it has encountered mixed user feedback on some of its more experimental AI-generated features. This experience has taught Spotify the importance of striking a balance between innovative technology and user expectations, ensuring that the algorithms not only delight but also broaden users' musical horizons.[42,43]

In addition to its internal innovations, Spotify has excelled in data storytelling and marketing. The annual Spotify Wrapped campaign transforms complex listening data into engaging, shareable narratives that capture each user's unique music journey over the past year. This creative presentation of data has evolved into a viral marketing phenomenon, further deepening user engagement and solidifying the brand's identity.[44]

Spotify's dedication to a data-driven approach has been instrumental in its transformation from a small startup into a dominant global force. Through continuous experimentation, thoughtful integration of user feedback, and compelling data storytelling, Spotify exemplifies how a strategic focus on data can drive innovation, enhance customer experiences, and create a sustainable competitive advantage.

Airbnb: Optimize Pricing and User Trust

Airbnb disrupted the hospitality industry by creating an innovative platform that connects travelers with hosts offering unique accommodations. From its inception, the company relied heavily on data to refine its business model. Early on, Airbnb embraced

---

[41] Sebastian Ankargren, "Experiment like Spotify: A/B Tests and Rollouts," Spotify Confidence Blog, accessed January 29, 2025, `https://confidence.spotify.com/blog/ab-tests-and-rollouts`.

[42] Erin Neil, "Hua Hsu on the Costs of Spotify," The New Yorker Newsletter, accessed February 5, 2025, `https://www.newyorker.com/newsletter/the-daily/hua-hsu-on-the-costs-of-spotify`.

[43] Kyndall Cunningham, "Spotify Wrapped misses the mark with this one joyless feature," Vox, accessed February 1, 2025, `https://www.vox.com/culture/389869/spotify-wrapped-google-ai-music-streaming`.

[44] Anna Kaufman, "Spotify Wrapped is out: How to see yours and what's new in 2024," USA Today, accessed January 30, 2025, `https://eu.usatoday.com/story/entertainment/music/2024/12/04/spotify-wrapped-2024-overview/76705879007/`.

a simple yet effective data infrastructure. Their first data scientist's job was just to query their production MySQL database and share very basic stats with the rest of the team. It may be hard to believe, but these were groundbreaking insights for their growth. The whole company was just seven people. It was really easy to keep everyone in the loop about all major metrics. Despite its humble beginnings, this data-centric approach proved to be invaluable in shaping the company's growth strategy.[45]

One of the most notable ways Airbnb harnessed data was through its dynamic pricing tool. By analyzing historical booking trends, local events, and competitor rates, the tool was able to recommend optimal prices for hosts. This real-time adjustment not only maximized revenue for hosts but also enhanced overall customer satisfaction by aligning prices with market demand. In the early days of Airbnb, such insights were gleaned from limited but focused datasets, proving that even imperfect data can yield powerful results when applied strategically.

Airbnb also used data to improve user trust, a critical factor for a platform built on personal interactions. By developing a robust review system and sophisticated fraud detection algorithms, the startup was able to create a safe and reliable environment for both hosts and guests. The company analyzed early transaction data and user feedback to design a double-blind review process that reduced bias and encouraged honest, constructive evaluations. These measures not only built trust but also played a significant role in scaling the platform rapidly.

In its formative stages, Airbnb discovered that visual appeal was a crucial determinant of user engagement. By closely reviewing user behavior data, the founders noticed that listings with low-quality photos received far fewer clicks and bookings compared to those with high-quality images. Acting on this insight, Airbnb launched a professional photography program in 2010, sending photographers to hosts' homes to capture appealing images. The results were dramatic: properties with professional photos saw bookings multiply by two to three times, a decision that underscored the value of leveraging data even for seemingly peripheral aspects of the business.[46]

---

[45] Riley Newman, "How we scaled data science to all sides of Airbnb over 5 years of hypergrowth," VentureBeat, accessed February 7, 2025, `https://venturebeat.com/dev/how-we-scaled-data-science-to-all-sides-of-airbnb-over-5-years-of-hypergrowth/`.
[46] Leigh Gallagher, *The Airbnb Story: How Three Guys Disrupted an Industry, Made Billions of Dollars ... and Plenty of Enemies* (Virgin UK, 2018).

Airbnb's journey was also marked by its keen ability to identify early product-market fit through data. When the platform was first launched during the 2008 Democratic National Convention in Denver, the founders observed a surge in demand as hotel rooms sold out. They meticulously tracked booking inquiries and user behavior during such events, identifying a clear niche for alternative accommodations during periods of high demand. This insight allowed Airbnb to refine its positioning and target marketing efforts toward major events and conferences, capturing a market segment that traditional hospitality services had overlooked.

Learning from competitors was another critical aspect of Airbnb's data-driven evolution. In its early years, the company scrutinized platforms like Craigslist to understand pricing trends, popular neighborhoods, and the types of properties that were most in demand. By extracting these insights, Airbnb not only fine-tuned its pricing strategy but also implemented creative tactics such as cross-posting listings to increase visibility and attract new users. This strategic analysis of competitor data helped the platform solidify its market presence and rapidly grow its user base.

Airbnb's relentless commitment to iteration further exemplified its data-driven ethos. The company continuously ran A/B tests on various elements of its user interface—from the placement of the search bar to the clarity of the booking process—to optimize conversions. Early tests revealed that users were abandoning the booking process due to unclear pricing or confusing navigation. By simplifying these elements based on data insights, Airbnb improved its conversion rates and overall user satisfaction, laying a solid foundation for scalable growth.[47]

The evolution of Airbnb's data infrastructure also played a pivotal role in its success. While the company started with simple tools for data collection and analysis, the exponential growth in user activity necessitated a shift to more sophisticated technologies. Over time, Airbnb adopted scalable solutions such as Hadoop and Apache

---

[47] Jan Overgoor, "Experiments at Airbnb," Airbnb Engineering, accessed January 26, 2025, https://medium.com/airbnb-engineering/experiments-at-airbnb-e2db3abf39e7#.miqyczkzb.

Spark to process vast amounts of data efficiently, ensuring that its insights remained timely and actionable as the platform expanded globally. Airbnb developed analytical tools such as Insights Report to provide hosts with tips on how to optimize listings and maximize revenue.[48,49]

At its core, Airbnb's success as a startup can be attributed to its unwavering focus on data. By systematically collecting, analyzing, and acting on data across various stages—from pricing optimization and visual enhancement to trust building and user experience refinement—Airbnb not only disrupted the hospitality industry but also set a benchmark for data-driven innovation. The lessons from its early data-driven decisions continue to serve as an inspiring blueprint for startups around the world, demonstrating that even with limited resources, a strategic approach to data can drive exponential growth and create a sustainable competitive advantage.[50]

Uber: Fueling the Ride from Startup to Global Giant

Uber's meteoric rise from a small San Francisco startup to a global ridesharing powerhouse is a story deeply rooted in its innovative use of data. In its early days, Uber faced the formidable challenge of matching supply and demand in real time. The company developed sophisticated algorithms that analyzed app activity, historical booking trends, weather conditions, and local events. This data-driven approach allowed Uber to predict where and when riders would need a ride, guiding drivers to high-demand zones. During large events like concerts or conferences, these insights proved invaluable, dramatically reducing wait times and maximizing earnings for drivers.[51]

Dynamic pricing, or surge pricing as it became known, was another groundbreaking innovation born from Uber's reliance on data. By monitoring real-time ride requests and driver availability, Uber's algorithms could automatically adjust fares during peak

---

[48] "What's the Airbnb Insights Report? A Guide to Listing Performance," Rankbreeze, accessed February 2, 2025, https://rankbreeze.com/airbnb-insights-report/.

[49] James Mayfield, Krishna Puttaswamy, Swaroop Jagadish, and Kevin Long, "Data Infrastructure at Airbnb," Airbnb Engineering, accessed February 6, 2025, https://medium.com/airbnb-engineering/data-infrastructure-at-airbnb-8adfb34f169c.

[50] Georgios Zervas, Davide Proserpio, and John Byers, "The Rise of the Sharing Economy: Estimating the Impact of Airbnb on the Hotel Industry," Journal of Marketing Research, vol. 54, no. 5, 2017, pp. 687–705.

[51] Mike Isaac, *Super Pumped: The Battle for Uber* (W. W. Norton & Company, 2019).

periods. Although surge pricing sparked controversy initially, it effectively incentivized drivers to work during high-demand periods, ensuring that riders received timely service. This adaptive pricing model not only balanced supply and demand but also reinforced the company's commitment to a market-responsive strategy.

Uber also harnessed the power of GPS and mapping data to enhance the rider experience. In its infancy, the startup leveraged Google Maps to provide basic navigation and driver matching. However, by continuously analyzing location data, Uber refined its ability to match riders with the nearest available drivers, which in turn reduced wait times and optimized routes. This focus on precise, real-time location analytics was critical for scaling the service in densely populated urban areas.

Data played a key role in Uber's expansion strategy as well. When entering new markets, the company relied on demographic data, local transportation statistics, and competitive analysis to identify optimal launch cities. This strategic approach ensured that sufficient drivers were onboarded before the platform went live, thereby providing a seamless experience for riders from day one. By methodically evaluating market opportunities through data, Uber was able to prioritize regions where its service could have the greatest impact.

Improving the overall experience for both riders and drivers was another facet of Uber's data-driven journey. The company implemented a rating system that collected and analyzed user feedback, enabling continuous refinement of the service. A/B testing was embedded into the culture; Uber routinely experimented with different user interface designs and booking flows to determine which variations led to higher conversion rates and increased satisfaction. These iterative improvements were pivotal in building a robust, user-centric platform.[52,53]

Safety and trust, essential for gaining widespread adoption, were also enhanced through data. Uber used real-time analytics to monitor rides and detect fraudulent activities. Features such as live tracking and trip sharing were introduced to ensure transparency and security, fostering trust among users. This commitment to safety, backed by rigorous data analysis, helped Uber overcome initial skepticism and build a loyal customer base.

---

[52] "Supercharging A/B Testing at Uber," Uber Blog, accessed January 28, 2025, `https://www.uber.com/en-PL/blog/supercharging-a-b-testing-at-uber/`.

[53] "Uber One A/B Test Fail," Real World Marketer, accessed February 10, 2025, `https://www.realworldmarketer.com/p/uber-one-ab-test-fail`.

Throughout its evolution, Uber's journey demonstrates that even a startup with limited resources can harness the power of data to drive innovation, optimize operations, and create a sustainable competitive advantage. From the rudimentary tools of its early days to the sophisticated data infrastructure that now underpins its global operations, Uber's story is a testament to the transformative potential of a data-driven approach.

Duolingo: The Power of Gamification

Duolingo, the language-learning app that transformed education into an engaging, game-like experience, stands as a shining example of how data can drive both growth and user retention. From its very inception, Duolingo embraced a data-driven strategy to make language learning accessible, fun, and effective for millions around the globe. Rather than relying solely on traditional teaching methods, the company integrated advanced analytics with gamification techniques, creating a platform where every lesson was an opportunity to learn through play.[54]

At the core of Duolingo's success lies its ability to use data to fuel gamification. By collecting and analyzing extensive user data—ranging from lesson completion rates and quiz responses to the frequency of daily logins—Duolingo identified the key drivers of engagement. For instance, the introduction of daily streaks, point systems, leaderboards, and badges was directly inspired by patterns observed in the data. The app's data scientists noticed that users who maintained regular practice through streaks were significantly more likely to stick with the program, so the team experimented with various streak mechanisms, such as streak freezes and celebratory animations, ultimately boosting retention rates.[55]

Duolingo's development process has always been characterized by continuous A/B testing. The team frequently experimented with different reward structures, user interface designs, and notification strategies. By comparing variations in real time and measuring their impact on metrics like lesson completion and user satisfaction,

---

[54] Daniel Pereira, "Duolingo Business Model," Business Model Analyst, accessed February 4, 2025, https://businessmodelanalyst.com/duolingo-business-model/?srsltid=AfmBOorqcdX6I65 Cyb_NcB6_IeT-Z-x8YmkOhTVyxKoWBTrYa-_MOH-m.

[55] Jasmine Bilham, "Case study: How Duolingo Utilises Gamification to Increase User Interest," Raw.Studio Blog, accessed February 8, 2025, https://raw.studio/blog/how-duolingo-utilises-gamification/.

Duolingo was able to fine-tune its features. This relentless experimentation ensured that the platform evolved in direct response to user behavior, leading to an ever-improving learning experience that resonated with its diverse user base.[56]

Personalization has been another critical component of Duolingo's approach. The platform's adaptive learning algorithms analyze detailed user performance data—such as error rates and the time spent on particular exercises—to dynamically adjust the difficulty of lessons. This means that if a learner struggles with a specific grammatical concept, the app automatically provides more exercises to reinforce that area. In addition, personalized notifications and streak reminders are tailored based on individual usage patterns, ensuring that users receive timely encouragement to continue their studies.

The power of habit formation has been central to Duolingo's strategy as well. Recognizing that consistent practice is essential for language acquisition, Duolingo has built features that turn learning into a daily ritual. Through subtle behavioral nudges like push notifications sent at optimal times—determined by analyzing usage data—the platform successfully ingrains the habit of regular study. Over time, these habits not only boost user engagement but also contribute to significant improvements in learning outcomes.

Duolingo's data-driven approach also extended to scaling its operations. By monitoring user acquisition costs and customer lifetime value, the company refined its freemium model to support sustainable growth. Data insights guided decisions on which new languages to introduce, resulting in an expansion to more than 40 language courses, including niche options that cater to both global and local markets. This strategic use of data ensured that Duolingo's growth was not random but carefully mapped out according to market demand and user preferences.[57]

Moreover, Duolingo continually harnessed artificial intelligence to enhance learning outcomes. The integration of AI-powered features—such as context-sensitive hints that help users correct mistakes—ensured that learners not only practiced

---

[56] Lavanya Aprameya, "Improving Duolingo One Experiment at a Time," Duolingo Blog, accessed February 11, 2025, `https://blog.duolingo.com/improving-duolingo-one-experiment-at-a-time/`.

[57] Erin Gustafson, "Meaningful metrics: How data sharpened the focus of product teams," Duolingo Blog, accessed February 2, 2025, `https://blog.duolingo.com/growth-model-duolingo/`.

but also understood their errors. The system's ability to learn from each interaction further refined its recommendations, making the entire process increasingly effective over time.[58]

Despite its many successes, Duolingo was not without its challenges. Early iterations of certain features, such as aggressive upsell tactics or poorly timed notifications, led to lower engagement levels. However, by rigorously analyzing user feedback and iterating on these features, Duolingo managed to overcome these obstacles and optimize the overall user experience. The company's willingness to embrace failure as a learning opportunity and pivot quickly based on data has been a crucial element of its enduring success.

Beyond operational improvements, Duolingo has used its wealth of data for compelling storytelling. The platform frequently shares user success stories and global learning trends, reinforcing its mission to make education free and accessible. These narratives, backed by concrete data, resonate with both users and investors, strengthening Duolingo's brand and driving further engagement.

Duolingo's journey from a small startup to a global leader in language education illustrates the transformative power of data when combined with gamification. By continuously analyzing user behavior, running experiments, and iterating its features, Duolingo not only created a highly engaging learning platform but also set a powerful example for startups across industries.

Stripe: Building Developer-Friendly Payment Infrastructure

In its early days, Stripe set out on an ambitious journey to simplify payment integrations for developers. The startup's founders realized that traditional payment systems were overly complex and cumbersome for the modern developer. With this insight, they began with local prototypes built around open source tools. This lean, almost rudimentary data infrastructure was intentionally simple—a deliberate choice that allowed them to iterate

---

[58] Bernard Marr, "The Amazing Ways Duolingo Is Using AI And GPT-4," Forbes, accessed January 29, 2025, https://www.forbes.com/sites/bernardmarr/2023/04/28/the-amazing-ways-duolingo-is-using-ai-and-gpt-4/.

quickly and validate core features without the burden of heavy enterprise solutions. Even though these early systems were imperfect, their simplicity was a catalyst for rapid experimentation and agile problem-solving.[59,60,61]

As the volume of transactions increased and the demands of their growing customer base evolved, Stripe's approach to data evolved as well. Recognizing the need for more scalable solutions, they transitioned to a cloud-based infrastructure while retaining many of the open source tools that had served them so well. The integration of technologies like Kafka for data streaming and Redis for caching exemplified how incremental scaling could be achieved without sacrificing the nimbleness that defined their early operations. This phase of their development was marked by constant learning—each technical challenge provided a new opportunity to refine their processes and optimize data flow, transforming early constraints into stepping stones for innovation.

One of Stripe's most transformative leaps came with its adoption of machine learning (ML) to combat payment fraud—a critical challenge for any financial platform. Stripe's Radar product leverages ML models trained on billions of transactions to detect anomalies in real time. By analyzing patterns such as unusual purchase locations, mismatched billing details, or atypical spending behavior, Radar flags high-risk transactions with remarkable accuracy.

The system employs adaptive learning, continuously retraining models on new fraud tactics to stay ahead of evolving threats. For instance, when fraudsters shifted to exploiting "card testing" attacks (using stolen cards to make small purchases), Stripe's ML models quickly identified these micro-transactions as outliers, even when they mimicked legitimate user behavior. Risk scoring algorithms further prioritize flagged transactions, enabling Stripe's partners to focus investigations on the most suspicious cases while reducing false positives that frustrate legitimate customers.

---

[59] Derek Andersen, "The Collison Brothers and Story Behind the Founding of Stripe," Startup Grind Blog, accessed February 1, 2025, `https://www.startupgrind.com/blog/the-collison-brothers-and-story-behind-the-founding-of-stripe/`.

[60] "When Stripe Was Young: The Early Years," Vator.tv, accessed February 5, 2025, `https://vator.tv/2016-12-29-when-stripe-was-young-the-early-years/`.

[61] Stephen Armstrong, "The untold story of Stripe, the secretive $20bn startup driving Apple, Amazon and Facebook," Wired, accessed January 27, 2025, `https://www.wired.com/story/stripe-payments-apple-amazon-facebook/`.

Beyond fraud prevention, Stripe recognized that its users—often startups themselves—lacked the resources to build custom analytics pipelines. Stripe Analytics solved this by embedding actionable insights directly into its platform. Users automatically gain visibility into key metrics such as revenue trends, retention cohorts, customer lifetime value and churn analysis. These tools eliminate the need for businesses to manually aggregate data or build dashboards, allowing even non-technical teams to make data-driven decisions. For example, a SaaS company using Stripe can instantly identify which pricing tier retains users longest or pinpoint seasonal dips in payment success rates—insights that once required weeks of SQL queries and engineering bandwidth.

Over time, data evolved from an operational tool to a core differentiator. By embedding ML into fraud detection and democratizing analytics for users, Stripe didn't just solve its own challenges—it turned data into a product feature that locks in customer loyalty. Their story exemplifies how startups can start lean, scale intelligently, and transform technical hurdles into strategic assets.[62]

Canva: Design Tool for Everyone

Founded in 2012, Canva emerged as a disruptor in the graphic design industry with a mission to "empower everyone to design." The platform's intuitive drag-and-drop interface, vast template library, and freemium model addressed a critical gap: making professional-grade design accessible to non-experts. By 2024, Canva boasted 180 million monthly active users and a $40 billion valuation—a trajectory fueled not just by product simplicity but by a relentless focus on data-driven decision-making.[63,64]

Canva's rise was underpinned by its ability to harness data at scale. Early on, the company recognized that user behavior analytics were critical to refining its product. For instance, Canva's engineering team built a real-time event tracking system capable of processing 25 billion daily events (800 billion monthly), capturing everything

---

[62] "How Machine Learning Works for Payment Fraud Detection and Prevention," Stripe Resources, accessed February 9, 2025, `https://stripe.com/ae/resources/more/how-machine-learning-works-for-payment-fraud-detection-and-prevention`.

[63] Andre Borczuk, "Canva's Billion Dollar Growth: A PLG Case Study," ProductMonk.io, accessed February 7, 2025, `https://www.productmonk.io/p/canva-s-billion-dollar-growth-a-plg-case-study`.

[64] "Canva Case Study," The Clueless Company, accessed February 3, 2025, `https://www.theclueless.company/canva-case-study/`.

from template usage to feature adoption. By structuring these events with Protobuf schemas and enforcing compatibility rules, Canva ensured data consistency across its microservices and analytics pipelines. This granular visibility allowed teams to identify friction points—such as underutilized tools or onboarding drop-offs—and iterate rapidly.[65]

One pivotal innovation was the development of Datumgen, a custom code generator that automated schema validation and generated TypeScript, Java, and SQL definitions. This tool reduced engineering overhead and ensured that every event—whether a design save or a collaboration action—could be analyzed for insights.

Canva's freemium model thrived by converting free users to paid subscribers, but scaling this required sophisticated analytics. The company migrated from third-party tools like ProfitWell to an in-house subscription data architecture. This shift enabled granular segmentation—tracking metrics like churn by country, device, and subscription tier. By modeling churn using the Fader-Hardie probabilistic framework, Canva predicted user lifetime value (LTV) with improved accuracy, even with limited data points. This allowed growth teams to target high-risk cohorts with personalized retention campaigns, significantly reducing voluntary churn.[66,67]

As Canva's user base exploded, its data infrastructure faced scaling challenges. The engineering team replaced daily database snapshots with change-data-capture (CDC) systems, reducing data replication times and enabling near-real-time analytics. By adopting a service-aligned data architecture, Canva decentralized ownership of data pipelines, empowering service teams to define their own schemas and transformations. This modular approach, combined with tools like Snowpipe for streaming ingestion, allowed the platform to handle petabytes of data without compromising performance.[68]

---

[65] Prachi Kothiyal, "Canva's Robust Product Analytics Pipeline: Collecting 25 Billion Events Daily," Talent500.co Blog, accessed January 26, 2025, `https://blog.talent500.co/ canva-records-billion-events-daily/`.

[66] Chuxin Huang, Paul Tune, and Grant Noble, "Subscription Analytics at Scale," Canva.dev Blog, accessed January 30, 2025, `https://www.canva.dev/blog/engineering/subscription- analytics-at-scale/`.

[67] Jaskirat Grover and Justin Ty, "Service-aligned Data Platform Architecture," Canva.dev Blog, accessed February 6, 2025, `https://www.canva.dev/blog/engineering/service-aligned- data-platform-architecture/`.

[68] "How do Canva's engineers and analysts scale data platforms to keep up with growth? — with Krishna Naidu," Firebolt.io Blog, accessed February 10, 2025, `https://www.firebolt.io/blog/ how-do-canvas-analysts-and-engineers-scale-data-platforms-to-keep-up-with-growth- with-krishna-naidu`.

The company employs a data-driven approach to continuously refine its user experience, focusing on key elements such as navigation design, customer targeting, and social proof. Canva's landing page features over 15 strategically placed buttons and hyperlinks, carefully balanced to encourage action without overwhelming visitors. The company uses a hidden customer journey strategy, guiding users through the product's capabilities without explicitly stating steps. By analyzing user behavior, Canva has implemented a versatile design that caters to multiple customer segments, from individuals to enterprise teams. The landing page showcases Canva's best features and includes social proof from recognizable brands early on to build trust. This data-informed optimization strategy has contributed to Canva's impressive growth and valuation, making it a prime example of effective conversion rate optimization in the digital product space.[69]

Canva's success exemplifies how startups can leverage data to balance agility with scalability. By embedding analytics into every layer of its operations—from real-time event tracking to predictive churn modeling—the company transformed a niche tool into a global powerhouse. This data-driven approach has enabled Canva to innovate rapidly, introducing cutting-edge AI features such as the text-to-image generation tool. This feature allows users to create unique images simply by describing them in text, showcasing Canva's commitment to staying at the forefront of technology. For data professionals and founders, Canva's journey underscores the importance of infrastructure flexibility, cross-functional collaboration, and a relentless focus on user-centric metrics, all while continuously integrating advanced technologies to enhance user experience and product capabilities.[70]

ElevenLabs: AI Voice Unicorn

ElevenLabs, a synthetic voice technology startup founded in 2022 by Mati Staniszewski and Piotr Dabkowski, emerged as a unicorn within two years, achieving a $3.3 billion valuation in 2025. The company's mission—to break language barriers and democratize audio content creation—was propelled by its cutting-edge AI models for real-time voice

---

[69] Manan Modi, "How Canva, a $40 Billion Design Startup, Optimizes its Landing Page to Increase Conversions," Medium, accessed February 12, 2025, `https://uxplanet.org/how-canva-a-40-billion-design-startup-optimizes-its-landing-page-to-increase-conversions-12faa82f6824`

[70] Cameron Adams, "Supercharging the Visual Suite," Canva Newsroom, accessed February 4, 2025, `https://www.canva.com/newsroom/news/supercharging-the-visual-suite/`.

cloning, multilingual dubbing, and emotionally resonant text-to-speech (TTS) synthesis. Born from the founders' frustration with poor dubbing quality in their native Poland, ElevenLabs leveraged data science to refine its core technology, ensuring voices retained the speaker's tone, pitch, and emotional nuances with uncanny accuracy.[71]

ElevenLabs' ascent hinged on iterative improvements guided by user data. Early adopters tested its speech-to-speech dubbing tool, translating scripts across languages while preserving vocal identity. Imagine running an experiment—recording yourself speaking in English and Spanish, then comparing ElevenLabs' AI-generated translations. Such tests revealed minor discrepancies but underscored the technology's realism. This feedback loop informed refinements in linguistic alignment and emotional inflection, critical for scaling adoption in global markets like entertainment and customer service.

The company also harnessed data to expand its Voice Library, a repository of customizable voice models. By analyzing usage patterns from hundreds of thousands users, ElevenLabs identified high-demand vocal profiles (e.g., narrative-friendly voices for audiobooks or energetic tones for commercials). This data-driven curation attracted diverse clients, from publishers like HarperCollins to gaming studios.[72]

As ElevenLabs continued to evolve, its impact on the global audio landscape became increasingly profound, driven by the powerful synergy of AI and data. The company's innovative approach not only revolutionized the way we consume and create audio content but also opened up new possibilities for cross-cultural communication and understanding. By leveraging machine learning algorithms trained on vast datasets of human speech, ElevenLabs broke down language barriers and enabled emotionally authentic voice replication, effectively shrinking the world and bringing diverse cultures and perspectives closer together. The journey from a startup born out of frustration with poor dubbing to a unicorn reshaping the audio industry serves as a testament to the transformative power of AI-driven innovation, data-centric development, and the relentless pursuit of perfection in the realm of synthetic voice technology.

---

[71] Ingrid Lunden and Marina Temkin, "ElevenLabs has raised a new round at $3B+ valuation led by ICONIQ Growth, sources say," TechCrunch, accessed February 2, 2025, `https://techcrunch.com/2025/01/24/elevenlabs-has-raised-a-new-round-at-3b-valuation-led-by-iconiq-growth-sources-say/`.

[72] "HarperCollins Publishers and ElevenLabs to Bring More Stories to Life Through Audio," ElevenLabs Blog, accessed January 25, 2025, `https://elevenlabs.io/blog/harpercollins-publishers`.

Lessons Learned

The journeys of Robinhood, Spotify, Airbnb, Uber, Duolingo, Stripe, Canva, and ElevenLabs offer compelling evidence of data's transformative power in the startup world. These companies, operating across diverse industries, demonstrate that a strategic approach to data can be a crucial differentiator, even with limited resources.

One consistent theme is the importance of starting small and iterating rapidly. Robinhood's initial landing page, devoid of a finished product, proves that even a simple, data-driven approach can generate significant traction and investor interest. Airbnb's early reliance on basic stats from their MySQL database highlights how groundbreaking insights can emerge from humble beginnings.

Innovation is another key takeaway. Spotify's Discover Weekly and Daily Mix features showcase the potential of data-driven personalization to enhance user engagement and loyalty. Uber's dynamic pricing model demonstrates how real-time data analysis can optimize operations and balance supply and demand. Canva's landing page optimization approach exemplifies how important it is to shape users' behavior before they even start using the product.

Experimentation is essential. Spotify's extensive A/B testing framework, with thousands of experiments run each year, underscores the value of continuous refinement. Duolingo's constant experimentation with reward structures and user interface designs illustrates how data can drive the evolution of a highly engaging learning platform.

Personalization emerges as a significant driver of success. Duolingo's adaptive learning algorithms, which tailor lessons to individual user performance, highlight the power of personalized content. ElevenLabs' data-driven curation of its Voice Library demonstrates the importance of catering to specific user needs and preferences.

Finally, these startups demonstrate the importance of leveraging data to build trust and ensure safety. Airbnb's robust review system and fraud detection algorithms, Uber's safety features, and Stripe's Radar product all showcase how data can create a secure and reliable environment for users. ElevenLabs' proactive AI safety protocols further underscore the ethical considerations in leveraging data, particularly in the age of deepfakes.

In essence, these startups prove that data is not just a tool for analysis but a strategic asset that can drive innovation, enhance customer experiences, and create a sustainable competitive advantage. By embracing a data-driven culture, startups can make informed decisions, optimize their operations, and ultimately achieve remarkable success.

Summary

From Charles Babbage's early insight that "errors using inadequate data are much less than those using no data at all" to the AI-driven innovations of today's unicorn startups, this chapter has demonstrated how data serves as the lifeblood of entrepreneurial success. In the chaotic and high-stakes environment of startups, where uncertainty and rapid iteration are constants, relying solely on intuition is a risk no founder can afford. Data is not just a tool—it is a mindset, a compass that guides decisions, validates assumptions, and uncovers transformative opportunities.

The role of data begins even before a product exists. Frameworks like the Lean Startup methodology and Customer Development emphasize hypothesis testing and market validation, showing how early-stage startups can use even limited or imperfect data to validate demand and attract investors. Examples like Airbnb's reliance on basic user behavior metrics or Robinhood's waitlist sign-ups illustrate how minimal yet strategic use of data can build momentum and confidence in an idea. As startups grow, data evolves into a strategic asset that optimizes operations, personalizes user experiences, and creates defensible advantages. Stripe's fraud detection algorithms, Uber's dynamic pricing models, and Canva's real-time analytics are prime examples of how data can fuel scalability and innovation.

A true data-driven culture goes beyond dashboards and metrics—it fosters curiosity, agility, and accountability. Spotify's and Airbnb's A/B testing ethos philosophy demonstrates how embedding data into decision-making democratizes insights across teams and transforms raw numbers into actionable narratives. Startups like Duolingo and ElevenLabs highlight how iterative experimentation informed by user behavior can create habit-forming products while addressing ethical considerations in AI development.

For investors, data bridges vision with viability. Metrics such as retention rates, unit economics, and network effects signal scalability and defensibility when applied effectively. Airbnb's pivot to professional photography or Stripe's usage analytics reveal how startups can use data to refine their offerings while building trust with stakeholders. However, the power of data comes with responsibility; startups must prioritize ethical governance to avoid algorithmic bias and maintain user trust.

The evolution from reactive analytics to predictive AI underscores the transformative potential of data. Startups that thrive will treat data as both shield and sword—navigating risks, anticipating trends, and crafting products that resonate deeply with their users. As Babbage foresaw with his mechanical computations, the true value of data lies in its ability to turn ambiguity into action. For startups, it is not just about survival; it is about fueling creativity, resilience, and lasting impact.

As we transition to Chapter 2, we will explore how startups can lay the groundwork for harnessing this power systematically. From setting up scalable infrastructure to defining actionable KPIs, this chapter will provide the blueprint for embedding a robust data-driven approach into every aspect of your business journey.

CHAPTER 2

Building a Data-Driven Strategy

Winning should be at the heart of every strategy.

—A.G. Lafley and Roger L. Martin,

Playing to Win: How Strategy Really Works[1]

Strategy, at its core, is about making choices to win, and this holds especially true for startups, where survival and rapid growth are paramount. In this context, a robust data strategy isn't just an advantage—it's a necessity. Without a clear plan for how to collect, analyze, and leverage data, startups risk falling into the "ad hoc trap," where their data teams are constantly bombarded with requests and reports but never truly contribute to the company's strategic goals.

This chapter delves into why a well-defined data strategy is critical for startup success, especially for companies seeking to operate without enterprise-scale resources. A clear strategy provides direction, empowers data teams, and moves the organization beyond simply reacting to data. Instead, it enables them to proactively generate insights, anticipate customer needs, and make data-backed decisions that fuel growth. To do this, we will first explore established frameworks like the Data Value Chain, the Data Flywheel, and the Lean Analytics Cycle to extract practical approaches for resource-constrained environments. The chapter then distills six essential components of an effective startup data strategy: business objective alignment, governance and quality framework, scalable architecture, analytics capabilities, a data-driven culture, and continuous learning mechanisms. Finally, a step-by-step methodology will guide

---

[1] A.G. Lafley and Roger L. Martin, *Playing to Win: How Strategy Really Works* (Harvard Business Review Press, 2013).

© Piotr Sidoruk 2026

P. Sidoruk, *From Data to Dollars*, https://doi.org/10.1007/979-8-8688-1898-1_2

you through assessment, planning, and implementation, highlighting how to avoid common pitfalls and transform data from a technical burden into a strategic asset that scales with your business. It's crucial to understand that a data strategy is not primarily about the data itself. Too often, startups fall into the trap of believing that the latest shiny technology will magically solve their problems. In reality, technical solutions are merely tools—and often expensive ones—to achieve business objectives. A truly effective data strategy begins with a deep understanding of the business: its goals, its challenges, and its unique operational context. It's about finding the right way to use data, not necessarily the perfect way nor the way many academics might recommend. The right way for a startup is the way that demonstrably helps it achieve its specific, often aggressive, growth targets. This chapter will explore how startups can avoid the common pitfall of tech-centric thinking and build a data strategy that truly aligns with their business needs, focusing on practical, actionable steps.

Defining Your Data Strategy

The concept of a data strategy might appear as a luxury reserved for established enterprises. This perception couldn't be further from the truth—the absence of intentional data planning during formative years creates compounding technical debt that stunts growth and limits strategic options as companies scale. A data strategy for startups isn't about building expensive infrastructure or hiring armies of analysts; it's about creating an intentional framework that aligns every byte of information collected with the organization's evolving business objectives while avoiding the fatal mistakes that derail 35% of startups through misaligned product-market fit according to CB Insights research.[2]

At its core, a startup data strategy answers three fundamental questions: What data matters most to achieving our immediate goals and long-term vision? How will we ensure this data flows reliably from source to insight? And what cultural practices will turn these insights into decisive action across all levels of the organization? Antoine Le Nel, Revolut's Chief Growth and Marketing Officer, crystallizes this mindset: "The growth is coming from the product. Full stop. If your product was that great, you wouldn't have an awareness problem." This product-centric view demands that startups instrument

---

[2] "The Top 12 Reasons Startups Fail," *CB Insights*, accessed March 8, 2025, `https://www.cbinsights.com/research/startup-failure-reasons-top/`.

their offerings to capture the behavioral signals that validate market fit while exposing operational inefficiencies—a process impossible without strategic data collection from the earliest development stages.[3]

A well-constructed data strategy serves as the backbone of a startup's ability to translate raw information into actionable insights, operational efficiency, and competitive differentiation. At its core, this strategy must transcend mere technical implementation, instead functioning as a dynamic blueprint that harmonizes data capabilities with overarching business objectives. For startups navigating the complexities of rapid growth and resource constraints, a deliberate approach to defining this strategy ensures that every data initiative directly contributes to survival, scalability, and market relevance.

The Foundation: Business Objectives As the North Star

A data strategy divorced from business goals is little more than an exercise in technical vanity. Startups must anchor their data initiatives in the specific outcomes they aim to achieve—whether that's increasing customer lifetime value, optimizing CAC (Customer Acquisition Cost), or accelerating product-market fit. As Gartner defines it, a data strategy is a "highly dynamic process" designed to support the acquisition, organization, analysis, and delivery of data in service of business objectives. This necessitates a reversal of traditional thinking: rather than starting with data availability or technological capabilities, startups should first articulate their strategic priorities and then identify how data can amplify those goals.[4]

For example, a fintech startup targeting underserved markets might prioritize data initiatives that enhance risk assessment models or improve financial inclusion metrics. By aligning data collection and analysis with these objectives, the startup avoids the common pitfall of accumulating disconnected datasets that fail to inform decision-making. This alignment requires close collaboration between technical teams and business leaders to ensure data projects address tangible pain points, such as reducing churn or identifying high-value customer segments. The process begins with

---

[3] "Advice for Startup Founders: Lessons from Top Entrepreneurs," Viva Technology, accessed March 6, 2025, https://vivatechnology.com/news/advice-for-startup-founders-lessons-from-top-entrepreneurs.

[4] "Data Strategy," Gartner, accessed March 4, 2025, https://www.gartner.com/en/information-technology/glossary/data-strategy.

a rigorous assessment of the current data landscape—understanding existing assets, gaps, and inefficiencies—before defining measurable, time-bound objectives that bridge these gaps.

The Inspiration: Popular Frameworks

Before dissecting the core components of a startup data strategy, it's critical to ground the discussion in established frameworks that have shaped modern data practices. These frameworks provide structured approaches that help startups avoid common pitfalls while maximizing the value extracted from their data assets. Each framework offers unique perspectives on how data can be transformed into actionable insights that drive business growth. Understanding these established models enables founders to adapt proven principles to their specific context rather than building from scratch, saving valuable time and resources during critical growth phases. The following sections explore six influential frameworks, highlighting their core components and relevance to startup environments.

The Data Value Chain

The Data Value Chain framework offers a powerful, **structured approach to maximizing value** from an organization's data assets. Developed by Open Data Watch, an international nonprofit organization of data experts, it is inspired by Michael Porter's value chain concept from his book *Competitive Advantage: Creating and Sustaining Superior Performance*. It visualizes the entire life cycle of data, from initial collection to its final impact on decision-making.[5,6,7]

---

[5] Michael E. Porter, *Competitive Advantage: Creating and Sustaining Superior Performance* (Free Press, 1998).

[6] "The Data Value Chain: Moving from Production to Impact ," Open Data Watch, accessed February 20, 2025, https://opendatawatch.com/publications/the-data-value-chain-moving-from-production-to-impact/.

[7] "About Us," Open Data Watch, accessed October 15, 2025, https://opendatawatch.com/about/.

The framework helps organizations identify gaps, prioritize efforts, and ensure that data doesn't just get collected but gets used effectively. The Data Value Chain comprises four core stages:

1. **Collection:** This initial stage involves identifying what data is needed, establishing a process for gathering it (through surveys, logs, or other methods), and then processing it to ensure it is clean, accurate, and correctly stored for future use.

2. **Publication:** Once collected and processed, data must be made accessible. This stage includes publishing the data with clear documentation, disseminating it through appropriate channels, and performing analysis to make it understandable.

3. **Uptake:** This stage bridges the gap between available data and its potential users. It involves actively connecting users to the data, incentivizing them to incorporate it into their workflows, and influencing a culture that values evidence-based decisions.

4. **Impact:** The final and most crucial stage is where data creates tangible value. This happens when data is actively used to inform a decision, which in turn leads to a positive change or improved outcome. It also includes the reuse of data for new analyses, generating further value over time.

Startups that successfully implement the Data Value Chain can develop a data-driven test-and-learn culture that dramatically improves operational efficiency and market responsiveness. By treating data as a product and building trust through transparent, data-backed decision-making, startups can transform historically siloed information into new data products that create sustainable competitive advantages and drive growth—even with limited infrastructure and without hiring armies of analysts.[8]

---

[8] Karan Dhawal, "Data Value Chain: Analysis-Enable Data Products," The Data Administration Newsletter, accessed February 6, 2025, https://tdan.com/data-value-chain-analysis-enable-data-products/29012.

The Data Flywheel

The data flywheel represents a dynamic, **self-reinforcing mechanism** that transforms data collection from a static activity into a continuous engine of business growth:

1. More product usage generates more data.

2. More data leads to a better product.

3. A better product drives more usage.

4. The cycle repeats, accelerating with each rotation.

The term *flywheel effect* was popularized by Jim Collins and described in his book *Good to Great: A Study of Management Strategies of Companies with Lasting Growth.* Unlike linear frameworks, the flywheel concept emphasizes how data initiatives can create powerful momentum through positive feedback loops that accelerate over time. Essentially, a data flywheel is a strategy that utilizes various data sources, including internal, customer, and external information, to foster business expansion and strengthen customer connections. This self-perpetuating system becomes increasingly valuable as more data accumulates, creating a compounding effect that can provide startups with sustainable competitive advantages in their markets.[9,10]

The fundamental principle driving data flywheels is that the more and better data a company collects, the more value it can provide. For resource-constrained startups, this mechanic offers particular appeal because initial investments in data infrastructure can yield disproportionately large returns as the system gains momentum. The flywheel operates by combining existing resources such as customer feedback, analytics tools, marketing automation systems, and service data with artificial intelligence algorithms to generate increasingly sophisticated customer insights. These insights enable startups to tailor their products and services more precisely to customer needs, creating better experiences that in turn generate more engagement and more data to fuel the cycle.

The ultimate goal for startups implementing a data flywheel is to reduce dependency on external sources for growth, instead leveraging internal data assets as primary drivers of business expansion. As customer data accumulates over time, startups gain deeper

---

[9] Jim Collins, *Good to Great: A Study of Management Strategies of Companies with Lasting Growth* (Harper Business, 2001).

[10] Bruce MacVarish, "Data Flywheels: Creating a Continuous Cycle of Business Growth," York IE Blog, accessed February 15, 2025, https://york.ie/blog/data-flywheels-creating-a-continuous-cycle-of-business-growth/.

understanding of customer priorities and behaviors, enabling more personalized offerings that strengthen customer relationships. This increased personalization typically leads to higher customer satisfaction, which generates more interactions and therefore more data points, continuing to accelerate the flywheel's momentum. For startups focused on rapid scaling, establishing this virtuous cycle early can create significant advantages over competitors who lack similar data-driven reinforcement mechanisms in their business models.

The Lean Analytics Cycle

The Lean Analytics Cycle, described in the book *Lean Analytics* by Alistair Croll and Benjamin Yoskovitz, provides startups with an **iterative framework for making data-driven decisions** while conserving resources and maintaining the agility needed in uncertain market conditions. Grounded in the principles of the broader Lean Startup methodology, this approach emphasizes learning and adaptation through systematic experimentation rather than exhaustive analysis. The framework articulates a step-by-step process that begins with defining clear business goals, forming testable hypotheses about how to achieve those goals, and selecting focused metrics to evaluate progress. This disciplined methodology prevents startups from drowning in vanity metrics or pursuing data initiatives disconnected from business outcomes.[11]

At its core, the Lean Analytics Cycle is driven by the principles of data-driven decision-making and validated learning, which systematically test assumptions to minimize wasted effort. To put these principles into practice, the cycle follows four distinct steps:

1. Choose a metric to improve.

2. Form a hypothesis.

3. Run an experiment.

4. Act on the results.

A key concept within this process is the One Metric That Matters (OMTM), which forces a startup to focus its limited resources on the single measurement most critical to its current goals. This focus on a single metric is particularly valuable for early-stage

---

[11] Alistair Croll and Benjamin Yoskovitz, *Lean Analytics: Use Data to Build a Better Startup Faster* (O'Reilly, 2013).

startups that need both organizational alignment and rapid, measurable progress. A practical application of this cycle is illustrated by Avinash Kaushik, an author of the influential book *Web Analytics 2.0*, using Airbnb's well-known experiment. On his popular blog, Kaushik described how Airbnb, with the goal of increasing the number of nights a property was rented, hypothesized that amateur photos were creating a trust gap. Their experiment—offering professional photography to a subset of listings—validated this assumption, leading to a clear lift in bookings and providing a data-driven path to improve a key business metric.[12,13]

This iterative approach allows startups to make incremental improvements while maintaining the flexibility to pivot when data suggests current strategies aren't yielding desired outcomes. For founders navigating market uncertainties, the Lean Analytics Cycle offers a pragmatic balance between data-informed decision-making and the speed required in startup environments. By emphasizing frequent experiments with clearly defined metrics, startups can optimize resource allocation while building a culture of continuous learning that becomes increasingly valuable as the organization matures and data volumes grow.

Data Maturity Models

The Data Maturity Model provides startups with a **diagnostic framework to assess their current data capabilities** and chart a strategic path toward more sophisticated data practices. Unlike prescriptive frameworks that dictate specific actions, maturity models offer a benchmarking system that helps organizations understand their relative position on the data evolution spectrum and identify appropriate next steps. Multiple established frameworks exist, including the Dell Data Maturity Model, Gartner Data Maturity Model, and CMMI maturity levels, each with distinct strengths and emphases that organizations can select based on their specific needs. Broadly, these frameworks serve as a means for organizations to assess their data management capabilities by examining their data

---

[12] Avinash Kaushik, "The Lean Analytics Cycle: Metrics > Hypothesis > Experiment > Act," Occam's Razor by Avinash Kaushik Blog, accessed March 2, 2025, https://www.kaushik.net/avinash/lean-analytics-cycle-metrics-hypothesis-experiment-act/.

[13] Avinash Kaushik, *Web Analytics 2.0: The Art of Online Accountability and Science of Customer Centricity* (Sybex, 2009).

management processes against predefined maturity levels, effectively benchmarking their data practices. This assessment approach proves particularly valuable for startups seeking to balance immediate operational needs with longer-term data infrastructure investments.[14]

Typically, data maturity frameworks define four to five progressive levels through which organizations evolve their data practices. The journey begins at Level 1 (Initial/ Ad Hoc), characterized by reactive processes and a general absence of data ownership or accountability. Many early-stage startups naturally begin here, with data collection happening haphazardly and analysis occurring primarily in response to immediate problems rather than strategic objectives. As organizations grow, they progress to Level 2 (Repeatable), where some processes and roles become defined and consistency emerges in tools and knowledge management. This transition often coincides with startups moving beyond founder-led analysis to establishing dedicated data roles and implementing basic data governance principles.[15]

The more advanced stages of maturity include Level 3 (Defined), marked by emerging centralized data management capabilities and coordinated data policies, and Level 4 (Managed), where data processes become systematically measured and controlled. For resource-constrained startups, these maturity models offer value by helping founders prioritize investments that match their current stage and business needs rather than attempting to implement enterprise-scale solutions prematurely. The assessment process itself delivers significant benefits by identifying specific gaps in current data processes, enabling targeted improvements that align with overall business strategy. While complete data maturity represents a long-term journey, startups can achieve substantial competitive advantages by systematically advancing their capabilities across key data management dimensions.[16]

---

[14] Keith D. Foote, "Creating a Data Maturity Model: What, Why, How," Dataversity, accessed March 4, 2025, https://www.dataversity.net/creating-a-data-maturity-model-what-why-how/.

[15] "What are Data Maturity models?: An extensive guide," Airbyte Data Engineering Resources, accessed March 4, 2025, https://airbyte.com/data-engineering-resources/what-are-data-maturity-models.

[16] "CMMI Levels of Capability and Performance," ISACA CMMI Performance Solutions, accessed March 4, 2025, https://cmmiinstitute.com/learning/appraisals/levels.

The DataOps Methodology

The DataOps Methodology adapts principles from software development's DevOps movement to address the unique challenges of managing data pipelines and analytics workflows in dynamic business environments. This framework emphasizes **collaboration, automation, and continuous improvement** to accelerate the delivery of high-quality data products while maintaining reliability and governance. Essentially, the DataOps framework, as understood within the industry, provides a collection of practices, procedures, and technological tools that empower organizations to enhance the pace, precision, and dependability of their data management and analytical operations. This methodology proves particularly relevant for startups experiencing rapid growth, where manual data processes quickly become bottlenecks that impede business agility.[17]

The core philosophy of DataOps treats data as a valuable asset requiring efficient management processes throughout its life cycle. In essence, DataOps stresses the need for cooperation among various teams, including data engineers, data scientists, and business analysts, to guarantee timely and appropriate data access for all involved. For startups with limited specialized personnel, this collaborative emphasis helps bridge organizational silos that naturally form as companies grow, ensuring that technical data management remains aligned with business needs. The methodology also establishes a culture of continuous improvement, with cross-functional teams regularly identifying and addressing inefficiencies in data pipelines through iterative refinement rather than major overhauls.

Automation represents a critical component of the DataOps framework, enabling startups to scale their data operations without proportional increases in headcount. By implementing automated processes for data ingestion, transformation, and analysis, organizations can reduce the potential for human error while freeing specialized talent to focus on higher-value activities such as developing new insights and strategic initiatives. For startups navigating the transition from ad hoc data practices to more structured approaches, DataOps provides a balanced methodology that improves data quality and accessibility without imposing overly rigid processes that might impede

---

[17] Eric Jones, "DataOps Framework: 4 Key Components and How to Implement Them," IBM Think, accessed March 4, 2025, `https://www.ibm.com/think/topics/dataops-framework`.

innovation. The framework's emphasis on measurable improvements in data delivery speed and quality aligns well with startup environments where demonstrable business impact remains the ultimate measure of successful data initiatives.

PwC's Data Strategy Archetypes

PwC's Data Strategy Archetypes offer companies a **categorical framework for understanding different approaches to data strategy implementation** based on organizational maturity and business context. This model recognizes that no single data strategy fits all situations, instead identifying distinct patterns or "archetypes" that organizations typically follow in their data evolution. According to PwC research, there are three data strategy archetypes that reflect an organization's stage in the data and analytics journey and its overall maturity level. While the PwC study addresses enterprises of various sizes and industries, startups can find valuable insights there to understand how a company's data needs and maturity evolve over time. [18,19]

The archetypes generally follow a three-step evolutionary path:

1. **Foundational and Decentralized:** It begins with uncoordinated data activities and opportunistic governance. In this initial stage, typical of many startups, teams remain scattered across functions with limited standardization or strategic alignment.

2. **Centralized:** As organizations mature, they progress toward greater centralization with standardized approaches to data management and governance, establishing more consistent practices across the enterprise.

---

[18] Moritz Mark, "Which data strategy archetype are you?," PwC Research and Insights, accessed March 4, 2025, https://www.pwc.ch/en/insights/digital/data-strategy-archetype.html.

[19] PwC Switzerland, *One data strategy to rule them all: A comparative perspective on data strategies*, 2020, https://www.pwc.de/de/digitale-transformation/one-data-strategy-to-rule-them-all-pwc.pdf.

3. **Embedded and Decentralized:** This is the most advanced stage, where a company gives data responsibilities back to its individual business departments. While everyone follows the same company-wide standards for handling data, each department has the freedom to manage its own projects to solve its unique problems. The goal is to get the best of both worlds: consistent, reliable data across the entire organization, plus the flexibility for each team to innovate and move quickly.

The framework considers five crucial elements that influence successful data strategies, which encompass aspects like a company's data and analytics structure, governance, platform technology, and operational model. For startups, understanding these factors helps focus limited resources on the aspects of data management that will deliver the greatest strategic value at their current stage.

PwC's methodology highlights that recognizing a company's existing archetype enables the generation of tailored insights, which can then be leveraged for strategic advantage. This practical orientation helps startups avoid the common pitfall of implementing unnecessarily complex or advanced data strategies before establishing fundamental capabilities. For founders navigating early data decisions, the archetypes provide valuable context for evaluating whether their current approach aligns with both immediate needs and longer-term strategic objectives. The framework acknowledges that organizations generally evolve through these archetypes sequentially, suggesting that startups should focus on mastering basic data capabilities before attempting more sophisticated approaches that might create technical debt or organizational friction without delivering proportional business value.

Core Components of a Startup Data Strategy

Drawing from the collective wisdom of established data frameworks, six essential components emerge as foundational elements of any effective startup data strategy. These components distill the most valuable aspects of various approaches into a comprehensive blueprint applicable to resource-constrained environments. By addressing each of these areas, startups can establish robust data capabilities aligned with their business objectives while maintaining the flexibility needed for rapid growth and adaptation (see Figure 2-1).

CORE COMPONENTS OF A STARTUP DATA STRATEGY

1  **Business Objective Alignment**
Ensure every data initiative directly supports survival and growth by answering key business questions and tracking metrics that matter.

2  **Data Governance & Quality Framework**
Establish lightweight rules for data ownership, quality, and security to ensure your data remains trustworthy and usable without creating bureaucracy.

3  **Scalable Data Architecture**
Build a flexible technical foundation—often using cloud services—that can handle increasing data volume and complexity as your startup grows.

4 **Analytics Capabilities & Insight Generation**
Develop streamlined processes to transform raw data into actionable insights, starting with simple metrics and progressing to more advanced analytics.

5 **Data-Driven Culture**
Foster a company-wide mindset where decisions are guided by evidence, data literacy is encouraged, and leadership visibly uses data to navigate challenges.

6  **Continuous Learning & Adaptation**
Create feedback loops to regularly review, test, and improve your data strategy, ensuring it evolves alongside your business.

Figure 2-1. *Core Components of a Startup Data Strategy*

Business Objective Alignment

The cornerstone of an effective data strategy is its alignment with specific business goals and challenges. Unlike enterprise organizations that might pursue data initiatives for their own sake, startups must ensure every data investment directly contributes to survival and growth objectives. This component involves identifying key business questions data should answer, defining metrics that measure progress toward goals, and prioritizing data initiatives based on potential business impact. For startups, this alignment means focusing on metrics directly tied to product-market fit, customer acquisition, retention, or operational efficiency rather than building elaborate data infrastructure without clear purpose.

Data Governance and Quality Framework

Even resource-constrained startups need foundational governance principles to ensure data remains trustworthy and usable. This component establishes clear ownership of data assets, defines quality standards, and implements appropriate security and compliance measures. For startups, lightweight governance approaches focus on establishing just enough structure to prevent chaos without creating bureaucratic overhead. This typically involves defining data ownership roles, implementing basic documentation practices, establishing data quality checks at collection points, and ensuring compliance with essential regulations relevant to the startup's industry and customer base.

Scalable Data Architecture

Startups must establish technical foundations that accommodate exponential growth without requiring complete rebuilds. This component involves designing systems that collect, store, and process data in ways that remain viable as volume, variety, and velocity increase. An effective startup data architecture balances immediate needs with future scalability, often leveraging cloud-based infrastructure to avoid heavy upfront investments. As highlighted in the frameworks, data architecture defines the structure of an organization's data assets and how they interact with each other. For early-stage companies, this might mean implementing simple but extensible data models, adopting standardized collection methods, and establishing clear integration patterns that prevent future data silos while avoiding premature technical complexity.

Analytics Capabilities and Insight Generation

The ability to transform raw data into actionable insights represents the core value proposition of any data strategy. This component focuses on establishing appropriate analytical methods, tools, and workflows that extract meaningful signals from collected data. The Lean Analytics framework emphasizes starting with "One Metric That Matters" (OMTM) before expanding to more complex analyses. For startups, this often means beginning with simple descriptive analytics focused on critical business questions before progressing to more sophisticated predictive or prescriptive approaches. The key is establishing streamlined processes that move quickly from data to insight to action, rather than creating analysis bottlenecks that delay critical business decisions during crucial growth phases.

Data-Driven Culture

Technical elements alone cannot create value without corresponding cultural practices that embrace data in decision-making processes. This component addresses the human aspects of data strategy, including organizational structure, skills development, and decision-making norms. A data-driven culture means cultivating a setting where data is appreciated, and decisions are made based on data. For startups, this means leadership visibly using data in their own decisions, establishing clear processes for integrating insights into product development and business operations, developing basic data literacy across all functions, and creating feedback loops that demonstrate the impact of data-informed choices on business outcomes.

Continuous Learning and Adaptation

Recognizing that both business needs and data capabilities evolve rapidly, effective startup data strategies incorporate mechanisms for systematic refinement over time. This component establishes processes for regular assessment, experimentation, and improvement in all aspects of data management and utilization. Drawing from the DataOps framework, this approach emphasizes collaboration, automation, and continuous improvement to accelerate the delivery of high-quality data products. For startups, this typically involves implementing regular review cycles for data strategy effectiveness, establishing experimental processes to test new data approaches, documenting learnings from both successes and failures, and creating clear pathways for incorporating these insights into strategy refinements as the organization matures.

By thoughtfully addressing these six core components, startups can establish solid data foundations that support current business priorities while building capabilities that scale with growth. Rather than attempting to implement enterprise-grade data systems prematurely, this modular approach enables founders to make targeted investments aligned with specific business objectives, gradually expanding their data capabilities as organizational maturity and resources increase. The integrated framework provides structure without sacrificing the agility essential to startup environments, helping founders extract maximum value from limited data resources while establishing sustainable practices that mature alongside the business.

From Assessment to Execution: A Stepwise Approach

When joining a startup as the first data hire, you face both a challenge and an opportunity: establishing the foundation that will shape how the company leverages data for years to come. Without a structured approach, it's easy to become overwhelmed by competing priorities or trapped in the "ad hoc trap" where your data initiatives lack strategic direction. The following stepwise methodology provides a practical road map for developing and implementing a data strategy that balances immediate business needs with long-term data capabilities.

To bring these concepts to life, let's imagine you've just joined FitScale, an innovative subscription-based fitness platform for fitness trainers and social media influencers. FitScale currently has around 20,000 active paid subscribers (90% monthly, 10% annual) and is preparing for Series A funding discussions. However, their existing data landscape is fragmented, inconsistent, and unreliable. Your task is clear: develop a high-level plan to create and implement FitScale's first coherent data strategy.

Step 1: Current State Analysis

Your journey begins with a comprehensive assessment of the existing data landscape. This crucial first step provides the contextual understanding needed to build an effective strategy aligned with business realities rather than theoretical ideals.

Start by conducting a thorough audit of existing data assets, identifying what data the company currently collects, where it resides, how it flows between systems, and—most importantly—who uses it and for what purposes. It's essential to understand where the founders are getting the data from for their PowerPoint decks and what operational tools are already in place. This pragmatic approach grounds your assessment in actual business practices rather than assumptions. At FitScale, you identify the following data assets:

- The SQL database storing app event data from native iOS and Android apps

- The NoSQL database aggregating payments data from Apple App Store

- Google Play Store and Stripe

- Salesforce CRM used by the Sales team for tracking Fitness Creator leads

- Google Analytics tracking individual trainer landing pages

- Product analytics software implemented early on (with unclear event definitions)

- Various Google Sheets and Looker Data Studio reports

Next, identify critical data gaps by mapping existing data against key business processes and decisions. Look specifically for missing data sources like key user behavior not being tracked at all—these gaps often represent significant blind spots in the company's understanding of its customers or operations. For startups in particular, missing product instrumentation data can severely limit understanding of product-market fit, so prioritize assessing whether appropriate usage tracking is in place. At FitScale, your audit quickly reveals critical gaps, such as incomplete workout tracking, which represents a significant blind spot in understanding how trainers use the platform.

Simultaneously, begin developing a data dictionary in collaboration with key stakeholders across the organization. According to Atlan's guidance, an effective data dictionary typically includes field name, data type, size/length, description, and relationships with other data assets. This collaborative process serves a dual purpose: documenting existing data while building relationships with the teams that will ultimately be your partners in data strategy implementation. As you work through this documentation, ask stakeholders questions like "What does each variable/element/ field/attribute within a dataset mean? What is it describing? Did the meaning change over time?" and "Who collected your data? Are they still the owners, or is it somebody else?" These conversations often reveal inconsistencies in how different teams define and interpret the same data elements—a critical issue to address in your strategy.[20]

At FitScale, you work closely with stakeholders from engineering, sales, marketing, and product teams to create a shared data dictionary. Clearly define key terms like "subscriber," "active user," "workout started," etc., ensuring consistency across teams. This collaborative exercise not only documents existing definitions but also builds relationships essential for future implementation phases.

Step 2: Objective Setting with Stakeholder Alignment

With a clear understanding of the current data landscape, the next step involves aligning your data strategy with business objectives through structured stakeholder engagement. This phase is crucial for ensuring your data initiatives directly support the company's strategic priorities.

You schedule one-on-one meetings with the CEO, the Head of Sales, and the Product Manager, probing for their most pressing business questions: "What decisions would be easier if you had the right data?" These conversations reveal a company-wide focus on preparing for Series A funding, which makes subscriber retention a top priority. You then organize a workshop to translate these high-level goals into formal, data-related Objectives and Key Results (OKRs), a framework we will explore in greater detail in Chapter 6. You guide the team to establish a clear, primary objective: **Improve subscriber retention**. To make this goal actionable, you define a measurable key result: **Increase the share of annual subscriptions from 10% to 20% within six months**. This process creates shared ownership and ensures your data initiatives are directly tied to what matters most to the business.[21]

Step 3: Governance and Architecture Blueprinting

With objectives established, the next step focuses on designing the structural foundation for your data strategy: governance policies and technical architecture that enable reliable, secure, and accessible data assets. For startups, lightweight governance is

---

[20] "Data Dictionary: Examples, Templates, and How to Create One in 2025," Atlan, accessed March 7, 2025, `https://atlan.com/what-is-a-data-dictionary/`.

[21] John Doerr, Measure What Matters: How Google, Bono, and the Gates Foundation Rock the World with OKRs (Portfolio, 2018).

essential—create clear policies that ensure data quality and compliance without imposing bureaucratic processes that impede agility. Do it without overwhelming the organization with complex procedures.

On the technical side, design a data architecture blueprint that balances immediate needs with scalability for future growth. At FitScale, your blueprint outlines a clear path: data from the app's SQL database, the NoSQL payment system, and Salesforce will be consolidated into a central cloud-based data warehouse. This will create a single source of truth and is the first step toward enabling future self-service analytics for the product and marketing teams, empowering them to answer their own questions.

Step 4: Iterative Implementation and Feedback Loops

The final step involves putting your strategy into action through a phased approach that delivers incremental value while building toward your longer-term vision. This iterative implementation is particularly important for startups where demonstrating concrete business impact quickly is essential for continued investment in data initiatives.

Begin by identifying and executing high-impact, low-complexity projects that can demonstrate the value of your data strategy. These "quick wins" build organizational confidence in data-driven approaches while providing learning opportunities for your implementation team. At FitScale, building a reliable subscription dashboard that unifies data from the fragmented payment sources could be such a project. This directly addresses a major pain point for the leadership team and provides a clear, trustworthy view of a critical metric.

As you select initial projects, leverage the Data Flywheel concept, having in mind that the more (and better) data your system collects, the more value it can provide. Projects that contribute to this virtuous cycle—where insights lead to product improvements that generate more valuable data—offer particularly strong returns on investment. Establish clear feedback mechanisms that capture learnings from each implementation phase. The DataOps Methodology promotes collaboration between different teams to ensure that everyone has access to the right data at the right time. Regular cross-functional reviews of ongoing data initiatives provide opportunities to refine your approach based on practical experience rather than theoretical plans. As your implementation progresses, continuously reassess your strategy against evolving business needs.

By following this stepwise approach—from comprehensive assessment through iterative implementation—you create a data strategy that delivers immediate business value while establishing the foundation for sustainable competitive advantage through data.

As the first data hire, your systematic approach to strategy development not only addresses current data needs but also positions the startup for data-driven success as it scales.

From Foundation to Future: Crafting a Long-Term Strategic Road Map

The four-step process provides a clear path for establishing a foundational data strategy. However, the journey does not end there. The long-term success of your data initiatives will depend on your ability to build a strategic road map and prepare for critical business milestones, such as fundraising. While the specifics will differ for every startup, the following road map for FitScale illustrates how you can transition from initial wins to building a mature, data-driven culture.

As you achieve early successes through iterative implementation, you can begin planning explicitly toward a longer-term vision of becoming a fully data-driven organization. This involves developing a clear road map that outlines the evolution of your data capabilities over time:

- **Short Term (0–2 Months):** Focus on auditing existing assets, standardizing key definitions, and delivering high-impact "quick wins." This phase is about building trust and demonstrating the immediate value of a structured data approach.

- **Medium Term (2–6 Months):** Begin implementing a centralized data warehouse and establishing basic ELT (Extract, Load, Transform) pipelines. The goal is to create a single source of truth and roll out initial dashboards for key business functions.

- **Long Term (6+ Months):** With a solid foundation in place, you can shift your focus to more advanced capabilities. This includes developing recommendation systems, enabling personalized user experiences, and scaling up A/B testing to drive product innovation.

Throughout this process, it is crucial to foster a strong data culture. This can be achieved through regular training sessions tailored to each department's needs, providing accessible documentation to empower self-service analytics, and actively encouraging a culture of experimentation.

As your startup approaches key funding milestones like a Series A round, your ability to articulate your data strategy and demonstrate progress becomes critical. Investors will want to see that you have a firm grasp of your key metrics and a clear plan for leveraging data to drive growth.

Be prepared to provide concise, data-backed answers to common investor questions:

- What are your key metrics, and how precisely are they calculated?

- How do you ensure accurate and reliable reporting across multiple data sources?

- Can you provide segmented insights about your most valuable customer cohorts?

- What is your long-term plan for leveraging advanced analytics and personalization to build a competitive advantage?

By following a structured approach—from initial assessment through iterative implementation—you can create a robust, high-level plan that positions your startup not only for a successful fundraising process but also for a sustainable, data-driven future. As the first data professional, you have the unique opportunity to shape how your organization leverages its most valuable asset: its own data.

Data Strategy with No Data

But how do you build your data strategy if you are just getting started with your startup and there is no data to work with? Data-driven decision-making begins even before you have a Minimum Viable Product (MVP). A well-defined data strategy in the pre-MVP stage lays the groundwork for understanding your customer, validating your assumptions, and, ultimately, achieving product-market fit. Many startups fail because they neglect this crucial step, collecting data haphazardly or postponing it until later. This often leads to misaligned data, hindering their ability to understand customer behavior and make sense of their information.[22]

---

[22] "Data Strategy for early-stage startups (that investors will love)," Paralect YouTube Channel, April 14, 2022, https://www.youtube.com/watch?v=8RXFblDO6mI.

A pre-MVP data strategy isn't about having vast amounts of data—it's about planning how you will collect and use data to inform your decisions. It's about establishing the right mindset and framework from the outset. Here's how to approach it:

1. **Define Your Customer Journey:** Even without a product, you should have a clear idea of how a customer will interact with your proposed solution. Map out the steps a potential customer will take, from initial awareness to becoming a loyal user. This customer journey map will be the backbone of your data strategy.

2. **Align Your Vision and Mission:** Clarify who you are building the product for, what problem it solves, and why your solution is unique. Articulate your vision and mission. This foundational work is crucial because it directly influences the type of data you'll need to collect. Without a clear understanding of your purpose, you won't know what data to track.

3. **Identify Key Metrics:** Based on your customer journey and business goals, determine the key metrics you'll need to track. What information will tell you if you're on the right track? These metrics will vary depending on your business, but they might include things like user engagement, conversion rates, or customer acquisition cost. Even in the design phase, start thinking about Time to Value. This metric measures how quickly a user experiences the core benefit of your product. Map out the steps a user takes to achieve that "aha!" moment and look for ways to reduce friction and accelerate the time to value. This can inform your design decisions and help you prioritize features that deliver value quickly. While you might not have real user data at this stage, you can still create a plan for measuring and optimizing TTV once your product is live.[23]

---

[23] Carlos Gonzalez de Villaumbrosia, "Time to Value: The Metric You Can't Afford to Ignore," Product School, October 16, 2025, https://productschool.com/blog/product-strategy/time-to-value.

4. **Plan Your Data Collection:** How will you gather the data you
    need? Consider what tools and technologies you'll use. Even in
    the pre-MVP stage, you can start thinking about how you'll track
    user interactions, gather feedback, and measure the effectiveness
    of your efforts.

5. **Validate Your Assumptions:** Your initial ideas about your target
    market and their needs are just hypotheses. Use the data you
    collect to validate these assumptions. Are you reaching the
    right people? Does your solution address their pain points? Be
    prepared to iterate and adjust your strategy based on what the
    data tells you.

By focusing on these key elements in the pre-MVP stage, you'll build a solid
foundation for data-driven decision-making, increasing your chances of success and
avoiding the pitfalls that many early-stage startups face. A well-defined data strategy is
not just a nice-to-have—it's a critical component of building a successful startup.

Establishing a Data-Driven Culture

Establishing a data-driven culture is not merely an operational upgrade—it's a
fundamental transformation in how decisions are made, risks are assessed, and
opportunities are seized. A data-driven culture transcends the mere presence of
analytics tools or dashboards; it requires a systemic shift in mindset, processes, and
organizational behavior. For startups, where agility and precision are survival traits,
embedding this culture early can mean the difference between scaling sustainably and
floundering in ambiguity.

The foundation of a data-driven culture lies in leadership commitment. Executives
and founders must model data-centric decision-making visibly and consistently, as
demonstrated by Piyush Gupta, CEO of DBS Bank, who famously rewarded employees
for calculated risks even when experiments failed, reinforcing the value of data-informed
experimentation over perfectionism. Startups lack the bureaucratic inertia of larger
organizations, making them uniquely positioned to instill this top-down cultural shift.
Leaders must articulate why data matters—whether to optimize customer acquisition
costs, personalize user experiences, or identify market gaps—and align every team's

objectives with measurable outcomes. This requires transparent communication about how data ties to the startup's mission, as vague directives like "become data driven" yield little traction without context.[24,25]

Equally critical is democratizing data access while maintaining rigor. Startups often fall into two traps: hoarding data within technical teams or flooding non-technical employees with unstructured information. The solution lies in balancing empowerment with governance. For instance, JPMorgan Chase's program gamified data literacy, enabling employees across functions to engage with machine learning concepts through collaborative competitions. Startups can adopt similar low-barrier entry points, such as self-service analytics platforms with guided workflows, ensuring that marketing, operations, and product teams derive insights without relying on data engineers for every query. However, accessibility must coexist with robust data governance—clear ownership, quality checks, data security to control who can access specific data, and ethical guidelines—to prevent misinterpretation or misuse.[26,27]

Cultural resistance remains a formidable barrier. Employees accustomed to intuition-driven decisions may perceive data as a threat to their expertise or creativity. Overcoming this requires reframing data as a collaborative tool rather than a critique. Airbnb's early adoption of A/B testing exemplifies this: by treating every design change as a hypothesis to be validated, they normalized data as a partner to creativity, not its antagonist. Startups should foster psychological safety, encouraging teams to share data-backed failures as learning opportunities.[28]

[24] "Creating a Data-Driven Culture," Barc, accessed March 6, 2025, https://barc.com/data-driven-culture/.

[25] Ganes Kesari, "Building a Data-Driven Culture: Four Key Elements," MIT Sloan Management Review, accessed March 6, 2025, https://sloanreview.mit.edu/article/building-a-data-driven-culture-four-key-elements/.

[26] Dexter Chu, "Building a Data-Driven Culture," Secoda, accessed March 6, 2025, https://www.secoda.co/learn/building-a-data-driven-culture.

[27] Ganes Kesari, "Building a Data-Driven Culture: Four Key Elements," MIT Sloan Management Review, accessed March 6, 2025, https://sloanreview.mit.edu/article/building-a-data-driven-culture-four-key-elements/.

[28] Santiago Pampillo, "Creating a Data-Driven Culture in Startups," TechDisrupt, accessed March 6, 2025, https://santiagopampillo.github.io/TechDisrupt/Articles/84-Startups-154-creating-a-data-driven-culture-in-startups.html.

Data literacy is the linchpin of this cultural shift. Founders often assume that hiring a data scientist absolves the rest of the team from engaging with data, but this siloed approach stifles scalability. Startups must prioritize literacy at all levels, tailoring training to role-specific needs. Sales teams might focus on CRM analytics, while product teams delve into user behavior metrics. Resources like Qlik's "Data Literacy Project" or Coursera's specialized courses can supplement internal training, but the key is contextual relevance—showing employees how data directly impacts their daily workflows and strategic goals.[29]

The integration of data into daily rituals further solidifies this culture. Stand-up meetings, sprint reviews, and board decks should routinely incorporate key metrics, fostering accountability and transparency. Startups like Canva have institutionalized this by embedding OKRs (Objectives and Key Results) into every team's workflow, ensuring alignment between individual contributions and company-wide data benchmarks.

Yet, data-driven cultures must guard against analysis paralysis. Early-stage startups, in particular, face the temptation to over-index on metrics at the expense of speed. The mantra "fail fast" applies equally to data initiatives: iterate on minimum viable analytics rather than pursuing exhaustive reports. As Antoine Latrille of Newfund observes, gut instinct often drives initial investment decisions, with data serving to validate or challenge those intuitions. Startups should adopt a "test and learn" approach using lightweight tools like Google Analytics for rapid insights while reserving advanced models for scalable use cases.[30]

Finally, storytelling with data ensures cultural adoption. Numbers alone rarely inspire action; narratives that connect metrics to human outcomes do. When Spotify analyzes user listening habits to curate personalized playlists, they're not just optimizing engagement—they're crafting stories of musical discovery. Founders should champion data storytellers within their teams, individuals who can translate churn rates into customer pain points or CAC trends into market opportunities. This bridges the gap between data professionals and decision-makers, embedding data into the organizational DNA.

---

[29] "About DLP," Data Literacy Project, accessed March 6, 2025, `https://thedataliteracyproject.org/about-dlp/`.

[30] Elena Ghinita, "By 2025, 75% of Deal Analysis Will Be Data-Driven—But Most VCs Still Trust Their Gut," The Recursive, accessed March 7, 2025, `https://therecursive.com/by-2025-75-of-deal-analysis-will-be-data-driven-but-most-vcs-still-trust-their-gut/`.

For startups, building this culture is neither a side project nor a one-time initiative. It's a continuous investment in people, processes, and tools, ensuring that every pivot, product launch, and partnership is grounded in evidence rather than ego. Those who succeed will find themselves not just surviving the startup gauntlet but redefining its rules.

Avoiding Common Pitfalls in Strategy

Despite the growing awareness of data's transformative potential, startup data strategies frequently fail to deliver their intended results. This is often due to several recurring pitfalls that founders and teams inadvertently fall into.

One common trap is the "Perfect Data" fallacy. Early-stage companies frequently delay critical decisions and actions until they've assembled comprehensive, flawless datasets—a perfectionist impulse that ignores the pragmatic wisdom of the 80/20 rule. Instead, startups should prioritize acquiring "good enough" data that directly addresses their most pressing business questions, enabling rapid iteration and informed decision-making without unnecessary delays.

Another frequent misstep is overinvestment in technology. While advanced tools and AI/ML platforms promise powerful insights and competitive advantages, they are ineffective without clear foundational alignment to business objectives. A useful rule of thumb is to invest in technology only when it demonstrably supports a specific, well-defined business outcome. Technology should always follow strategy—not precede it.

Startups also commonly suffer from siloed ownership of their data strategy. When responsibility for data initiatives is confined solely to engineering teams, misalignment inevitably arises. Successful data strategies require collaborative ownership across functions: product managers should articulate clear use cases, marketers define meaningful KPIs, and engineers select tools that effectively serve both. Regular cross-functional synchronization—such as monthly data review boards or joint planning sessions—can prevent fragmentation and ensure cohesive progress.

Neglecting data quality is another widespread pitfall. Startups often deprioritize validation and testing in their rush to ship products quickly. Yet, implementing rigorous testing and validation processes during development is crucial; developers must feel accountable for the quality of data produced by their software. Without reliable input from source systems, even the most talented data teams will struggle to generate meaningful insights.

Similarly problematic is the "metrics-first" myopia—when startups begin analysis by aimlessly exploring metrics rather than systematically testing hypotheses. Data exploration without clear hypotheses can lead to confusion, wasted resources, and missed opportunities. Instead, startups should start from clearly defined hypotheses tied to strategic goals using metrics as tools for validation rather than as ends in themselves.[31]

Compliance blind spots also frequently undermine startup data strategies. Many startups encounter significant regulatory requirements within two years of founding; privacy simply cannot be treated as an afterthought. Privacy considerations must be embedded into a startup's architecture from day one—through practices such as pseudonymization at data ingestion points, role-based access controls, and automated consent management systems—to proactively mitigate risk and build user trust.

Another common pitfall is the capability-goal mismatch: startups often overestimate their readiness for complex data initiatives, launching ambitious projects prematurely without sufficient maturity or resources. This mismatch between aspiration and capability leads to frustration, wasted effort, and disillusionment with data-driven approaches. Instead, startups should honestly assess their current capabilities and gradually scale complexity as their maturity grows.

Finally—and perhaps most critically—startups frequently underestimate the importance of cultural adoption ("cultural osmosis failure"). Even the best technology stack and highest-quality datasets will fail without genuine buy-in from the people who use them daily. Effective change management is essential: teams must clearly understand the rationale behind the data strategy—how it benefits them personally and professionally—and leadership must actively support adoption efforts by addressing concerns openly and transparently. Data strategy requires ongoing optimization; engaging people throughout this iterative process is vital for sustained success.[32]

By anticipating these pitfalls through proactive design—prioritizing governance before scaling up, quality before quantity, alignment before investment—startups can transform their data strategy from a costly distraction into a powerful engine of growth.

---

[31] "The Four Cringe-Worthy Mistakes Too Many Startups Make with Data," First Round Review, accessed October 16, 2025, https://review.firstround.com/the-four-cringe-worthy-mistakes-too-many-startups-make-with-data/.

[32] Randy Bean, "Why Culture Is the Greatest Barrier to Data Success," MIT Sloan Management Review, September 30, 2020, https://sloanreview.mit.edu/article/why-culture-is-the-greatest-barrier-to-data-success/.

The key lies not in completely avoiding mistakes (an impossible goal) but rather in building resilient systems that make failures visible early on, contained in scope, and instructive for future improvement.

Summary

This chapter presents a comprehensive framework for building and implementing data strategies in startup environments. Beginning with the fundamentals of why data strategy matters for early-stage companies, it explores how intentional data planning can prevent technical debt while driving business outcomes. The chapter examines six established data strategy frameworks—Data Value Chain, Data Flywheel, Lean Analytics Cycle, Data Maturity Models, DataOps Methodology, and PwC's Data Strategy Archetypes—extracting practical principles applicable to resource-constrained startups.

From these frameworks emerge six core components essential to any effective startup data strategy: business objective alignment, governance and quality framework, scalable data architecture, analytics capabilities, data-driven culture, and continuous learning mechanisms. The chapter provides a stepwise approach to implementation, from conducting a current state analysis to iterative execution, while addressing the unique challenges of pre-MVP data planning.

Special attention is given to establishing data-driven culture and avoiding common pitfalls that lead to strategy failure, particularly the "Perfect Data" fallacy, overinvestment in technology, siloed ownership, neglecting data quality, metrics-first myopia, compliance blind spots, capability-goal mismatch, and cultural adoption failure. The chapter concludes with a practical exercise demonstrating how to apply these principles in a real-world startup scenario using the fictional fitness platform FitScale as a case study. By following this structured approach, startups can establish solid data foundations that support current business priorities while building capabilities that scale with growth. Next, we'll dive into foundational concepts around building your startup's data infrastructure.

CHAPTER 3

Data Infrastructure Basics

As your startup gains momentum and begins attracting industry attention, a curious phenomenon unfolds: your inbox becomes a battleground for consultants and tech vendors claiming to have the solution perfect for your data needs. They'll dazzle you with sleek dashboards, cutting-edge AI tools, and promises of scalability—all at a premium price. But here's the hard truth: no external party understands your business like you do. I've witnessed startups hemorrhage cash on over-engineered systems that ignored their unique constraints, team skills, or industry realities. Worse, I've been on the other side—working for a tech giant, selling analytics software with the urgency of a sidewalk salesman. A skilled seller can convince you that your startup needs tools it absolutely doesn't, leaving you with costly, inflexible baggage.

Your advantage? You know your business environment, budget, and team capabilities better than any vendor. Lean into that. Before committing to a solution, talk to peers in your industry. Ask: What stack do they use? What surprised them post-implementation? How do hidden costs (maintenance, scaling) compare to initial quotes? Seek insights from founders who've navigated this maze, not just vendors with glossy brochures. Dive into resources authored by hands-on experts—think VC advisors or engineers who've scaled startups, not articles sponsored by Big Tech.

This chapter equips you to make informed, pragmatic decisions about your startup's data infrastructure. We'll cover

- **Core Components:** Understanding the five foundational elements of a data stack—Data Sources, Ingestion, Storage, Transformation, and Output

- **Cloud vs. On-Premise:** When to leverage cloud solutions vs. on-premise or hybrid approaches

- **Step-by-Step Guidance:** How to centralize diverse data sources into a cohesive "single source of truth" using modern tools

© Piotr Sidoruk 2026
P. Sidoruk, *From Data to Dollars*, https://doi.org/10.1007/979-8-8688-1898-1_3

   – **Practical Considerations:** Quality control, security, compliance, and cost-efficiency tailored specifically for startups

   – **Real-World Scenario:** Choosing the right data stack for a hypothetical startup

By the end of this chapter, you'll understand how to build a flexible data infrastructure that evolves alongside your business—without overspending or over-engineering.

Stack Selection

A startup's data stack typically revolves around five core components presented in Figure 3-1.

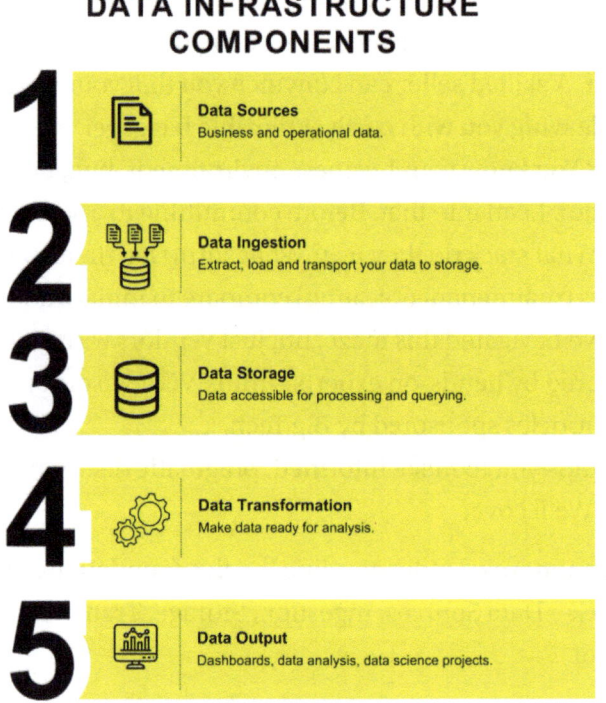

Figure 3-1. *Data Infrastructure Components*

The complexity of each component depends entirely on your business's stage and goals. A solo Instagram travel influencer selling e-books, for example, might thrive with basic tools: Instagram analytics, Google Analytics for website traffic, and sales dashboards from their payment processor. They could even claim they're "doing data science" by using Mailchimp's built-in A/B testing—no custom infrastructure required. If their business stays small, investing in a dedicated data engineer or warehouse would be overkill.

But scale changes everything. Suppose that influencer launches a global travel app, hires a team, and needs to merge data from dozens of sources—user behavior, payment systems, third-party APIs. Suddenly, consistency, transformation, and scalable storage become critical. This isn't hypothetical: Airbnb's first data hire didn't start with a complex stack. They built a culture of data-driven decisions using minimal tools, which later laid the groundwork for sophisticated infrastructure as the company grew.[1]

The lesson? Avoid overbuilding early. If your startup doesn't yet handle massive datasets, frequent analyses, or tangled data sources, focus on lightweight solutions. A complex stack demands heavy investment—not just in tools but in engineers to maintain them and processes to ensure quality. Ask yourself:

- Do I need deeper capabilities in this area right now?

- What's the ROI of investing time/money here?

- Could simpler tools suffice until we hit inflection points?

Startups evolve, and so should their stacks. Build with intentionality, not because a vendor insists you're "behind." The right infrastructure isn't about size—it's about aligning with your trajectory. Most of this book is written from a perspective of a cloud-based startup. Let's examine why most startups choose cloud computing and what the specific cases are where on-premise solutions might be preferred.

Cloud vs. On-Premise

When establishing your data infrastructure, one of the most fundamental decisions you'll face is whether to build on cloud services or establish on-premise systems. For the vast majority of startups, cloud computing represents the clear winner—and for good reason. Cloud solutions eliminate the hefty upfront investments in hardware, software licenses, and data centers that can quickly drain your limited funding.

---

[1] Riley Newman, "At Airbnb, Data Science Belongs Everywhere," Airbnb Engineering & Data Science Blog, July 7, 2015, https://medium.com/airbnb-engineering/at-airbnb-data-science-belongs-everywhere-917250c6beba.

With pay-as-you-go models, you're only paying for the computational resources you actually use, allowing you to allocate precious capital toward product development or customer acquisition instead. The cloud's inherent scalability is equally compelling; your infrastructure can grow alongside your user base without the planning, delays, and costs associated with purchasing and configuring physical servers. Most cloud providers also handle security updates, maintenance, and backup protocols, freeing your team from infrastructure management to focus on your core product.[2]

Beyond cost efficiency and scalability, cloud computing provides startups with unprecedented agility. In the fast-paced startup environment where pivoting is common, cloud services offer the flexibility to quickly experiment with new features or business models. Developers can spin up new instances in seconds rather than waiting weeks for hardware procurement, accelerating development cycles and innovation. This accessibility also supports distributed teams—increasingly common in today's startup ecosystem—by enabling secure access to company data and applications from anywhere with an internet connection.[3]

When On-Premise Makes Sense

That said, there exist specific scenarios where startups might opt for on-premise solutions. Companies working in military contracts, advanced cybersecurity, or handling extremely sensitive regulated data (like certain healthcare or financial information) might require the complete control that on-premise infrastructure provides. The need for data sovereignty—ensuring data remains within specific geographical boundaries to meet compliance requirements—can also drive on-premise decisions for startups in highly regulated industries.[4]

[2] Karen Rogers, "The Benefits of Cloud Computing for Small Businesses," *LinkedIn*, March 19, 2024, https://www.linkedin.com/pulse/benefits-cloud-computing-small-businesses-karen-rogers-syxje/.

[3] Google Cloud, "Advantages of Cloud Computing," Google Cloud, accessed March 6, 2025, https://cloud.google.com/learn/advantages-of-cloud-computing.

[4] Justin Burns, "Cloud vs. On-Premises: Which is the Best Fit for Startups," LinkedIn, accessed March 8, 2025, https://www.linkedin.com/pulse/cloud-vs-on-premises-which-best-fit-startups-justin-burns-7cfjf/.

Some startups have also found that once they reach a certain scale with predictable workloads, owning their infrastructure can be more cost-effective long-term than paying premium cloud rates—though this inflection point typically comes much later than most founders anticipate. However, such reverse migrations from cloud to on-premise remain relatively rare and typically apply to larger companies with stable, predictable workloads rather than early-stage ventures.

Hybrid Approaches

For startups with specific security requirements but still wanting cloud benefits, hybrid cloud approaches offer a middle ground. These solutions combine public cloud, private cloud, and on-premises infrastructure to create a unified IT environment. This approach allows startups to keep sensitive data on-premises while leveraging the cloud for scalable application development and less sensitive operations. A fintech startup, for instance, might store customer financial data on-premises for regulatory compliance while using cloud services for application development and analytics.[5]

Making the Right Choice

For most early-stage companies, the flexibility, reduced overhead, and ability to quickly iterate that cloud computing offers align perfectly with the dynamic nature of startup growth. Unless your specific industry or security requirements dictate otherwise, cloud infrastructure provides the most practical foundation for building your data stack. Remember that your choice isn't permanent—as your startup evolves, so too can your infrastructure strategy.

To illustrate, let's revisit our hypothetical global travel app. It might begin with a simple, low-cost cloud database to manage initial user sign-ups and bookings. As the app scales and the business expands, however, the challenge evolves. A need emerges to unify data from a growing number of data sources to create a single source of truth. This holistic "big picture" view becomes essential for running the comprehensive, multidimensional analysis required to make strategic decisions. This could mean adopting a scalable cloud data warehouse and more robust data ingestion tools. The initial choice of a cloud environment is what makes this seamless evolution possible.

---

[5] Ida Ożarowska, "What to choose: on-premise, private, public, hybrid, or multi-cloud?" FOTC Blog, October 27, 2022, https://fotc.com/blog/infrastructure-types/.

The key is selecting an approach that supports your current needs while providing the agility to adapt as your business grows and changes. Let's dive in and explore the five main components of the typical data infrastructure: Data Sources, Ingestion, Storage, Transformation, and Output.

Data Sources

At the foundation of any data stack are the raw inputs—your data sources. Let's revisit our hypothetical global travel app to see how varied these can become:

- Product analytics platforms (like Mixpanel or Amplitude) tracking user behavior across iOS and Android apps

- Google Analytics capturing website traffic patterns

- Payment systems (Stripe, Apple App Store, Google Play Store) logging subscription renewals, one-time purchases for e-books, or travel product sales

- CRM tools (e.g., Salesforce) storing data on partnerships with travel influencers

- Marketing automation platforms (Braze, Mailchimp) housing responses to emails, push notifications, or SMS campaigns

- External APIs (weather data APIs predicting ideal travel conditions for specific regions and seasons)

- Social media platforms (Instagram, TikTok, Facebook, X) offering insights into influencer performance, follower engagement, or identifying rising creators who could join the platform

- Internal spreadsheets (Google Sheets, Excel) where business teams manually track metadata about destinations, influencer collaborations, or ad hoc analyses

- Advertising platforms (Google Ads, Meta Ads) detailing campaign performance and ROI

As your startup scales, these sources multiply. Data pours in from every corner—user interactions, payments, partnerships, marketing, external tools, and even manual inputs. Without a strategy to centralize and harmonize this data, it becomes noise. You'll drown in spreadsheets, inconsistent formats, and siloed insights, making it impossible to answer basic questions like "Which marketing channel drives the highest lifetime value?" or "How do weather trends impact sales of seasonal travel packages?"

Worse still, when you ask different team members about the same metrics, you'll get different answers. Each person might be basing their response on a different data source, a different version of a spreadsheet, or even different definitions of what constitutes a user interaction. For instance, one team member might define a "user click" as when a button is pressed, while another counts it only if the action leads to a specific outcome. This lack of standardization leads to confusion and mistrust in the data itself. Without a unified view, decision-making becomes a guessing game, where everyone's insights are valid only within their own silo.

In general, startups should prioritize tools that offer pre-built connectors to common data sources, reducing the need for custom engineering work. Solutions like Fivetran, Stitch, or Airbyte provide automated data movement with minimal configuration, allowing even non-technical team members to set up and monitor data pipelines. This approach addresses a key startup challenge: limited engineering resources that shouldn't be diverted to building custom data integrations.

Table 3-1 describes the most common types of data sources for startups, along with examples and tools of each category. Understanding these sources is crucial for building a comprehensive data stack that supports informed decision-making.

Table 3-1. *Common Data Source Categories for Startups*

Category	Examples	Tools/Platforms
User Behavior	App clicks, page views, feature usage, session duration, funnel analysis, A/B test results	Mixpanel, Amplitude, CleverTap, Google Analytics, Heap, PostHog
Payments & Transactions	Subscription renewals, one-time purchases, refunds, chargebacks, transaction logs	Stripe, PayPal, App Store/Play Store, Braintree, Recurly
CRM & Sales	Lead tracking, partnership details, influencer collaborations, sales pipeline, customer interactions	Salesforce, HubSpot, Airtable, Pipedrive, Zoho CRM
Marketing	Email open rates, campaign ROI, ad performance, social media ad data, SEO data	Mailchimp, Braze, Meta Ads, Google Ads, SEMrush, Ahrefs
External APIs	Weather data, third-party integrations, financial data, geolocation data	OpenWeatherMap, Google Maps API
Internal Documentation & Logs	Business metadata, error logs, application logs, spreadsheets created manually	Google Sheets, Excel, Notion, Confluence, Datadog, Sentry, ELK Stack
Social Media	Follower growth, engagement rates, influencer performance metrics, sentiment analysis	Instagram, TikTok, X, Sprout Social, Brandwatch, Hootsuite
Operational Databases	User profiles, product inventory, order details, application data, settings, configurations, system event logs	PostgreSQL, MySQL, MongoDB, Redis, Cassandra, DynamoDB
Customer Support	Ticket volume, resolution time, customer feedback, chat logs	Zendesk, Intercom, Freshdesk, Help Scout
Product Usage Telemetry	Device information, performance metrics, crash reports, resource consumption	Firebase Crashlytics, New Relic, Datadog APM, Prometheus

(continued)

Table 3-1. (*continued*)

Category	Examples	Tools/Platforms
Financial Systems	Accounting data, payroll, expense reports	QuickBooks, Xero, NetSuite, SAP, Oracle Financials
Supply Chain/ Logistics (if applicable)	Inventory levels, shipping data, delivery tracking	ShipStation, Shippo, in-house logistics systems, ERP modules
Security & Compliance	Audit logs, security alerts, access controls, compliance reports	AWS CloudTrail, Google Cloud Audit Logs, Splunk, GuardDuty

The takeaway? Startups rarely lack data—they lack cohesion. Your job isn't to collect everything but to identify which sources align with your key metrics today while designing a system flexible enough to absorb new inputs as you grow.

For example, our hypothetical travel app might initially focus on combining data from its core **Operational Database** (like PostgreSQL, which stores all user profiles and booking information) with a few key third-party sources shown in Table 3-1: **User Behavior** data from Mixpanel and Google Analytics (we will elaborate more on these kinds of tools in Chapter 10), **Marketing** data from Mailchimp, and **Payments & Transactions** from Stripe. This targeted approach allows the team to answer critical early-stage questions like "Which marketing campaigns are driving the most high-value bookings?" As the startup matures and launches a partner program, it can then expand its data stack to include numerous additional data sources, such as **CRM & Sales** data from Salesforce.

This approach aligns with startup methodology by embracing iterative development. Rather than attempting to build a complex data infrastructure from day one, early-stage startups can start with basic capabilities addressing their most pressing analytical needs. Then expand based on feedback and evolving business requirements. At the very beginning, most startups don't really need investments in huge data systems and sophisticated tools. This preserves precious resources for core business activities. Now, let's dive into how to funnel this chaos into something actionable by exploring data ingestion.

Data Ingestion

Imagine you have a multitude of data sources—each with its own format and frequency of updates. The first step is to assess which sources are crucial for your analytics needs. Then, you need to figure out how to ingest all that data and transfer it to one place—your single source of truth. There are several ways to achieve this. For some data sources, especially those within the same ecosystem (like Google Analytics and BigQuery), native integrations make the process seamless. However, not all platforms offer such convenience.

When native integrations aren't available, data integration platforms like Fivetran, Airbyte, Stitch, or Hevo become invaluable. These tools can handle a wide range of common data sources, reducing the need for custom engineering work. Their advantages include minimal configuration efforts, allowing even non-technical team members to set up and monitor data pipelines. This approach addresses a key startup challenge: limited engineering resources that shouldn't be diverted to building custom data integrations. However, there are drawbacks. These solutions can become costly as your data volume grows, and the quality of the data transferred might not be perfect. While they might be good enough for non-critical datasets like user engagement metrics, using them for financial data where accuracy is paramount can be risky.

If you need 100% control over the logic of data ingestion and transfer, building your own custom data pipeline might be necessary. This approach requires more time upfront and ongoing maintenance, which demands dedicated engineering resources. For small startups with limited resources, it's crucial to balance ease of use, time required, cost, and engineering allocation. You need to be strategic about when to use off-the-shelf solutions and when to invest in custom infrastructure. This decision will significantly impact your ability to scale efficiently and make data-driven decisions.

Let's explore various tools and strategies available to startups and compare their features, costs, and suitability for different stages of growth. Before diving into specific tools, it's essential to understand the fundamental difference between two primary approaches to data processing: ETL and ELT.

Understanding ETL vs. ELT

The difference between Extract, Transform, Load (ETL) and Extract, Load, Transform (ELT) is more than just the order of operations. This distinction significantly impacts how startups should approach their data architecture. The simple illustration in Figure 3-2 presents the flow of both ELT and ETL processes.

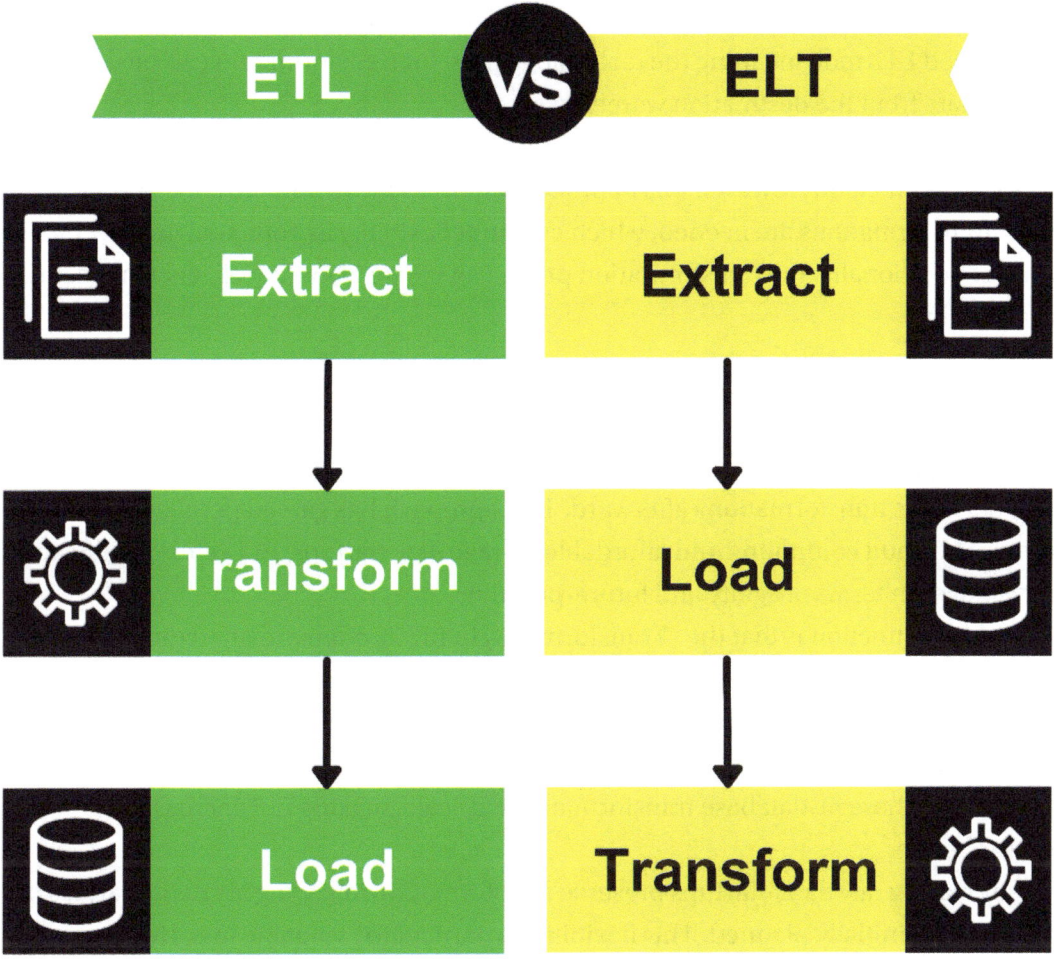

Figure 3-2. *ETL vs. ELT*

ETL: The Traditional Approach

In ETL, data is transformed before loading into your target system. This approach has been the industry standard for decades and offers certain advantages for startups. Data is cleaned, structured, and transformed before it reaches your data warehouse, ensuring that only high-quality, relevant information is stored. This reduces storage requirements and can make analytics queries faster since data is already formatted appropriately.

In this model, the "Transform" (T) step is handled by a separate processing engine or a specialized ETL tool, meaning the technology used for transformation can be thought of as separate from the destination warehouse.[6]

For startups with well-defined data requirements and limited storage budgets, ETL can be efficient. However, this approach requires knowing in advance exactly what transformations are needed, which can limit flexibility as your analytics needs evolve. Additionally, the transformation phase can create a bottleneck, delaying data availability.

ELT: The Modern Alternative

The ELT approach, by contrast, loads raw data directly into your data warehouse or lake and performs transformations afterward. This approach has gained popularity with the rise of cloud computing and affordable storage. For startups, ELT offers significant advantages in terms of agility and future-proofing.

A key distinction is that the "Transform" (T) in ELT is often executed directly within the data warehouse rather than by an intermediary processing engine. This means the transformation technology is intrinsically tied to the "Load" (L) destination. Modern cloud data warehouses like Snowflake, BigQuery, and Redshift are optimized for performing these in-database transformations at scale, making ELT increasingly cost-effective.[7]

By storing raw data, startups preserve all information, allowing for new analyses that weren't initially planned. This flexibility is particularly valuable in early-stage companies where data requirements often change rapidly. For most startups, especially those with cloud-based infrastructure, ELT is becoming the preferred approach due to its flexibility, scalability, and alignment with modern data tools. However, the right choice ultimately depends on your specific requirements, especially regarding compliance, data sensitivity, and analysis needs.

---

[6] Joe Reis and Matt Housley, *Fundamentals of Data Engineering: Plan and Build Robust Data Systems* (Sebastopol, CA: O'Reilly Media, 2022).

[7] DataCamp, "ETL and ELT in Python," DataCamp, accessed November 5, 2025, `https://www.datacamp.com/courses/etl-and-elt-in-python`.

Key Data Ingestion Tools for Startups

Let's examine four critical categories of data ingestion tools and compare the leading options in each category, focusing on factors most relevant to startups: ease of implementation, cost, maintenance requirements, and scalability.

Data Replication Tools

Data replication tools are the workhorses of modern data stacks, moving data from source systems to your data warehouse with minimal configuration. Let's take a look at three popular tools: Airbyte, Fivetran, and Stitch (see Table 3-2). As mentioned in this book's introduction, the tool comparisons presented in this chapter are a subjective snapshot designed to provide a high-level overview from a startup perspective. For a full disclaimer on the methodology, please see the introduction.

Table 3-2. *Comparison of Popular Data Replication Solutions*

Feature	Airbyte	Fivetran	Stitch
Approach	Open source with cloud option	Fully managed service	Managed service with simple UI
Maintenance Burden	Moderate (self-hosted) to Low (cloud)	Very low	Low
Community Support	Growing community, active development	Enterprise support	Established support
Customization	High	Limited	Moderate
Number of Connectors[8]	600+	700+	130+
Best For	Teams needing flexibility and cost control	Teams prioritizing reliability over customization	Teams seeking balance of simplicity and capabilities

---

[8] As of November 2025.

Airbyte stands out as an open source platform that has gained significant traction for its flexibility and extensive customization options. Airbyte offers an intuitive user interface and relatively quick setup process. Its open source nature allows organizations to inspect, modify, and contribute to the codebase, providing transparency and potential cost savings for budget-conscious teams. The platform offers both free open source and paid cloud options, making it accessible for early-stage startups. However, users should be aware that managing an open source deployment requires some technical expertise and maintenance overhead. The community-driven development model means that while some connectors are extremely robust, others may require additional attention or customization.[9]

Fivetran emphasizes fully managed, automated data pipelines with minimal maintenance requirements. With comprehensive connector support and enterprise-grade security features, it provides a reliable solution for organizations prioritizing seamless operation over complete control. Fivetran's fully managed approach means that technical teams can focus on data analysis rather than pipeline maintenance. This convenience comes at a cost, as Fivetran operates on a paid model that can become significant as data volumes grow. However, for startups with limited engineering resources seeking reliability and simplicity, the investment may be justified by reduced internal overhead and faster time-to-insight.[10]

Stitch focuses on simplicity and ease of use, offering a straightforward user interface. With an affordable entry point, it appeals particularly to early-stage startups with straightforward data integration needs. The platform's strength lies in its accessibility for non-technical users and quick implementation timeline. While Stitch provides reliable service for standard use cases, some startups report that its pricing model doesn't scale as favorably as alternatives when data volumes increase significantly. Additionally, it offers somewhat more limited customization options compared to Airbyte or enterprise-focused solutions.[11]

---

[9] "The Open Standard for Data Movement," Airbyte, accessed November 5, 2025, `https://airbyte.com/`.

[10] "Fivetran," Fivetran, accessed November 5, 2025, `https://www.fivetran.com/`.

[11] "Stitch," Stitch, accessed November 5, 2025, `https://www.stitchdata.com/`.

Custom vs. Off-the-Shelf: Making the Right Choice

Ready-made solutions can experience bugs or limitations that impact critical data workflows. For startups handling mission-critical data, developing custom replication code offers better control over transformations, independence from vendor pricing changes, and tailored error handling. However, custom development requires significant engineering resources and may delay time-to-market.

When working with critical datasets where high accuracy and reliability are essential, thorough due diligence is crucial. Even if a replication tool lists a connector for your data source (such as Stripe for payment data), verify that it supports all object types your business requires. Dive deeper into the documentation to confirm that the integration covers your specific needs comprehensively, not just partially. Data sources frequently update their APIs, and replication tools must keep pace with these changes to maintain data accuracy. If particular object types aren't supported, your data integrity could be compromised.

Alternatively, if you choose to develop a custom solution, you'll need to commit resources to monitor all planned changes by data source providers and ensure your integration code is updated accordingly. This responsibility requires vigilance but provides maximum control over your data pipeline.

When making your decision, consider: the criticality of your data operations, available engineering resources, uniqueness of your requirements, and your tolerance for occasional issues vs. development costs. It's worth noting that the three tools discussed here represent just a small sample of the available options in the market. Numerous other replication tools exist with varying features, pricing models, and specializations. The landscape continues to evolve rapidly, so evaluating tools based on your specific requirements rather than brand recognition is essential. Remember that no solution is perfect, and your optimal approach will likely evolve as your organization grows and matures.

Workflow Orchestration Tools

Workflow orchestration tools help coordinate and schedule your data pipelines, ensuring that dependencies are managed correctly and failures are handled appropriately. Let's take a look at several popular tools in Table 3-3. Apache Airflow remains the most widely adopted orchestration tool, but its setup and maintenance can be challenging for small teams. We will explore this topic in much greater detail in Chapter 8. Newer tools like Dagster offer more intuitive interfaces while maintaining the flexibility of code-based definitions. For startups already committed to a specific cloud provider, the native offerings (Step Functions, Cloud Composer, Data Factory) can provide a more seamless experience with less maintenance overhead.

Table 3-3. *Workflow Orchestration Tools*

Feature	Apache Airflow	Dagster	AWS Step Functions	Google Cloud Composer	Azure Data Factory
Approach	Open source, Python-based, extensive ecosystem	Intuitive interface, well-structured environment, Airflow integration	Serverless, visual workflow builder, AWS integration	Built on Airflow, GCP integration, managed service	No-code/ low-code options, Azure integration
Pricing	Free (open source)	Open source core, cloud pricing varies	Paid	Paid	Paid
Maintenance	High	Medium	Low	Low	Low
Scalability	High	High	High	High	High
Best For	Technical teams comfortable with Python	Teams needing better visibility into data flows	AWS-centric startups	GCP-centric startups	Azure-centric startups

As you can see in Table 3-3, the choice of an orchestration tool often depends on your team's existing technical expertise and cloud environment. For example, **Apache Airflow**, an open source tool, offers high scalability but comes with high maintenance, making it best for technical teams comfortable with Python. On the other hand, solutions like **AWS Step Functions** or **Google Cloud Composer** are deeply integrated into their respective cloud ecosystems, offering lower maintenance for startups already invested in those platforms.[12,13,14,15,16]

Event Streaming Tools

Event streaming platforms handle real-time data flows, enabling applications to react to events as they occur rather than processing data in batches. For a startup, this can power such features as real-time analytics dashboards or instant fraud detection. Table 3-4 compares several popular event streaming solutions and elucidates how their core characteristics—such as being open source or fully managed—make them suitable for different startup scenarios and levels of engineering resources.

Table 3-4. *Event Streaming Solutions*

Feature	Apache Kafka	Amazon Kinesis	Google Pub/Sub	Azure Event Hubs
Startup-Friendly Features	Open source, extensive ecosystem	Fully managed, easy setup, AWS integration	Fully managed, serverless, GCP integration	Fully managed, Azure integration
Pricing	Free (open source), managed options available	Paid	Paid	Paid
Maintenance	High	Low	Low	Low
Scalability	Very high	High	Very high	High
Best For	Data-intensive applications requiring custom control	AWS-centric startups	GCP-centric startups	Azure-centric startups

---

[12] "Apache Airflow," Apache Airflow, accessed November 5, 2025, `https://airflow.apache.org/`.

[13] "Dagster," Dagster, accessed November 5, 2025, `https://dagster.io/`.

[14] "AWS Step Functions," AWS Step Functions, accessed November 5, 2025, `https://aws.amazon.com/step-functions/`.

[15] "Cloud Composer," Google Cloud: Cloud Composer, accessed November 5, 2025, `https://cloud.google.com/composer`.

[16] "Azure Data Factory," Azure Data Factory, accessed November 5, 2025, `https://azure.microsoft.com/products/data-factory`.

For most early-stage startups, managed services like Amazon Kinesis provide the best balance of capabilities and operational overhead. As Table 3-4 illustrates, the high maintenance burden of a tool like Apache Kafka makes it better suited for teams with dedicated infrastructure engineers or specific high-throughput needs that can't be met by managed alternatives.[17,18,19,20]

However, startup data architecture experts emphasize that event streaming tools adoption should be prioritized only if there are clear and well-defined real-time analytics or application requirements, as premature implementation risks introducing unnecessary complexity into data infrastructure.

Reverse ETL Tools

Reverse ETL tools move processed data from your warehouse back to operational systems, enabling data-driven actions in your business applications. Let's take a look at several popular reverse ETL solutions in Table 3-5.

Table 3-5. *Reverse ETL Solutions*

Tool	Hightouch	Census	RudderStack
Startup-Friendly Features	Broad integrations, simple configuration, quick setup	Simplicity and good documentation	Open source option, combined Customer Data Platform and reverse ETL
Pricing	Paid with a free tier	Paid with a free tier	Free (self-hosted), Paid (cloud) with a free tier
Maintenance	Low	Low	Low (cloud) to High (open source)
Scalability	High	High	High

[17] "Apache Kafka," Apache Kafka, accessed November 5, 2025, https://kafka.apache.org/.

[18] "Amazon Kinesis," Amazon Kinesis, accessed November 5, 2025, https://aws.amazon.com/pm/kinesis/.

[19] "Pub/Sub," Google Pub/Sub, accessed November 5, 2025, https://cloud.google.com/pubsub.

[20] "Event Hubs," Azure Event Hubs, accessed November 5, 2025, https://azure.microsoft.com/products/event-hubs.

Reverse ETL is a relatively new category but increasingly important as startups seek to operationalize their analytics. These tools enable teams to close the loop between analytics and operations, pushing insights back into the tools where work happens.

As you can see in Table 3-5, the landscape is led by tools like Hightouch and Census, which are particularly startup-friendly due to their low maintenance overhead and pricing models that can scale with usage. For an early-stage company, a key differentiator is the availability of a free or low-cost entry point. Both **Hightouch** and **Census** offer free tiers to get started, while **RudderStack** provides both a cloud-based Free Tier and a completely separate open source option for teams with the technical resources to self-host. This allows startups to experiment with reverse ETL and demonstrate its value before committing to a significant expense.[21,22,23]

Strategic Considerations for Startups

When designing a data ingestion strategy, startups must prioritize four key factors. Engineering resources are often the scarcest asset—tools like Airbyte, Fivetran, or Stitch with no-code/low-code interfaces empower non-technical teams to manage pipelines, preserving developer bandwidth for core product work. Cost structures require balancing immediate expenses with future scalability: open source solutions (e.g., Airbyte) minimize upfront costs but demand maintenance, while managed services (e.g., Fivetran) reduce operational effort but scale in price with data volume. Data complexity dictates tool selection—simple SaaS integrations work with any replication tool, whereas other data sources or streaming events may require specialized platforms. Finally, future-proofing is critical: adopt flexible architectures like ELT (preserving raw data) and tools with strong ecosystems (Airflow, Fivetran) to avoid technological dead ends.

---

[21] "Pricing," Hightouch, accessed November 4, 2025, `https://hightouch.com/pricing`.

[22] "RudderStack Open Source," RudderStack, accessed November 4, 2025, `https://www.rudderstack.com/docs/get-started/rudderstack-open-source/`.

[23] "Pricing," Census, accessed November 5, 2025, `https://www.getcensus.com/pricing`.

Drawing Your Conclusion: Key Principles for a Startup Data Ingestion Strategy

You have now navigated the complex world of data ingestion, from the fundamental choice between ETL and ELT to the diverse landscape of tools that can move your data. The goal of this section was not to crown a single "perfect" tool but to arm you with a strategic framework for making choices that align with your startup's unique constraints and ambitions. As you move forward, your decisions should be guided by a few core principles:

- **An ELT-First Approach Gives You Flexibility:** For most startups, the ELT (Extract, Load, Transform) model is the superior choice. By loading raw data directly into your warehouse, you preserve all data, allowing for future analyses that you can't even anticipate today. This agility is critical in a dynamic environment where business questions and data requirements are constantly evolving.

- **A "Perfect" Stack Does Not Exist:** Prioritize pragmatism. The ideal data stack is not the one with the most advanced tools; it's the one that solves your current problems while conserving your most valuable resources: time, money, and engineering talent. Start with managed, low-maintenance replication tools for common tasks and avoid the temptation to over-engineer your infrastructure early. You can always add complexity later when the need is clear and justified.

- **Choose the Right Tool for the Right Task:** A modern data ingestion layer is rarely a single tool; it's a combination of specialized components. Use simple data replication tools for standard SaaS data, a workflow orchestrator to manage dependencies, and event streaming platforms only when you have a well-defined need for real-time data. Don't try to force one tool to do everything; build your stack incrementally based on specific business cases.

- **Keep Trade-Offs in Mind When Building Your Stack:** Every ingestion tool decision involves a balance. The most common trade-off for a startup is between cost and convenience. Managed services offer low maintenance but can become expensive as your data volume grows, while open source tools may have lower upfront costs but require significant engineering resources to implement and maintain. Be conscious of these trade-offs and choose the path that best aligns with your team's current capacity and budget.

Ultimately, your data ingestion strategy should be a living part of your data infrastructure, adapting as you grow. By making intentional, pragmatic choices based on these principles, you will build a foundation that not only centralizes your data but also empowers your entire team to make smarter, faster decisions.

Data Storage

At the heart of a typical startup's data infrastructure lies the data warehouse—the single source of truth. While on-premises database systems were once the norm, today's startups overwhelmingly lean on cloud-based solutions. These aren't just trendy; they address critical startup needs: scalability, cost-efficiency, and minimal upfront engineering. Cloud platforms let small teams set up infrastructure quickly, often through free tiers and user-friendly interfaces, without deep server expertise. Consumption-based pricing models mean you pay as you grow—starting cheap and scaling costs alongside your business. For most startups, avoiding the cloud is the exception, not the rule (think highly regulated industries or niche use cases), so we'll focus on cloud-first approaches here.

When choosing your first data warehouse, prioritize immediate value over grand plans. Months spent building a "perfect" infrastructure are months wasted in the startup world, where speed and iteration are survival skills. Cloud warehouses like BigQuery, Snowflake, or Redshift let you start small, test cheaply, and adapt as you learn what works. If your startup already uses tools from a specific provider (e.g., Google Cloud for other services), sticking with their warehouse solution might streamline integration and reduce vendor sprawl.

Don't get lost in the hype of new tools. Ask practical questions:

- Do you need real-time analytics (millisecond-level speed), or is batch processing sufficient?

- Are you storing historical data for periodic analysis or querying constantly?

- How critical is cost predictability vs. scalability?

For most startups, a warehouse's job is straightforward: serve as a central repository for combining and querying data. Merging user behavior with payment data, for example, becomes as simple as joining tables with SQL—no complex scripts pulling from fragmented sources. This efficiency is why warehouses exist.

Wait—how is this different from a database?

- – Databases handle day-to-day operations: storing and managing transactional data (e.g., processing orders).

- – Data warehouses are built for analysis: aggregating, transforming, and querying data across sources to uncover insights.

Your data warehouse isn't about chasing the latest tech trends. It's about solving today's problems while staying flexible enough to evolve tomorrow.

Storage Solutions in Modern Data Infrastructure

To understand how storage fits into the broader data ecosystem, let's revisit the a16z architecture diagram (see Table 3-6) from their influential article "Emerging Architectures for Modern Data Infrastructure." This diagram isn't just a static blueprint— it's a living framework that maps the evolving landscape of data tools. Developed in collaboration with data leaders from organizations like LinkedIn, Databricks, Fivetran, and Stanford, it synthesizes patterns seen across startups, scale-ups, and enterprises.[24]

---

[24] Matt Bornstein, Jennifer Li, and Martin Casado, "Emerging Architectures for Modern Data Infrastructure," Andreessen Horowitz, accessed August 3, 2025, `https://a16z.com/emerging-architectures-for-modern-data-infrastructure/`.

Table 3-6. *Unified Data Architecture 2.0 (a16z) Summary Table*[25]

Infrastructure Component	Representative Tools & Technologies
Data Sources	**OLTP Databases**, **ERP & Operational Apps** (Salesforce, Oracle, HubSpot), **Event Collectors** (Snowplow, Segment), **Logs**, **3rd Party APIs** (Stripe), **File & Object Storage**
Data Ingestion & Transport	**Replication** (Fivetran, Stitch, Airbyte), **Event Streaming** (Confluent/Kafka, Pulsar, AWS Kinesis), **Workflow Manager** (Airflow, Dagster, Prefect), **Reverse ETL** (Census, Hightouch)

Data Storage, Query & Processing	**Data Warehouse:** Snowflake, BigQuery, Redshift
	Data Lakehouse Storage:
	  — Data Lake Table Formats: Delta, Iceberg, Hudi
	  — File Formats: Parquet, ORC, Avro
	  — Storage: S3, GCS, ABS, HDFS
	Data Lakehouse Query & Processing:
	  — Spark Platform: Databricks, EMR
	  — Query Engine: Starbust/Presto/Trino, Hive, Dremio, Databricks Photon
	  — DS/ML Platforms: Pandas, PyTorch
	Real-Time Analytics Database: Imply/Druid, ClickHouse, Pinot
	Stream Processing: Databricks, Confluent, Flink, Upsolver

Data Transformation	**Metrics Layer & Data Modeling** (LookML, dbt), **Workflow Manager** (Airflow, Prefect, Dagster)
Data Output & Analysis	**Dashboards:** Looker, Superset, Tableau **Embedded Analytics:** Sisense, Looker, cube.js **Augmented Analytics:** ThoughtSpot, Outlier, Anodot, Sisu **Data Workspace:** Mode, Hex, Deepnote **DS/ML Tooling:** Databricks, Sagemaker **App Frameworks:** Plotly Dash, Streamlit **Custom Apps**

---

[25] Matt Bornstein, Jennifer Li, and Martin Casado, "Emerging Architectures for Modern Data Infrastructure," Andreessen Horowitz, accessed August 3, 2025, https://a16z.com/emerging-architectures-for-modern-data-infrastructure/.

The diagram organizes data infrastructure into several major areas: Sources, Ingestion and Transport, Storage, Query and Processing, Transformation, and Analysis and Output. While it includes different kinds of tools and advanced systems, it's important to remember that no single startup uses all these components. For startups, the diagram serves as a compass, not a checklist. It acknowledges the rapid pace of innovation in data tech—new tools emerge monthly, and best practices shift—but underscores a timeless goal: getting data from scattered sources into a usable state for the business.

As Table 3-6 illustrates, a modern data infrastructure is composed of several interconnected layers, each with specialized tools. For instance, in the "Data Storage, Query & Processing" component, you can see a distinction between the Data Warehouse (like Snowflake), the Data Lakehouse, and Real-Time Analytics Databases (like ClickHouse). This shows how different storage solutions are optimized for different tasks, from business intelligence to machine learning.

Focusing on the Storage section, the diagram highlights three core components: data warehouses, lakehouses (with data lakes), and real-time analytics databases. Each serves distinct purposes, and understanding their differences helps startups choose the right tools for their evolving needs. Let's break them down.

Data Warehouse

The Data Warehouse is what we've been discussing so far—structured storage optimized for analytical queries, business intelligence, and reporting. It enables the data team to start building unified dashboards, reports, and queries that describe what has happened or is currently happening in your business. Think of it as your structured single source of truth where data is organized, cleaned, and ready for business analysis.[26]

---

[26] Amberle McKee, "What Is a Data Warehouse?," DataCamp Blog, July 29, 2024, https://www. datacamp.com/blog/data-warehouse.

Lakehouse and Data Lake

In contrast, a Data Lake stores raw, unstructured, or semi-structured data—all those messy logs, images, text files, and other data that doesn't fit neatly into tables. Data lakes emerged to handle the massive volumes of varied data that modern businesses generate. They're particularly valuable for data science teams building models that need access to raw data, not just the curated views that warehouses provide.[27]

The Lakehouse is a newer concept that aims to combine the best of both worlds. It provides the structure, performance, and governance of data warehouses with the flexibility and scale of data lakes. For startups scaling quickly and wanting to avoid maintaining separate systems, lakehouses can be appealing—they enable both traditional business intelligence and advanced data science from a single platform. For instance, top tech companies like Uber and Netflix leverage data lakehouse architectures to handle their petabyte-scale data challenges, enabling everything from real-time analytics to advanced machine learning on a unified data foundation.[28,29,30,31]

---

[27] Matt Bornstein, Jennifer Li, and Martin Casado, "Emerging Architectures for Modern Data Infrastructure," Andreessen Horowitz, accessed August 3, 2025, `https://a16z.com/emerging-architectures-for-modern-data-infrastructure/`.

[28] Xinli Shang, Kai Jiang, Huicheng Song, Jianchun Xu, and Mohammad Islam, "Fast Copy-On-Write within Apache Parquet for Data Lakehouse ACID Upserts," Uber Engineering Blog, June 29, 2023, `https://www.uber.com/blog/fast-copy-on-write-within-apache-parquet/`.

[29] Nishith Agarwal, "Building a Large-scale Transactional Data Lake at Uber Using Apache Hudi," Uber Engineering Blog, June 9, 2020, `https://www.uber.com/blog/apache-hudi-graduation/`.

[30] Dao Mi, Pablo Delgado, Ryan Berti, Amanuel Kahsay, Obi-Ike Nwoke, Christopher Thrailkill, and Patricio Garza, "From Facts & Metrics to Media Machine Learning: Evolving the Data Engineering Function at Netflix," Netflix Technology Blog, August 21, 2025, `https://netflixtechblog.com/from-facts-metrics-to-media-machine-learning-evolving-the-data-engineering-function-at-netflix-6dcc91058d8d`.

[31] David Myriel, "Netflix's Media Data Lake and the Rise of the Multimodal Lakehouse," August 22, 2025, `https://lancedb.com/blog/case-study-netflix/#where-the-multimodal-lakehouse-shines`.

Real-Time Analytics Database

Finally, Real-Time Analytics Databases address a specific need: analyzing data as it's generated. While traditional warehouses process data in batches (perhaps hourly or daily), real-time systems handle streaming data for immediate analysis. If your startup needs to make decisions based on what's happening right now—think fraud detection, recommendation engines, or dynamic pricing—these systems become relevant.[32]

What's fascinating about the a16z diagram is that it reflects how the lines between these different Storage and Query and Processing systems are blurring. We're seeing rapid convergence, with each approach borrowing features from the others.

For your startup, the choice isn't necessarily about picking one perfect system but understanding which capabilities matter most right now while keeping flexibility for the future. If immediate insights from structured data drive your decisions, start with a cloud data warehouse. If your data scientists need raw data for modeling, a data lake might be crucial. If you're building real-time features, streaming analytics capabilities become important.

Key Differences at a Glance

Most early-stage companies start with a cloud warehouse—it's cheap and scalable and requires minimal setup. But as they grow, hybrid needs emerge. A16z emphasizes that heterogeneity is here to stay: a startup might use Snowflake for BI, a lakehouse for ML pipelines, and a real-time database for customer-facing features. The key is to avoid over-engineering early. Begin with what solves today's problems (e.g., a warehouse for basic reporting) and layer in specialized tools only when needed. As your startup evolves, so should your storage strategy. See Table 3-7 to compare these solutions.

---

[32] Jacob Jensen, "Choosing between Lakehouse, Data Warehouse and Real-time Analytics in Microsoft Fabric," LinkedIn, accessed March 4, 2025, https://www.linkedin.com/pulse/lakehouse-data-warehouse-real-time-analytics-microsoft-jensen/.

Table 3-7. *Data Storage Solutions Comparison*

Feature	Data Warehouse	Data Lake	Lakehouse	Real-Time Analytics DB
Startup Fit	Best for defined analytics needs	Good for storing diverse raw data	Balanced solution for growing needs	Ideal for real-time data needs
Primary Use Cases	BI reporting, historical analysis	Data archiving, ML/AI experimentation	Combined BI, ML/AI in unified platform	Immediate insights, monitoring, alerts
Cost	Higher maintenance, expensive scaling	Low storage cost, separate processing costs	Moderate, cost-effective scaling	Can be expensive for high performance
Data Structure	Primarily structured, relational	Raw data in any format	Any format with unified management	Can handle varied data, but optimized for streaming data
Schema Approach	Schema-on-write, strict schema	Schema-on-read, flexible	Schema-on-write with flexible evolution	Flexible schema for query performance

The data landscape is dynamic, but foundational principles remain: centralization, accessibility, and usability. As a startup, it's crucial to avoid the temptation of building an overly complex data stack from the beginning. Instead, focus on what solves your immediate problems and allow your infrastructure to evolve as you grow. The Data Storage Solutions Comparison (Table 3-7) provides a clear road map for this journey.

Startups don't need every tool—focus on what solves immediate problems. A **Data Warehouse** is often the best starting point. As you can see in the table, its "Startup Fit" is described as *best for defined analytics needs*, with "Primary Use Cases" focused on *BI reporting and historical analysis*. This is because a warehouse is optimized for structured, relational data, making it perfect for creating a single source of truth to answer fundamental business questions like "Which marketing campaigns are most effective?" or "What is our customer lifetime value?"

As your startup scales, new challenges will emerge that require more advanced tools. For example, if you start collecting large volumes of raw, unstructured data—such as images, user logs, or social media feeds—to experiment with machine learning and AI models, a **Data Lake** becomes essential. Unlike a warehouse that requires data to be structured before it's loaded, a lake stores information in its native format. This "schema-on-read" approach gives data scientists the flexibility to explore diverse datasets and train complex models without being limited by a predefined structure. As noted in the table, its primary use case is *data archiving and ML/AI experimentation* precisely because of this flexibility. For teams looking to combine the structured analytics of a warehouse with the versatility of a lake, a **Lakehouse** offers a balanced solution for growing needs, unifying BI and ML on a single platform.

Finally, specialized tools like a **Real-Time Analytics DB** should only be adopted when a specific business requirement justifies the investment. The table highlights its fit for ideal for real-time data needs, with use cases like *immediate insights, monitoring, and alerts.* This becomes relevant only when you need to power customer-facing features like fraud detection or dynamic pricing.

The a16z framework reminds us that while tools change, the mission doesn't: turning raw data into actionable insights. By starting with a data warehouse for your core analytics and layering in more specialized tools like lakehouses or real-time databases only when a clear business need arises, you can build a powerful and cost-effective data infrastructure that grows with you.

Analytic or Operational Systems?

One of the most persistent architectural patterns in data infrastructure that a16z presents is the distinction between analytic and operational systems. Analytic systems primarily support data-driven decision-making, providing insights that inform strategic and tactical choices. Operational systems, by contrast, power data-driven products and services, enabling real-time interactions and automated processes. Despite predictions of convergence, these architectural patterns continue to exist in parallel, serving distinct but complementary purposes within the broader data ecosystem.[33]

This duality manifests in the continued growth of both cloud data warehouses (like Snowflake) focused on SQL users and business intelligence, alongside data lakehouses (like Databricks) that support a broader range of use cases. The persistence of these

---

[33] Matt Bornstein, Jennifer Li, and Martin Casado, "Emerging Architectures for Modern Data Infrastructure," Andreessen Horowitz, accessed August 3, 2025, `https://a16z.com/emerging-architectures-for-modern-data-infrastructure/`.

parallel systems reflects the diverse nature of data requirements within organizations, where different stakeholders may require different capabilities from the same underlying data. For startups, understanding this duality helps in making architectural choices that align with both immediate needs and long-term vision.

When designing their data stack, startup founders and data professionals need to consider this fundamental duality. Analytical systems typically prioritize read-heavy workloads, historical data analysis, and aggregation capabilities. Operational systems focus on write-heavy workloads, real-time data access, and transaction processing. Many startups find they need elements of both, implemented in ways that allow appropriate data flow between systems while maintaining performance and data integrity.

Data Transformation

Raw data rarely provides immediate value without some level of processing. The transformation layer prepares data for analysis by cleaning, structuring, and applying business logic. This step ensures that data is standardized and aligned with business requirements before reaching end users.

For startups, transformation tools like dbt (data build tool) have gained popularity due to their SQL-based approach that lowers the technical barrier to entry. Rather than requiring complex data engineering skills, dbt enables analysts with SQL knowledge to define transformations, significantly expanding the pool of talent that can contribute to data modeling. This addresses another common startup challenge: the difficulty of hiring specialized data engineering talent on limited budgets.[34]

The transformation layer represents a crucial inflection point in your data stack where technical decisions can either unlock agility or create overwhelming maintenance burdens. Startups should prioritize solutions that are intuitive enough for cross-functional teams to understand, require minimal custom documentation, and can be modified or migrated without massive rewrites. The transformation code itself should serve as documentation through clear structure and naming conventions, reducing the need for extensive external documentation that quickly becomes outdated.

---

[34] Mike Shakhomirov, "What is dbt? A Hands-On Introduction for Data Engineers," DataCamp Blog, October 30, 2024, `https://www.datacamp.com/tutorial/what-is-dbt`.

Transformation in the a16z Modern Data Infrastructure

Looking at the a16z unified data infrastructure architecture diagram, the Transformation area is divided into three critical components: Metrics Layer, Data Modeling, and Workflow Management. These components work together to create a cohesive transformation experience that bridges raw data and downstream analytics. Let's explore each component and its relevance for startups.[35]

Data Modeling

Data modeling forms the foundation of your transformation layer—creating business-oriented logical data concepts mapped directly to physical data structures in your warehouse. For startups, this means translating complex raw data into approachable models that business users can understand and query.

Tools like dbt (data build tool) and LookML (Looker Modeling Language, a language used within the Looker platform) have become staples for startups because they lower the barrier to entry. With dbt, anyone proficient in SQL can build robust transformation pipelines using version control and modular design principles. This "SQL-first" approach is powerful, and we cover the essential coding abilities for modern startup data roles in Chapter 7.

This approach enables startups to implement best practices like

- **Progressive Refinement:** Starting with simple models and iteratively improving them as business needs evolve

- **Self-Documenting Code:** Using naming conventions and inline documentation that makes the transformation logic transparent

- **Reusable Patterns:** Creating modular components that can be assembled to solve new business questions without starting from scratch

For early-stage startups, this means you can start building a solid data foundation without highly specialized engineering talent—a crucial advantage when resources are limited.

---

[35] Matt Bornstein, Jennifer Li, and Martin Casado, "Emerging Architectures for Modern Data Infrastructure," Andreessen Horowitz, accessed August 3, 2025, `https://a16z.com/emerging-architectures-for-modern-data-infrastructure/`.

Metrics Layer

The metrics layer serves as a centralized repository of standardized definitions. As highlighted in a16z's reference architecture, this layer becomes the "single source of truth" for metrics across the organization, addressing the common startup challenge of inconsistent definitions.[36]

Without a metrics layer, startups often experience "metric sprawl" where different teams calculate and report the same metrics in different ways. For instance, "Monthly Active Users" might be defined differently across marketing, product, and finance teams. A metrics layer establishes standard definitions for every critical business metric, ensuring that decisions are made using consistent measurements.

Workflow Management

As mentioned before, workflow management tools are widely used when talking about data ingestion and transport. But these tools can also orchestrate and automate your transformation processes, ensuring they run reliably and cost-effectively. For startups, this component is crucial for maintaining data freshness without requiring manual intervention. Modern workflow tools like Dagster, Prefect, or Airflow allow startups to

- Schedule regular transformation runs (daily, hourly, or real time)
- Handle dependencies between transformation steps
- Automatically retry failed processes
- Monitor performance and send alerts when issues arise[37]

Cloud vendors now offer managed workflow services like Google Cloud Composer or AWS Step Functions that can further reduce operational overhead for startups. The key for startups is balancing performance and cost—determining which transformations need to be materialized as tables vs. what can be calculated on demand. As noted by a16z experts, effective workflow management automates these decisions, optimizing cloud spending while maintaining acceptable query response times.

---

[36] Matt Bornstein, Jennifer Li, and Martin Casado, "Emerging Architectures for Modern Data Infrastructure," Andreessen Horowitz, accessed August 3, 2025, `https://a16z.com/emerging-architectures-for-modern-data-infrastructure/`.

[37] Jake Roach, "Dagster vs Airflow: Comparing Top Data Orchestration Tools for Modern Data Stacks," DataCamp Blog, September 15, 2024, `https://www.datacamp.com/blog/dagster-vs-airflow`.

SQL-First Data Modeling Tools: dbt Core vs. dbt Cloud vs. Alternatives

Let's get back to our hypothetical scenario of a global travel app. Imagine that to unify its growing data sources, the team has decided to adopt a scalable cloud data warehouse and has chosen Google BigQuery due to its pay-as-you-go pricing and minimal setup. With their data now centralized, the next crucial step is to select a tool for data modeling. Three choices come to mind: dbt (dbt Core vs. dbt Cloud) or a cloud-native alternative like Dataform, which is particularly appealing given their existing investment in the Google Cloud ecosystem.[38]

The dbt ecosystem presents two main options: dbt Core (open source) and dbt Cloud (managed service). Both offer powerful transformation capabilities, but with different trade-offs relevant to startups.

> **dbt Core** provides a free, command-line interface that allows complete control over your transformation environment. It uses SQL (and Python) to define models, with version control integration for tracking changes. While powerful, it requires more technical setup—you'll need to configure your own IDE, schedule jobs through external orchestrators like Airflow or GitHub Actions, and handle documentation hosting yourself. For technically proficient teams with tight budgets, this approach offers maximum flexibility without subscription costs.

> **dbt Cloud**, by contrast, is a subscription-based service that abstracts away much of the infrastructure management. It provides a browser-based IDE, built-in scheduling, documentation hosting, and continuous integration features.[39]

Major cloud providers have also developed native transformation tools. For our hypothetical travel app, which already uses Google Cloud and BigQuery, Google's offerings are a natural consideration. Google Cloud offers BigQuery integrated

---

[38] "Dataform," Google Dataform, accessed November 5, 2025, `https://cloud.google.com/dataform`.

[39] Jason Ganz and Jeremy Cohen, "How we think about dbt Core and dbt Cloud," dbt Labs Blog, last updated on October 15, 2024, `https://www.getdbt.com/blog/how-we-think-about-dbt-core-and-dbt-cloud`.

transformation capabilities through several options. You can use BigQuery scheduled queries that enable direct SQL transformations within BigQuery itself. It's easy to get started with and offers some scheduling features. But if you are looking for a Google Cloud service similar to dbt that allows version-controlled SQL transformations, you can use Dataform. These cloud-native options can reduce integration complexity for startups already committed to a particular cloud ecosystem, though they may increase vendor lock-in concerns. To make an informed decision, let's compare these three SQL-first data modeling tools directly (see Table 3-8).

Table 3-8. *Comparison of SQL-First Data Modeling Tools*

Feature	dbt Core	dbt Cloud	Google BigQuery + Dataform
Startup Applicability	Ideal for budget-conscious startups with technical talent	Good for growth-stage startups needing collaboration features	Best for startups already on Google Cloud
Cost Structure	Infrastructure costs only	Paid	Dataform is free, other services paid based on usage
Technical Expertise Required	Moderate-High	Low-Moderate	Low-Moderate
Maintenance Burden	Higher (self-hosted)	Low (fully managed)	Low (fully managed)
Integration Flexibility	High (works with any data warehouse)	High (works with major warehouses)	Medium (Google ecosystem focus)

As Table 3-8 demonstrates, the decision between these SQL-first modeling tools involves trade-offs between cost, convenience, and flexibility. For the global travel app, the choice depends entirely on its current resources and strategic priorities:

- If the startup has a technically skilled team, **dbt Core** is the most flexible option. It offers complete control and works with any data warehouse, avoiding vendor lock-in. However, this path requires a higher investment in setup and ongoing maintenance.

- If the startup is scaling quickly and needs to empower a growing team with collaborative features, **dbt Cloud** is a strong contender. The subscription cost is offset by the low maintenance burden and built-in features like scheduling and documentation, which frees up valuable engineering resources to focus on other priorities.

- Given the app is already on Google Cloud, **Dataform** presents the most seamless path forward. Its low maintenance, minimal technical barrier, and tight integration with BigQuery make it an attractive choice for teams that want to move quickly without managing additional infrastructure. The main trade-off is reduced flexibility and a deeper commitment to the Google ecosystem.

Ultimately, startups should begin with simple, well-documented data modeling and evolve incrementally. By first establishing consistent transformations and then layering on workflow automation and a metrics layer as the business grows, you can build a robust transformation pipeline that maximizes resources and supports scalable growth.

An Incremental Data Transformation Strategy for Startups

When implementing these transformation components, startups should start simple and evolve incrementally. They could

1. **Begin with Data Modeling:** This is your foundation. Before you worry about complex automation, focus on creating consistent, well-documented data transformations. Use approachable, SQL-first tools like dbt to build clean, logical models. This stage is about translating raw data into reliable building blocks that the entire business can understand and trust. The goal is to create a single source of truth for your core business entities.

2. **Add Workflow Automation:** Once your core models are stable and providing value, it's time to make them reliable and timely. Implement a workflow orchestrator to automate refreshes, monitor for failures, and manage dependencies. This step moves your data practice from ad hoc analysis to a dependable, automated system, freeing up your team from manual refreshes and ensuring fresh data is always available for decision-making.

3. **Introduce a Metrics Layer:** As the team grows, you'll face a new challenge: metric inconsistency. The marketing team's definition of an "active user" may differ from the product team's. A centralized metrics layer solves this by creating a single, undisputed source for all key business definitions. This ensures everyone, from the CEO to a marketing analyst, is speaking the same language and making decisions based on the same numbers.

This phased approach allows startups to build a robust transformation layer that grows with their business needs while maximizing the impact of limited engineering resources.

By thoughtfully implementing these three transformation components, startups can create data infrastructure that delivers consistent insights, adapts to changing requirements, and scales efficiently—all without requiring specialized data engineering teams from day one.

A very common and powerful setup you'll see in many startups is the combination of dbt for transformations and Apache Airflow for orchestration. In Chapter 8, we'll dive deep into this specific pairing, discussing implementation patterns, why it's so popular, and the trade-offs you should consider.

Data Output

At the culmination of your data stack lies what the a16z architecture refers to as the "Analysis and Output" layer—the interfaces through which insights are delivered to users. According to the a16z unified data infrastructure diagram, this layer encompasses several critical components that transform data into actionable intelligence. Let's explore each element and understand its relevance for startups at different growth stages.[40]

---

[40] Matt Bornstein, Jennifer Li, and Martin Casado, "Emerging Architectures for Modern Data Infrastructure," Andreessen Horowitz, accessed August 3, 2025, `https://a16z.com/emerging-architectures-for-modern-data-infrastructure/`.

Dashboards

Dashboards serve as the primary visual interface for business intelligence, presenting key metrics and insights through intuitive visualizations. While we explore the landscape of popular BI tools like Looker, Metabase, Power BI, QuickSight, and Tableau in greater detail in Chapter 9, the key for any startup is choosing a tool that fits its current stage and technical capabilities.

For early-stage startups, dashboards provide the quickest path to data democratization—enabling team members across functions to access insights without technical expertise. As one data leader at a Series A startup shared with me, "We spent four months building complex data integrations, but it was our first unified and automated dashboard that actually changed the way our founders made decisions."

However, a dashboard is only as valuable as the metrics it displays. The most effective ones focus on actionable metrics tied directly to business objectives rather than vanity metrics. Chapters 4–6 together offer a comprehensive framework for selecting these vital metrics. These chapters provide guidance that is relevant at each stage, covering everything from foundational KPIs to advanced analytical techniques that can drive strategy from day one. The best practice is to start with a minimal viable dashboard tracking these core KPIs before expanding to more specialized views as your team grows.

Embedded Analytics

Embedded analytics integrates data visualizations directly into your product or service interfaces, allowing users to access insights within their existing workflows. Unlike stand-alone dashboards, embedded analytics becomes part of your product experience.

In the Creator Industry, it's now standard practice to embed analytics into creator admin panels, allowing influencers to track performance metrics for their content without leaving your platform. This capability can transform from a nice-to-have feature into a competitive advantage, particularly for B2B startups where data visibility creates stickiness. Implementation typically follows a tiered approach, starting with basic filtered views before progressing to more advanced customization options as your customer base matures.[41]

---

[41] "Embedded Data Analytics," Cleartelligence, September 2024, https://www.cleartelligence.com/wp-content/uploads/2024/09/Embedded-Data-Analytics_0.pdf.

Augmented Analytics

Augmented analytics represents the next evolution in data interpretation, applying machine learning to automatically detect patterns, anomalies, and insights that might otherwise go unnoticed. The a16z architecture article identifies tools like ThoughtSpot and Anodot as leading examples in this category, enhancing human analysis through automated discovery. It's worth noting that this landscape is evolving rapidly—some tools mentioned in foundational analyses, such as Outlier, have since ceased operations, while others, like Sisu, have been acquired and integrated into larger platforms. We will also discuss this in Chapter 9.[42,43,44]

For resource-constrained startups, augmented analytics can function as a force multiplier, surfacing critical business insights without requiring a team of data scientists. This becomes particularly valuable in domains with vast datasets where manual analysis becomes impractical—such as monitoring user behavior across millions of interactions or detecting subtle market shifts.

Data Science and Machine Learning Tooling

As startups mature, many transition from descriptive analytics ("what happened?") to predictive modeling ("what will happen?") and prescriptive insights ("what should we do?"). The DS/ML tooling category encompasses platforms like Databricks, DataRobot, Sagemaker, and Dataiku that facilitate this evolution.[45]

These platforms streamline the development, deployment, and monitoring of machine learning models, helping startups overcome the notorious "last mile" problem in operationalizing ML. While early-stage startups may find these tools premature, growth-stage companies often discover competitive advantages through timely adoption of ML capabilities—whether for personalizing user experiences, optimizing operations, or forecasting business trends.

---

[42] Matt Bornstein, Jennifer Li, and Martin Casado, "Emerging Architectures for Modern Data Infrastructure," Andreessen Horowitz, accessed August 3, 2025, https://a16z.com/emerging-architectures-for-modern-data-infrastructure/.

[43] "Outlier AI," LinkedIn, accessed August 3, 2025, https://www.linkedin.com/company/outlier-ai/?originalSubdomain=br.

[44] "Sisu Data LinkedIn Post," LinkedIn, accessed August 3, 2025, https://www.linkedin.com/posts/sisu-data_weve-spent-the-last-five-years-working-on-activity-7119690491054485505-v2vL/.

[45] "AI and machine learning on Databricks," Databricks documentation, accessed November 5, 2025, https://docs.databricks.com/aws/en/machine-learning/.

The leap from descriptive analytics to predictive insights is as much about methodology as it is about technology. Before adopting advanced platforms, it's crucial to understand the fundamentals of scientific testing and model building. We provide the essential knowledge for getting started in later sections of this book. In Chapter 11, we cover how to validate your ideas with statistical rigor, and in Chapter 13, we delve into the core machine learning concepts that are most impactful for startups.

App Frameworks

App frameworks like Plotly Dash and Streamlit have dramatically lowered the barrier to creating interactive data applications. These tools enable startups to rapidly prototype and deploy data products without extensive front-end development resources.[46,47] For startups, these frameworks offer a middle ground between static dashboards and fully custom applications—allowing technical teams to create interactive tools for internal stakeholders or even customer-facing features.[48]

Custom Applications

As data capabilities mature, many startups eventually develop custom applications tailored to specific workflows or user needs. These purpose-built tools integrate deeply with your data stack while providing precisely tailored functionality beyond what off-the-shelf solutions can offer.

The rise of what a16z terms "warehouse-native" architecture has accelerated this trend, enabling developers to build applications directly on top of the core data platform. This approach reduces data duplication and ensures applications stay synchronized with your analytical systems.[49]

---

[46] "A faster way to build and share data apps," Streamlit, accessed November 5, 2025, https://streamlit.io/.

[47] "Dash Python User Guide," Plotly Dash, accessed November 5, 2025, https://dash.plotly.com/.

[48] "Tracking Assists App," Heroku, accessed March 6, 2025, https://tracking-dashboard-app.herokuapp.com/dashboard.

[49] Matt Bornstein, Jennifer Li, and Martin Casado, "Emerging Architectures for Modern Data Infrastructure," Andreessen Horowitz, accessed August 3, 2025, https://a16z.com/emerging-architectures-for-modern-data-infrastructure/.

Startup Implementation Strategy

Rather than attempting to implement all output components simultaneously, startups should develop an implementation strategy that aligns with their specific business model, product type, and growth stage. The three-stage approach outlined below represents just one possible implementation path:

1. **Foundation** (Seed to Series A): Implement core dashboards for executive metrics and team KPIs, focusing on actionable insights that drive decision-making.

2. **Differentiation** (Series A to B): Add embedded analytics capabilities to your product and explore automated insights through augmented analytics.

3. **Sophistication** (Series B+): Develop predictive capabilities through ML tooling and custom applications that create sustainable competitive advantages.

However, this framework must be customized to your startup's unique circumstances. A B2C marketplace might prioritize embedded customer-facing analytics early, while a deep-tech startup might invest in data science tooling from day one. The implementation strategy for a fintech startup handling sensitive financial data will necessarily differ from that of a content platform focused on engagement metrics.

As highlighted in research from the Tampere University, effective data strategies for startups should focus on a single business issue rather than attempting to solve everything at once. Similarly, Shopify's former Head of Data notes that startups have a significant advantage when implementing data strategies because they have less data debt and fewer established processes, allowing for more tailored approaches.[50,51]

---

[50] Tiia Wallin, "Planning a Data Strategy for a Startup Company," Tampere University Repository, accessed March 5, 2025, https://trepo.tuni.fi/bitstream/handle/10024/142676/WallinTiia.pdf?sequence=2.

[51] Bessemer Venture Partners, "How to get more out of your startup's data strategy," BVP Atlas, accessed March 4, 2025, https://www.bvp.com/atlas/how-to-get-more-out-of-your-startups-data-strategy.

The key is aligning your output strategy with your specific product needs, available resources, and business priorities. A more effective approach might involve asking: What critical decisions cannot be made without data? Which metrics directly impact revenue or user retention? How can analytics become an extension of your product's value proposition?

Data Warehouse Setup

The notion that setting up a data warehouse requires an extensive, complex process can be intimidating for startups. In reality, modern cloud providers have dramatically simplified this journey, making data warehousing accessible even to organizations with limited resources and technical expertise. The traditional view of warehouse setup as a series of obligatory technical hurdles deserves reconsideration in light of today's user-friendly cloud offerings and free tier options.

Let's return to our global travel app. The team is small, and every dollar counts. They're collecting a rapidly growing stream of data from bookings, user searches, and in-app behavior. This data is their key to understanding their users and growing the business, but it's scattered across different systems. They know they need a central hub—a single source of truth—but they can't afford to spend months and a fortune building it. The engineering team is already stretched thin developing new app features; they don't have time to manage complex infrastructure.

Earlier, we mentioned that this startup chose Google BigQuery as their data warehouse. Now, let's take a look at the data warehouse solutions landscape and explore why.

The Accessibility of Modern Cloud Data Warehouses

Cloud data warehouses have evolved to prioritize ease of use without sacrificing analytical power. Getting started with modern data warehouse platforms no longer requires significant upfront investment or complex configuration. Rather than treating setup steps as impediments, startups can view them as optional refinements to be implemented gradually as their data needs mature.

Many leading providers offer sandbox or free tier environments that allow users to experiment with features without financial commitment. These environments enable technical teams to evaluate platform capabilities before allocating resources. Free tier offerings typically include sufficient capacity for early-stage startups to begin deriving value from their data immediately, representing a significant democratization of technology that was previously accessible only to enterprises with substantial resources.

The technical barriers to entry have also been substantially lowered. Cloud data warehouses now feature intuitive interfaces and simplified setup procedures that don't require specialized data warehouse administration knowledge. Modern web interfaces enable users to create data warehouses in minutes, with integrated tools for data loading, transformation, and analysis that business users can leverage without extensive IT support. This usability-focused design means startups can begin extracting insights from their data almost immediately after signing up.

Key Considerations for Data Warehouse Selection

For the travel app team, the choice of a data warehouse isn't just a technical decision; it's a strategic one. Rather than focusing on precise pricing details that will likely be outdated by the time you implement your solution, consider these fundamental aspects when selecting and setting up a data warehouse:

- **Architectural Approach:** Do we prefer a serverless option like BigQuery that eliminates infrastructure management or a provisioned solution like Amazon Redshift that offers more configurability but requires more expertise?

- **Cost Structure Mechanics:** How does the pricing model align with our expected usage? A warehouse that charges by data scanned might be cost-effective for our selective queries but expensive for full-table scans.

- **Technical Expertise Requirements:** Does our team have the knowledge to manage a more complex system, or do we need a fully autonomous option that minimizes administration?

- **Integration Capabilities:** How easily will this warehouse integrate with the tools we already use, like Google Analytics and our future analytics tools?

- **Growth Path:** How easily will this solution scale as our data volume and complexity grow? Can we transition from a free to a paid tier without a painful migration?

- **Security and Compliance:** Does the solution provide the security controls we'll need as our data becomes more valuable, like column-level security and row-level access?

Comparative Analysis of Data Warehouse Solutions

When evaluating data warehouse options, startups should consider several factors beyond just current pricing, which tends to change quite frequently. Table 3-9 provides a general comparison of leading solutions based on their architectural approach and startup-friendliness.

Table 3-9. *Data Warehouse Solutions*

Data Warehouse Solution	Architecture Type	Technical Expertise Required	Setup Complexity	Maintenance Effort
Google BigQuery	Serverless	Low-Medium	Low	Low
Databricks (SQL + Unity Catalog)	Lakehouse	Medium	Low-Medium	Low-Medium
Oracle Autonomous DW	Autonomous	Low-Medium	Low	Low
SAP Data Warehouse Cloud	Multi-tenant	Medium	Medium	Medium
Amazon Redshift	Hybrid (Serverless & Provisioned)	Medium-High	Medium	Medium-High
Snowflake	Multi-cluster	Low-Medium	Low	Low
Microsoft Fabric	Unified SaaS	Low-Medium	Low	Low
Azure Synapse	Hybrid	Medium	Medium	Medium

As you can see in Table 3-9, the landscape of data warehouse solutions is diverse, offering different architectures and trade-offs in complexity and maintenance. For startups prioritizing speed and minimal overhead, fully managed solutions like **Google BigQuery** and **Snowflake** are often the top choices. Their architecture separates storage from compute, allowing you to scale resources independently and pay only for what you use, which is ideal for managing costs in a growing business. In a similar vein, **Databricks** offers a unique Lakehouse architecture, which merges the capabilities of a data warehouse with the flexibility of a data lake, allowing for both business intelligence and machine learning workloads on the same platform.[52,53,54,55]

Table 3-9 also includes two key offerings from Microsoft: **Azure Synapse Analytics** and **Microsoft Fabric**. It is important to understand their relationship, as Microsoft is heavily prioritizing Fabric as the future of its analytics platform. Think of Azure Synapse as a powerful Platform as a Service (PaaS) tool that integrates various analytics services but requires more hands-on management of its components. Microsoft Fabric, on the other hand, is a more recent Software as a Service (SaaS) offering that bundles Synapse's core capabilities with Power BI and Data Factory into a single, unified product. For a startup, this key distinction means Fabric aims to simplify management by handling the underlying infrastructure, while Synapse offers more granular control for teams that need it.[56]

Architecturally, Fabric is built around a central data lake called OneLake, whereas Synapse relies on integrating with Azure Data Lake Storage (ADLS) Gen2. While Fabric represents the clear future direction for Microsoft, Azure Synapse remains a relevant and powerful platform for many large-scale enterprise deployments. Both solutions are included here to provide a complete picture of the market, though startups beginning their journey today should strongly evaluate Fabric first.

Most major cloud data warehouse providers offer some form of free tier or trial program designed to lower the barrier to entry for startups and small teams. These typically include limited amounts of free storage, query processing, and compute

---

[52] "BigQuery: From data warehouse to autonomous data and AI platform," Google BigQuery, accessed November 5, 2025, `https://cloud.google.com/bigquery`.

[53] "Data warehousing on Databricks," Databricks Documentation, accessed November 5, 2025, `https://docs.databricks.com/aws/en/sql/`.

[54] "SAP Data Warehouse Cloud," SAP, accessed November 5, 2025, `https://api.sap.com/package/sapdatawarehousecloud/overview`.

[55] "Snowflake," Snowflake Documentation, accessed November 5, 2025, `https://docs.snowflake.com/`.

[56] "Microsoft Fabric vs Azure Synapse: Architecture, Features & FAQs (2025)," Atlan, last updated on December 9, 2024, `https://atlan.com/microsoft-fabric-vs-azure-synapse/`.

resources. While the specific allocations vary between providers and change over time, they generally provide sufficient capacity for initial exploration and proof-of-concept development.[57,58,59,60]

Justifying the Choice: Why the Travel App Chose BigQuery

When implementing your chosen data warehouse, consider adopting an incremental approach that prioritizes delivering value quickly while establishing patterns for future growth.

Begin with a minimal configuration focused on solving specific analytical needs rather than building comprehensive infrastructure. As your understanding of both the platform and your data requirements matures, you can gradually implement more sophisticated features like separate development environments, advanced security controls, and optimization techniques.

Most cloud data warehouses support this evolutionary approach by providing sensible defaults and simplified interfaces for basic operations while also offering advanced features that can be adopted as needed. This allows startups to begin deriving insights from their data almost immediately while establishing a foundation that can scale with their growth.

By focusing on these enduring considerations, startups can select and implement data warehouse solutions that will continue to meet their needs as both their requirements and the market evolve. The specific details of free tier offerings and pricing models will change, but the fundamental value proposition of cloud data warehouses— making powerful analytics accessible with minimal upfront investment—remains constant.

---

[57] "Get Started with Oracle Autonomous Data Warehouse for Free," Oracle, accessed November 5, 2025, https://www.oracle.com/pl/autonomous-database/autonomous-data-warehouse/get-started/.

[58] Amazon Web Services, "Amazon Redshift free trial," AWS Redshift, accessed March 6, 2025, https://aws.amazon.com/redshift/free-trial/.

[59] Oracle, "Always Free Autonomous Database," Oracle Cloud Documentation, accessed March 6, 2025, https://docs.oracle.com/en/cloud/paas/autonomous-database/serverless/adbsb/autonomous-always-free.html.

[60] Google Cloud, "Enable the BigQuery sandbox," Google Cloud Documentation, accessed March 6, 2025, https://cloud.google.com/bigquery/docs/sandbox.

After weighing these options, the travel app team decides to use Google BigQuery. This choice directly addresses their primary constraints of a small team and a tight budget.

- **Consumption-Based Pricing and a Generous Free Tier:** BigQuery's pricing model is a game-changer for them. With a generous free tier, they can get started without spending a dime. This consumption-based model makes costs predictable and proportional to their usage.

- **Minimal Setup and Management:** BigQuery's serverless architecture is a perfect fit for their small engineering team. There are no servers to manage, allowing them to set up their data warehouse in minutes and focus on product development.

- **Automatic and Seamless Scalability:** As a startup with unpredictable growth, the travel app needs a system that can handle sudden surges in users. BigQuery automatically scales storage and compute resources without any manual intervention or downtime.

- **Ease of Use and a Rich Ecosystem:** The team is already using other Google products like Google Analytics. BigQuery integrates natively with these tools, which simplifies their data pipeline and analysis workflows.

By choosing BigQuery, the travel app team makes a strategic decision that aligns perfectly with their startup reality. Their story shows that you don't need a massive budget or a dedicated data engineering team to build a solid data foundation. By adopting an incremental approach—starting with a minimal configuration to solve immediate analytical needs—they can deliver value quickly while establishing a pattern for future growth.

Quality Control and Access Security

As your data infrastructure grows with increasing volumes of information and expanding user access, quality control and security become essential to maintain trust in your analytical environment. Even early-stage startups must implement foundational practices that ensure data remains accurate, protected, and appropriately accessible. This section explores practical approaches to quality control and security that balance rigorous protection with the agility startups require.

The Startup Approach to Data Governance

Data governance doesn't need to be intimidating for startups. Rather than implementing enterprise-level frameworks immediately, focus on establishing flexible foundations that can grow with your company. The key is starting with minimal viable governance—just enough structure to ensure data reliability and security without creating bureaucratic overhead.[61]

When implementing data governance in a startup environment, follow the "crawl, walk, run" approach—a metaphor for an iterative, experimental problem-solving methodology. Begin with small wins that demonstrate immediate value before gradually expanding your governance framework:

1. **Crawl:** Identify your most critical data assets and establish basic quality and security controls around them.

2. **Walk:** Extend these controls to secondary data assets and formalize processes as your team grows.

3. **Run:** Implement comprehensive governance as your company matures and faces increased regulatory scrutiny.

This staged approach prevents overwhelming your team with processes while ensuring critical data remains protected and reliable throughout your growth journey.[62]

Quality Control Across the Data Life Cycle

Quality control must be implemented at each stage of your data life cycle, from collection to consumption. Let's examine how startups can approach this across their data infrastructure.

---

[61] "What is data governance?," Google Cloud Learn, accessed November 5, 2025, https://cloud.google.com/learn/what-is-data-governance.

[62] Willem Koenders, "Data Governance for Startups," Medium, June 24, 2024, https://medium.com/@willemkoenders/data-governance-for-startups-c372fe610889.

Data Sources: Establishing Quality at the Origin

The quality of your data ecosystem begins with your sources. Whether you're collecting data through your product, third-party APIs, or manual processes, establishing standards at this stage prevents downstream issues.

Developer Responsibility and Testing

Developers play a crucial role in ensuring data quality at the source. Encourage them to

1. Implement comprehensive validation rules in data creation processes

2. Document data structures and their intended meaning

3. Create automated tests that verify data integrity

4. Establish clear processes for communicating changes to data definitions

When developers understand that they are data producers, not just application builders, they become more conscious of how their work affects the overall data ecosystem. This cultural shift is essential for maintaining quality without excessive formal processes.

Minimizing Technical Debt

Changes to data structures without proper documentation create technical debt that eventually impacts data quality. Implement lightweight change management processes that don't burden developers but ensure data changes are communicated to analytics teams. This might be as simple as required data definition comments in code or a shared document of data structures that must be updated with each change.

Data Ingestion: Validating What Enters Your System

The ingestion layer represents your first opportunity to systematically control data quality before it enters your warehouse or lake.

Early Validation Mechanisms

Implement basic validation rules during ingestion to catch obvious quality issues:

1. Schema validation to ensure expected fields are present

2. Type checking to prevent format inconsistencies

3. Range validation for numeric values

4. Pattern matching for structured text fields

5. Deduplication checks to prevent multiple ingestion of the same data

These validations should focus on critical data elements first, particularly those affecting financial reporting, customer experience, or regulatory compliance.

Logging and Monitoring

Establish monitoring for your ingestion processes with alerts for significant data quality issues. Monitoring data quality allows testers to identify issues quickly and take actions to fix errors if needed. For startups, this monitoring can start simply—perhaps tracking failed ingestion attempts or validation errors—and become more sophisticated as your data needs grow.

Data Storage: Protecting Your Data Assets

How you store data affects both its quality and security. Modern data storage options offer various controls that startups can leverage without significant overhead.

Data Classification and Protection

Begin by classifying your data based on sensitivity and importance:

1. **Critical Business Data:** Financial, proprietary algorithms, strategic plans

2. **Sensitive Personal Data:** Customer PII, payment information, health records

3. **Operational Data:** Logs, metrics, non-sensitive application data

This classification guides your security measures, allowing you to focus resources where they matter most. You can start by identifying your most sensitive data and implementing strict access controls in select areas. Then, you can extend these controls further step by step.

Storage Optimization for Quality

Consider these approaches to maintain data quality in storage:

1. Implement appropriate partitioning strategies for time-based data.

2. Use data organization principles (bronze, silver, gold layers) to separate raw from processed data. The so-called Medallion architecture organizes data by progressively refining it: starting with the original, raw data (Bronze), then cleaning and standardizing it (Silver), and finally preparing it for business use through aggregation (Gold).[63]

3. Consider immutable storage for critical data to maintain audit trails.

4. Document storage structures and access patterns.

Data Transformation: Maintaining Quality Through Processing

Data transformations present both quality risks and opportunities. Each transformation step can introduce errors but also allows for data enrichment and standardization.

Testing Transformation Logic

For startups, you can start by implementing essential tests for your most critical transformations:

1. Row count validations before and after transformations

2. Sum checks for numeric columns

---

[63] "Medallion Architecture," Databricks Glossary, accessed November 5, 2025, https://www.databricks.com/glossary/medallion-architecture.

3.   Referential integrity checks between related datasets

4.   Null rate monitoring for required fields

5.   Distribution checks for key metrics

These tests can be implemented as simple assertions in your transformation code or using lightweight data quality frameworks.

Metadata Management

As your data transforms through various stages, maintaining metadata becomes increasingly important. Start with basic documentation of transformation logic and gradually implement more formal metadata management:

1.   Document business definitions of derived metrics.

2.   Record transformation lineage (what source data created which outputs).[64]

3.   Track when transformations were last validated or updated.

4.   Note known quality issues or limitations of specific datasets.

Data Output: Ensuring Reliable Analytics and Reporting

The final stage of your data life cycle—where insights are generated and consumed—requires particular attention to quality and access controls.

Access Controls and User Management

Consider these principles:

1.   Implement role-based access control (RBAC) for your BI tools and data warehouse.

2.   Apply the principle of least privilege—users should only access data necessary for their role.

3.   Create separate access policies for sensitive data elements.

---

[64] "What is data lineage?," IBM Think, accessed November 5, 2025, `https://www.ibm.com/think/topics/data-lineage`.

4. Establish a simple approval process for new data access requests.

5. Regularly audit and review access permissions as your team grows.

Quality Verification for Critical Reports

For reports and dashboards that drive key business decisions, implement additional quality verification:

1. Compare automated reports against manual calculations periodically

2. Implement anomaly detection for critical metrics

3. Create documentation explaining how key metrics are calculated

4. Establish clear owners for important dashboards and reports

Building a Data-Driven Culture with Quality at Its Core

Beyond specific technical measures, fostering a data-driven culture that values quality is essential for startups. Making data quality everyone's responsibility is crucial. Quality cannot be the sole responsibility of a data team. Instead

1. Train all employees on basic data concepts relevant to their roles

2. Recognize and reward good data practices across teams

3. Share examples of how data quality impacts business outcomes

4. Create simple escalation paths for reporting data issues

Regular Data Quality Reviews

Schedule lightweight data quality reviews focusing on your most critical data. These need not be formal audits—even a monthly meeting to discuss data issues can significantly improve quality awareness.

Security and Compliance for Startups in Regulated Industries

Startups in regulated industries face additional challenges balancing innovation with compliance requirements. However, compliance need not stifle growth if approached pragmatically.

Risk-Based Security Implementation

Security implementation should be prioritized based on risk:

1. Identify regulatory requirements specific to your industry and data types.

2. Implement essential security controls for high-risk data first.

3. Document your security approach and future plans for investors and auditors.

4. Consider compliance automation tools to simplify assessment and reporting.

As noted by startup security advisors, "Start by identifying your most sensitive data and implementing stricter access controls. Then, gradually extend these controls across your organisation."[65]

Data Compliance Automation

Use compliance software for an initial system check to see what security measures you have and what you lack. These tools can help you[66]

1. Assess your current compliance status against various frameworks

2. Generate required documentation automatically

---

[65] Emre Baran, "The Importance of Data Security for Startups," Startups Magazine, accessed March 7, 2025, https://startupsmagazine.co.uk/article-importance-data-security-startups.

[66] Matt Cooper, "The startup guide to making your first security hire," Vanta Resources, accessed March 7, 2025, https://www.vanta.com/resources/first-security-hire.

3. Monitor ongoing compliance with security controls

4. Prepare for formal audits when necessary

Practical First Steps for Startups

To begin implementing quality control and access security in your startup's data infrastructure, consider these practical first steps:

1. **Identify Critical Data:** Determine which data assets are most essential to your business operations and focus initial efforts there.

2. **Implement Basic Access Control:** Start with a simple role-based access system for your data warehouse and analytics tools.

3. **Establish Data Ownership:** Assign clear ownership for key datasets to ensure someone is responsible for their quality and security.

4. **Create Minimal Documentation:** Document your most important data definitions, sources, and transformation logic in a central location.

5. **Implement Basic Monitoring:** Set up alerts for critical data pipeline failures and quality issues in your most important datasets.

6. **Train Your Team:** Ensure everyone understands their role in maintaining data quality and security.

Data governance professionals point out that a lack of oversight will lead to system failure, even for the most well-designed platforms. These basic measures establish the necessary oversight without creating burdensome processes.[67]

---

[67] Ron Schmelzer and Kathleen Walch, "Top 9 AI Data Governance Best Practices for Security, Compliance, and Quality," PMI Blog, accessed March 6, 2025, `https://www.pmi.org/blog/ ai-data-governance-best-practices`.

Conclusion

Startups can handle quality control and access security without feeling overwhelmed. By focusing on critical data assets, implementing controls gradually, and fostering a culture that values data quality, you can build a robust data infrastructure that supports growth rather than hinders it. Remember that perfect is the enemy of good—start with essential practices, demonstrate their value, and expand as your company matures and your data needs evolve.

The investments you make in data quality and security today will pay dividends as your company grows, helping you avoid costly data remediation efforts, security incidents, and compliance challenges in the future. As your startup scales, these foundations will enable you to leverage your data as a strategic asset rather than managing it as a liability.

Summary

In this chapter, we've explored the fundamentals of building effective data infrastructure for startups. We discussed how critical it is to avoid over-engineering early on—focusing instead on lightweight solutions that match your current scale and analytical needs. We examined the decision between cloud and on-premise solutions, highlighting that cloud computing typically offers superior flexibility and cost-efficiency for most startups.

We then detailed the five core components of a typical startup data stack:

- **Data Sources:** Identifying and prioritizing relevant inputs while ensuring cohesion across diverse sources

- **Data Ingestion:** Comparing ETL vs. ELT approaches and reviewing popular ingestion tools that balance ease-of-use with scalability

- **Data Storage:** Evaluating cloud warehouses, lakehouses, and real-time databases based on your analytical requirements

- **Data Transformation:** Implementing tools like dbt to standardize raw data into actionable insights without heavy engineering overhead

- **Data Output:** Delivering insights through dashboards, embedded analytics, augmented analytics tools, or even custom-built applications

Finally, we emphasized the importance of incremental implementation—starting simple and evolving your infrastructure as your startup grows. By making pragmatic choices guided by analysis rather than vendor pitches, you ensure that your data infrastructure remains an asset rather than a liability.

My goal isn't to crown Redshift over BigQuery or vice versa. I'd like to equip you with a landscape of tools across key areas: What problems do they genuinely solve? How easily can components scale or adapt as your startup evolves? Startups thrive on experimentation—your data stack should too. You might adopt a tool today and replace it in six months, and that's okay. The key is architecting with flexibility, ensuring your infrastructure shifts as swiftly as your business does.

Stay pragmatic. Let analysis—not sales pitches—guide your decisions. After all, the best data infrastructure isn't the shiniest; it's the one that silently fuels your startup's agility, letting you move faster, smarter, and with confidence. In the next chapter, we will dive deeper into identifying and tracking metrics that matter most for your startup.

PART II

Metrics That Matter

PART II

CHAPTER 4

The Founder's Metrics Toolkit

The Airbnb guys used to draw a graph they wanted to hit—a forward-looking projected graph—and then they would put it up on the mirror in their bathroom and on their fridge door and everywhere. Every week they were just trying to grow 10%. Our best startups do this, and it's so important. It is this function that shapes the company and it's one metric to optimize for.

—Sam Altman,

CEO of OpenAI, former president of Y Combinator[1]

The story of Airbnb's obsession with metrics illustrates a fundamental truth about successful startups: they don't just track numbers—they live by them. Brian Chesky and his co-founders didn't merely analyze their growth data; they surrounded themselves with it, making their key metrics an inescapable part of their daily lives. This intense focus on measurable growth helped transform a simple idea of renting air mattresses into a global hospitality giant worth billions.[2]

Metrics matter because they convert the abstract concept of "progress" into concrete, actionable insights. For startups operating with limited resources and facing existential pressure to show traction, choosing the right metrics isn't just an academic exercise—it's a matter of survival. As Sam Altman learned through his experience at Y Combinator

---

[1] "Sam Altman | How to Get Funded by Y Combinator," Startup Grind YouTube channel, accessed June 20, 2025, `https://www.youtube.com/watch?v=kVKaMHgOTP4`.

[2] "Sam Altman | How to Get Funded by Y Combinator," Startup Grind YouTube channel, accessed June 20, 2025, `https://www.youtube.com/watch?v=kVKaMHgOTP4`.

© Piotr Sidoruk 2026

P. Sidoruk, *From Data to Dollars*, https://doi.org/10.1007/979-8-8688-1898-1_4

before leading OpenAI, the startups that succeed are those that identify, track, and relentlessly pursue improvement in metrics that genuinely reflect business health and growth potential.

However, the world of analytics presents a paradox for new ventures: data overabundance is common, yet actionable insight remains scarce. Many founders find themselves drowning in dashboards filled with vanity metrics that look impressive but fail to guide critical decisions. Others track too many numbers, creating "analysis paralysis" that slows decision-making rather than accelerating it.

What to Expect from This Chapter

Metrics should be the heartbeat of every successful startup. This chapter is designed to be your definitive guide to understanding, selecting, and leveraging the foundational metrics that truly drive your startup forward.

Here's what you will gain from this chapter:

- **A Clear, Step-by-Step Process:** You'll learn how to cut through the noise of vanity metrics and focus on numbers that reveal real progress and business health.

- **A Deep Dive into the North Star Metric:** Discover how to define the single metric that best captures your core value to customers and aligns your entire organization. You'll see how leading companies— from Airbnb to Facebook—have used this approach to drive growth and clarity.

- **Mastery of Essential Frameworks:** Explore proven concepts like Key Performance Indicators (KPIs) and the AARRR (Acquisition, Activation, Retention, Referral, Revenue) framework, complete with detailed examples and visualizations for each stage.

First, we will explore the foundations of measurement, covering why metrics matter, what makes a metric truly useful, and how to avoid the most common mistakes startups make when tracking progress. We will then build on that foundation with a detailed look at core metrics and guiding frameworks, providing you with the practical tools and models needed to turn data into a strategic advantage.

The Foundations of Measurement

If you dive deeper into the topic of metrics, you can easily feel overwhelmed. There are dozens of different metrics companies could potentially track, but the most important thing is understanding which ones are truly useful for your specific business. At the beginning, you should focus on just a few metrics that help you stay focused on what really matters.

Why Startups Need Metrics

Tom Blomfield, Y Combinator Group Partner, compares launching without metrics to "flying blind." As he explains, metrics give founders insight and oversight into how their business is performing, enabling them to monitor progress and base their decisions on data. The best startups in their early days focus on achieving fast, consistent weekly growth—usually targeting a 5–10% increase in their main metric—to drive momentum and quickly learn from results. This approach to exponential progress is a key indicator of a startup's potential for significant success.[3,4]

However, simplicity is crucial. You shouldn't overwhelm yourself with too many metrics. Divya Bhat, YC Visiting Partner, advises that, as a startup, you have limited time and infinite tasks. Prioritization helps you focus on what truly moves your business toward product-market fit. She recommends having one primary KPI (typically revenue growth for launched startups) and 3-5 secondary KPIs that support it.[5]

What Makes a Metric Great?

Every startup knows the obsession: the late-night search for that one perfect number—a single metric so elegant that it captures the very essence of the business. We dive into spreadsheets, layer on formulas, and chase complexity, believing that the more sophisticated the metric, the more profound its insight. This is a trap. And it's one that has stalled countless promising startups.

---

[3] "Adora Cheung - How to Set KPIs and Goals," Y Combinator YouTube channel, accessed June 29, 2025, https://www.youtube.com/watch?v=lL6GdUHIBsM&t=3s.

[4] "B2B Startup Metrics," Y Combinator Podcast Archive, accessed June 21, 2025, https://www.podcastarchive.wiki/podcasts/y-combinator/b2b-startup-metrics.

[5] "Setting KPIs and Goals | Startup School," Y Combinator YouTube channel, accessed June 29, 2025, https://www.youtube.com/watch?v=6DTK9yDP6pO&list=PLQ-uHSnFig5M9fW16o2l35jrfdsxGknNB.

The most powerful metrics aren't found in complexity; they are forged in simplicity. The only numbers that matter are the ones that your entire team can understand, rally behind, and influence. The ultimate test is effortless communication: **can you explain your key metric instantly, without overthinking it?**[6]

If you have to pause, it's a sign of a deeper problem, and the metric is destined to fail for several key reasons:

- **It Creates Friction, Not Focus:** When a metric is complex, team members spend more time debating its definition than improving it. This friction grinds progress to a halt.

- **It Disconnects Work from Results:** If an engineer, designer, or marketer can't see the direct line between their work and the movement of the key metric, they can't be expected to influence it. The metric becomes an abstract number that belongs to someone else.

- **It Undermines True Alignment:** A shared goal must be a clearly understood shared goal. A complex metric prevents your team from rowing in the same direction because they can't agree on what the destination looks like.

Confusion is the enemy of progress. When a team argues about what a metric means, they aren't working to improve it. To transform a number into your most meaningful metric, it must possess three core traits:

- **Instantly Understandable:** A great metric feels intuitive. It's easy to remember, easy to discuss, and easy to build a culture around. If it's not simple, it won't stick.

- **Decidedly Actionable:** A metric must demand a response. If it moves, and you do nothing, it's a headline, not a tool. Ask your team: "What would we do if this number doubled? What if it got cut in half?" If you don't have an answer, find a better metric.

---

[6] Jamie Cuffe, Avanika Narayan, Chandra Narayanan, Hem Wadhar, and Jenny Wang, "Defining Product Success: Metrics and Goals," Sequoia Capital Articles, accessed July 02, 2025, https://articles.sequoiacap.com/defining-product-success-metrics-and-goals.

- **Naturally Comparative:** Numbers need context to tell a story. "We had 100 new sign-ups" is data. "We had 100 new sign-ups, up 30% from last week and driven entirely by our new blog post" is intelligence. A great metric allows you to look back at the past, across different user groups, and toward the competition.

The Secret Weapon

- **Ratios:** The authors of *Lean Analytics* highlight a powerful feature of many world-class metrics: they are often ratios or rates. A ratio forces you to see the bigger picture by comparing two critical forces. It's the difference between celebrating 1,000 new users and realizing you paid twice their lifetime value to acquire them. Ratios reveal the health and sustainability behind the headline numbers.[7]

Whether you choose a simple count or a powerful ratio, the principle remains the same. A great metric isn't about complex math; it's about creating a shared understanding that drives focused, intelligent action.

A Word of Caution: When Good Metrics Go Bad

While well-defined metrics are powerful, it's crucial to be aware of the potential downsides of an overemphasis on goal achievement. The Harvard Business School paper titled "Goals Gone Wild" highlights that goals, when poorly implemented, can have systematic negative side effects. Startups must watch for these dangers:[8]

- **An Excessively Narrow Focus:** When a team is judged solely on one metric, they may neglect other critical areas of the business. For example, a relentless focus on new user acquisition could lead a team to ignore the retention of existing customers, creating a "leaky bucket" that dooms long-term growth.

---

[7] Alistair Croll and Benjamin Yoskovitz, *Lean Analytics: Use Data to Build a Better Startup Faster* (O'Reilly Media, 2013).

[8] Ordonez, Lisa D., Maurice E. Schweitzer, Adam D. Galinsky, and Max H. Bazerman. "Goals Gone Wild: The Systematic Side Effects of Over-Prescribing Goal Setting. "Harvard Business School Working Paper, No. 09-083, January 2009.

- **Encouraging Unethical Behavior:** Intense pressure to hit a specific target can lead to "gaming the system." To hit a "daily active users" goal, a team might be tempted to send excessive, low-value notifications. While this may temporarily boost the metric, it annoys users and erodes long-term trust.

- **Inhibiting Learning and Risk-Taking:** Rigid adherence to a metric can stifle innovation. If a team is afraid of a short-term dip in a key number, they may avoid running a bold experiment that could lead to a significant, long-term breakthrough. This is particularly dangerous for startups, where learning and adaptation are paramount.

- **Corroding Intrinsic Motivation:** Tying every action to an extrinsic reward or target can diminish a team's intrinsic passion for their work and for solving user problems.

Ultimately, metrics are meant to be guides, not tyrants. The goal isn't to find one perfect metric to rule them all but to use a balanced set of well-chosen indicators to facilitate intelligent conversation and drive strategic decisions. The best metrics illuminate the path forward and empower your team to act with clarity and confidence.

Avoiding Vanity Metrics

Vanity metrics are metrics that may appear impressive but do not provide actionable insights into a business's health or future success. They often produce large numbers that can be misleading and can distract from the metrics that truly matter for growth and sustainability.[9]

According to Andrew Chen of a16z, who was introduced earlier in the book, entrepreneurs often present a mix of "good" and "bad" metrics in their investor decks. While some metrics are genuinely indicative of product-market fit, others are vanity metrics that don't accurately reflect the company's performance. Chen advises focusing on metrics that demonstrate user engagement and retention rather than just top-line numbers.[10]

---

[9] "B2B Startup Metrics," Y Combinator Podcast Archive, accessed July 04, 2025, https://www.podcastarchive.wiki/podcasts/y-combinator/b2b-startup-metrics.

[10] Andrew Chen, "The red flags and magic numbers that investors look for in your startup's metrics – 80 slide deck included!," Andrew Chen Blog, accessed June 24, 2025, https://andrewchen.com/investor-metrics-deck/.

Tren Griffin from Microsoft, the author of *Charlie Munger: The Complete Investor*, emphasizes the importance of distinguishing between metrics that are easily manipulated and those that genuinely reflect customer value. For instance, in his "12 Things About Product-Market Fit" article, one key point is that metrics can be misleading. Startups can easily generate impressive-looking numbers through marketing spend, but these often don't translate to a sustainable business model if the underlying product doesn't retain users.[11,12]

Andreessen Horowitz's "16 Startup Metrics" further breaks down which metrics matter and which can be deceptive. The guide highlights that while a metric like "app downloads" might look good, it is far less important than "daily active users." A large number of downloads and sign-ups means little if those users do not engage with the product regularly. Similarly, metrics like page views can be easily inflated and do not necessarily correlate with user satisfaction or value creation.[13]

Here's an example of such a vanity metric—Cumulative Registered Users by Month (see Figure 4-1). The solid line represents the accelerating total sign-ups, while the dashed line shows the much flatter trend of active users. While the platform consistently adds new registered users, a significant churn problem prevents the active user base from growing at a similar rate.

---

[11] Tren Griffin, "12 Things About Product-Market Fit," Andreessen Horowitz, accessed June 24, 2025, `https://a16z.com/12-things-about-product-market-fit/`.

[12] Tren Griffin, *Charlie Munger: The Complete Investor* (Columbia Business School Publishing, 2015).

[13] Jeff Jordan, Anu Hariharan, Frank Chen, and Preethi Kasireddy, "16 Startup Metrics," Andreessen Horowitz, accessed June 30, 2025, `https://a16z.com/16-startup-metrics/`.

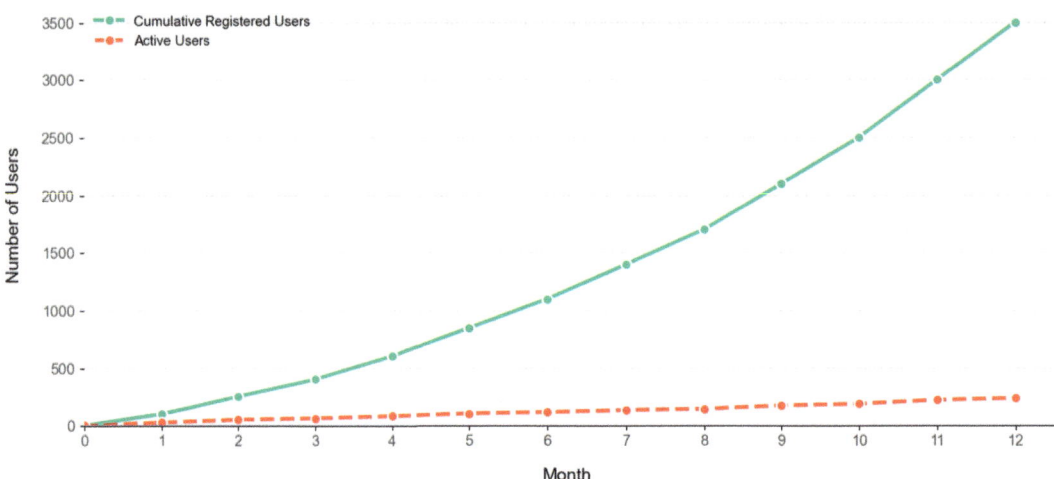

Figure 4-1. *Vanity Metrics: Cumulative Registered Users vs. Active Users*

Instead of focusing on these superficial numbers, startups should prioritize actionable metrics. As Andrew Chen points out, investors are more interested in cohorts and retention curves, which show how many users continue to engage with a product over time. These metrics provide a much clearer picture of product-market fit. For example, a strong indicator of product-market fit is when a significant percentage of users are using the product daily or weekly, and this engagement is sustained over time. Another powerful metric is the Net Promoter Score (NPS), as it measures user satisfaction and the likelihood of organic growth through word-of-mouth referrals.[14]

Ultimately, the goal is to identify and track metrics that drive strategic decisions and reflect real business value. As Marc Andreessen, another a16z co-founder, suggests, the life of any startup can be divided into two parts: before product-market fit and after. The right metrics help a startup understand where it is in that journey and what actions it needs to take to succeed. Relying on vanity metrics can create a false sense of security and lead to poor decision-making, while focusing on engagement, retention,

---

[14] Andrew Chen, "The red flags and magic numbers that investors look for in your startup's metrics – 80 slide deck included!," Andrew Chen Blog, accessed June 24, 2025, `https://andrewchen.com/investor-metrics-deck/`.

and customer value provides a clear path forward. Table 4-1 illustrates common vanity metrics in startups and provides an explanation for why each can be misleading when assessing a company's health and progress.[15,16]

Table 4-1. *Examples of Vanity Metrics*

Vanity Metric	Why It's a Vanity Metric
Registered Users	This metric tracks the total number of sign-ups but fails to distinguish active users from those who signed up and never returned. It does not measure engagement, user value, or retention.
Page Views/ Impressions	High numbers can be easily inflated through paid advertising (or bots) and do not necessarily correlate with user satisfaction or conversion. They fail to provide insight into whether users found value in the content.
App Downloads	Similar to registered users, this metric does not indicate actual usage. Many users may download an app and never open it, making it a poor indicator of product-market fit or user engagement.
Social Media Followers/Likes	A large following may appear impressive but does not automatically translate into revenue, leads, or active product usage. It is more a measure of potential brand reach than of actual business performance.
Gross Merchandise Value (GMV)	GMV reflects the total value of goods sold through a platform but is not the company's revenue. It often ignores crucial factors like the company's take rate, returns, and cancellations and therefore does not represent profitability.
Funding Amount Raised	While raising capital is a critical milestone, the amount of funding is not a direct indicator of a company's revenue, profitability, or sustainable business model. It reflects investor confidence but not the operational health of the startup.

---

[15] "Vanity Metrics vs. Proxy Metrics: Necessary Evil or Necessity?," volta.ventures, accessed June 29, 2025, https://www.volta.ventures/blog/vanity-metrics-vs-proxy-metric/.

[16] Marc Andreessen, "The only thing that matters," Pmarchive, accessed July 4, 2025, https://pmarchive.com/guide_to_startups_part4.html.

Core Metrics and Guiding Frameworks

With a clear understanding of what makes a metric meaningful, the next challenge is organization. Without a coherent framework, even the best metrics can get lost in a sea of data. Let's explore the most effective models for structuring your startup's measurement strategy, starting with the **North Star Metric**.

The North Star Metric

For centuries, sailors navigated vast, uncertain oceans by keeping their eyes on the North Star. It wasn't the destination itself, but a constant, reliable beacon that provided direction and ensured every adjustment kept them on the right course. Similarly, your **North Star Metric** is the guiding light for your startup. It's the single measurement that cuts through complexity, aligns every team toward a shared vision of customer success, and illuminates the path to sustainable growth.[17]

Every department should understand it and how they influence it—either directly or indirectly. It's something you should monitor regularly and prioritize over other measurements. What makes a good North Star Metric? It should

- Define the value you deliver to customers

- Measure company progress

- Focus on long-term success

- Directly or indirectly lead to revenue

Airbnb's North Star Metric is "nights booked." This single measure is shaped by a variety of factors, both quantitative—such as available listings and customer demand—and qualitative, like the overall user experience. The adoption of "nights booked" brought a newfound clarity and strategic focus to Airbnb, allowing leadership to unify the company's direction and streamline decision-making. By rallying teams around this shared metric, Airbnb was able to ensure that every project contributed directly to its core mission.

---

[17] Jack Chen, "What Is Your Startup's North Star Metric?" forbes.com, accessed June 25, 2025, https://www.forbes.com/councils/forbesbusinesscouncil/2022/11/11/what-is-your-startups-north-star-metric/.

As Airbnb grew and diversified its offerings, this metric evolved into "nights and experiences booked," reflecting the addition of locally hosted activities and tours to its platform. This evolution allowed Airbnb to continue aligning its teams and investments with the broader value it delivers to travelers worldwide.[18]

Figure 4-2 illustrates Airbnb's growth over time using only their North Star Metric. Trends highlighted by the chart:

- **Pre-pandemic Climb** (2015-2019): Bookings more than quadrupled, reflecting Airbnb's rapid international expansion and the mainstreaming of home sharing.

- **Pandemic Shock** (2020): Global lockdowns cut bookings by 41% vs. 2019.

- **Swift Rebound** (2021-2023): Pent-up travel demand, remote-work flexibility, and product improvements pushed 2022 and 2023 volumes well above the 2019 peak.[19,20]

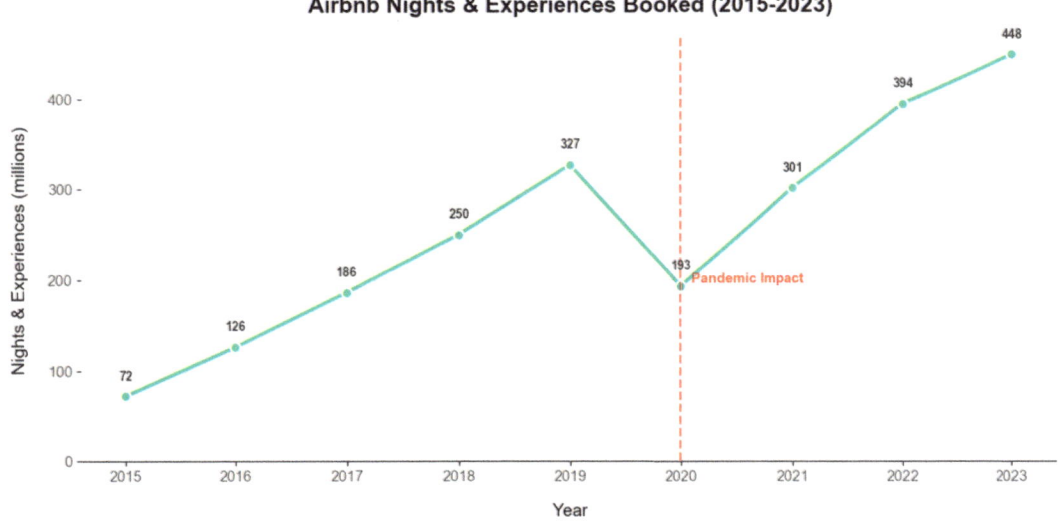

Figure 4-2. *North Star Metric Chart: From Airbnb Data*

[18] "Developing A Culture Of Experimentation At Airbnb," firstprinciples.ventures, accessed June 25, 2025, https://www.firstprinciples.ventures/insights/developing-a-culture-of-experimentation-at-airbnb.

[19] "Financials," investors.airbnb.com, accessed June 30, 2025, https://investors.airbnb.com/financials/default.aspx#sec.

[20] Backlinko Team, "Airbnb Stats," backlinko.com, accessed July 02, 2025, https://backlinko.com/airbnb-stats.

The nature of the product also dictates the ideal metric. Ad-driven businesses like Snap, Pinterest, or Facebook tend to base their North Star Metric on engagement. The main question for them is whether to focus on daily, weekly, or monthly active users (DAU, WAU, MAU). For freemium B2B products like Slack or Coda, the focus might be on engagement (like WAU) in the early stages, shifting to customer growth (e.g., "Number of Paid Teams") as the business matures and becomes more sales driven. Similarly, consumer subscription apps like Duolingo focus on engagement to build a large free user base that eventually converts to paid plans. Others, like Tinder, focus directly on customer growth—specifically, the percentage of paid accounts.[21]

While numerous companies use revenue (like ARR or GMV) as their North Star, many successful firms deliberately avoid it. A revenue-centric NSM can be volatile and difficult for teams to influence directly, as it is affected by external factors like currency rates. More importantly, focusing on revenue too early can lead to poor long-term decisions, such as optimizing pricing at the expense of user growth. A metric that is a leading indicator of revenue, such as "number of paid customers" or "nights booked," is often more inspiring for a team and a more direct measure of the value they are creating.[22]

To make a noticeable impact, most companies align around a single North Star Metric. This singular focus creates a cohesive strategy and clarifies decision-making across the organization. While some critics worry this leads to over-rotating on one aspect of the business, multiple NSMs are typically only used when a company needs to layer on a quality metric (like a Net Promoter Score) or when it operates distinct products with different goals, such as Spotify's subscription music and ad-based podcast businesses. Table 4-2 presents popular examples of North Star Metrics.[23,24,24,26,27]

[21] Lenny Rachitsky, "Choosing Your North Star Metric," future.com, accessed June 28, 2025, https://future.com/north-star-metrics/.

[22] Lenny Rachitsky, "Choosing Your North Star Metric," future.com, accessed June 28, 2025, https://future.com/north-star-metrics/.

[23] Lenny Rachitsky, "Choosing Your North Star Metric," future.com, accessed June 28, 2025, https://future.com/north-star-metrics/.

[24] Ward van Gasteren, "North Star Metric," growwithward.com, accessed June 22, 2025, https://growwithward.com/north-star-metric/.

[25] Ward van Gasteren, "North Star Metric," growwithward.com, accessed June 22, 2025, https://growwithward.com/north-star-metric/.

[26] Jory MacKay, "How to use North Star metrics to superpower your product," plan.io, accessed June 26, 2025, https://plan.io/blog/north-star-metrics/.

[27] Jory MacKay, "How to use North Star metrics to superpower your product," plan.io, accessed June 26, 2025, https://plan.io/blog/north-star-metrics/.

Table 4-2. *North Star Metrics Examples*

Company	North Star Metric
HubSpot	Weekly active teams
Uber	Weekly trips
LinkedIn	Monthly Active Users
Netflix	Watch time (median view hours per month)
Ecommerce companies	The number of paying users
SaaS companies	Monthly Recurring Revenue (MRR)

A North Star Metric should evolve as your startup grows. What makes sense for an early-stage startup won't be the same for an established business. At the seed stage, you might focus on customer acquisition; in early stages, revenue generation becomes more important; during growth, scaling the business takes priority; and at maturity, maintaining market position becomes crucial.

Checklist: Is Your North Star Metric Right?

– Does it measure the core value delivered to customers?

– Is it a leading indicator of long-term success (not just short-term revenue)?

– Can every team influence it, directly or indirectly?

– Is it easy to measure and communicate?

Key Performance Indicators

Key Performance Indicators (KPIs) are the specific metrics that directly contribute to and influence your North Star Metric (NSM), acting as the essential inputs that drive your company's overarching growth. These indicators play a crucial role in helping startups to measure their overall performance, identify specific areas where improvements are needed, and ultimately make data-driven decisions that contribute to their growth. It is important to distinguish KPIs from general business metrics. Table 4-3 clearly presents the differences between KPIs, metrics, and NSM.[28]

Table 4-3. *KPIs, Metrics, and NSM Comparison*

Characteristic	Metrics	KPIs	NSM
Definition	Any number you can track	The few, key numbers that show if you are on track	The single most important metric for company growth
Purpose	To monitor everything happening in the business	To measure performance against specific goals	To align the entire company on one ultimate goal
Example	Website clicks, social media likes	New Customers per Month, Customer Churn Rate	For Airbnb: "Nights and Experiences Booked"

While a startup might track a wide array of data points, KPIs are specifically chosen for their direct link to the company's overarching strategic objectives. KPIs can encompass various aspects of a startup's operations, including financial performance, customer satisfaction and engagement, operational efficiency, and overall growth. The selection of the most relevant KPIs is a critical step and is highly dependent on the startup's unique goals, the specific industry it operates in, and its current stage of development.[29]

A metric that is considered a key indicator of success for one startup might hold little to no relevance for another. Therefore, startups need to be discerning in their choice of KPIs, avoiding the temptation to track an excessive number of metrics and instead focusing on those that will provide the most actionable insights into their progress toward achieving their most important objectives.[30]

---

[28] "Are You Setting Your Sights On The Right North Star Metric?," thedrg.com, accessed June 26, 2025, https://thedrg.com/are-you-setting-your-sights-on-the-right-north-star-metric/.

[29] Emanuel Camilleri, *Key Performance Indicators: The Complete Guide to KPIs for Business Success* (Routledge, 2024).

[30] Matt Blumberg, Peter M. Birkeland, and Scott Dorsey, *Startup CXO* (Wiley, 2021).

Examples of Key Performance Indicators for Startups

The following KPIs are illustrative examples to help you get oriented—later in this chapter, we'll explore in detail how to calculate the foundational metrics and clarify what they truly measure in the context of your startup. The engine of your business is its financial health, tracked by financial KPIs. These metrics tell you about your viability, profitability, and runway. They are the undeniable numbers that investors and leadership watch closest. Key examples include

- Monthly Recurring Revenue (MRR) and Annual Recurring Revenue (ARR)

- Customer Acquisition Cost (CAC) and Customer Lifetime Value (CLV also known as LTV and CLTV)[31]

- Gross Margin and Burn Rate

- Cash Runway (How many months you can survive)

- Average Order Value (AOV)

While financials are critical, a business is nothing without its customers. Customer KPIs measure the pulse of your user base, revealing how well you attract, delight, and retain them. A sudden drop here is often a leading indicator of future financial trouble. Look to metrics like

- Customer Retention Rate vs. Churn Rate

- Net Promoter Score (NPS) and Customer Satisfaction (CSAT)

- Lead-to-Customer Conversion Rate

- Activation Rate (the percentage of new users who perform a key action)

For startups with a digital product, like a SaaS platform or mobile app, you must understand how people are interacting with what you've built. Engagement KPIs track this active usage and are crucial for identifying friction points and beloved features.[32]

---

[31] Rand Fishkin, *Lost and Founder: A Painfully Honest Field Guide to the Startup World* (Portfolio, 2018).

[32] Leandro Faria, *The essential SaaS metrics guide: How to grow your subscription business by measuring it the right way* (Leo Faria, 2015).

These include

- Daily Active Users (DAU) and Monthly Active Users (MAU)

- Session Length and Session Interval

- Feature Adoption Rate

Finally, you need to measure your momentum. Growth KPIs monitor your overall expansion and market penetration, showing how effectively you are scaling your presence and attracting new audiences. These often form the basis of your growth story.[33],[34]

Essential indicators are

- Month-on-Month (MoM) Growth Rate

- Funnel Conversion Rates

- Viral Coefficient (how many new users each existing user generates)

No single number on this list can tell you the whole story. The real power comes from seeing how they influence each other—how a dip in your Net Promoter Score might predict a rise in churn next month, or how a successful marketing campaign impacts both your Customer Acquisition Cost and Monthly Recurring Revenue. By tracking a balanced set of KPIs, you move from simply collecting data to building a comprehensive narrative of your startup's journey.[35]

Leading vs. Lagging KPIs

When considering KPIs, it is helpful to categorize them into leading and lagging indicators. Leading indicators are metrics that can predict future performance and provide early signals that can inform adjustments to strategy. Examples of leading KPIs for startups include the number of trial sign-ups, website visitor traffic, daily active users, and the Net Promoter Score.[36]

---

[33] Gustaf Alströmer, "Growth For Startups," ycombinator.com, accessed June 29, 2025, https://www.ycombinator.com/library/6k-growth-for-startups.

[34] Aaron Ross and Jason Lemkin, *From Impossible to Inevitable: How SaaS and Other Hyper-Growth Companies Create Predictable Revenue* (Wiley, 2019).

[35] Ash Maurya, *Scaling Lean: Mastering the Key Metrics for Startup Growth* (Portfolio, 2016).

[36] John Lovett, *The *NEW* Big Book of KPIs: (Key Performance Indicators)* (Independently published, 2025).

These metrics can give an indication of future trends in customer acquisition, engagement, and satisfaction. On the other hand, lagging indicators measure past performance and reflect outcomes that have already occurred. Common lagging KPIs include revenue generated, customer churn rate, overall customer satisfaction scores, and the number of paid clients. While lagging indicators provide valuable insights into the results of past efforts, they do not necessarily predict future success. It is crucial for startups to monitor both types of KPIs to gain a comprehensive understanding of their performance and to make well-informed decisions.

Leading indicators are particularly valuable in the fast-paced environment of a startup because they allow for proactive course correction. By identifying potential issues or positive trends early on, startups can make timely adjustments to their strategies before these factors significantly impact the lagging metrics. Lagging indicators, in turn, serve to validate the effectiveness of past strategies and provide a clear measure of the outcomes achieved. The interplay between leading and lagging KPIs provides a more complete picture of a startup's health and its trajectory for future growth.

The AARRR Framework: Pirate Metrics

The AARRR framework, also known as "Pirate Metrics," is a model that tracks five key stages of the customer life cycle: Acquisition, Activation, Retention, Referral, and Revenue. Created in 2007 by Silicon Valley investor and 500 Startups founder Dave McClure, AARRR was designed to help startups cut through the noise of superficial data and focus on actionable metrics that directly impact business health. Its memorable acronym, which sounds like a pirate's cry, makes it easy to remember.[37,38]

The framework is especially popular with product-led and Software-as-a-Service (SaaS) companies because it aligns perfectly with how users discover, engage with, and ultimately pay for digital products. It functions like a funnel, allowing you to identify where users are dropping off and where to focus your optimization efforts.

---

[37] Shoin Wolfe, "Pirate Metrics (AARRR)," Growth Analytics, accessed July 1, 2025, `https://www.growthanalyticsbook.com/frameworks/pirate-metrics-aarr`.

[38] Dave McClure, "Startup Metrics for Pirates" (PowerPoint presentation, SlideShare, uploaded August 21, 2007), `https://www.slideshare.net/slideshow/startup-metrics-for-pirates-long-version/89026`.

To use the AARRR framework, you identify key metrics for each stage, set up tools to track them, run experiments to improve them, and use the resulting data to inform your product and marketing strategies.

Each stage represents a milestone in transforming a prospect into a loyal, paying customer. Understanding these stages helps startups focus their efforts on the right metrics at the right time, enabling systematic growth and optimization. In the following sections, we will explore each stage in detail, highlighting key metrics and practical examples to guide your startup's growth strategy.

Acquisition: Generate Awareness and Attract a Relevant Audience

This first stage is about how you attract new users or customers to your product. It measures the volume and cost of bringing people into your ecosystem. It focuses on building awareness and attracting potential customers through channels like SEO, social media, advertising, and content marketing.

One of the simplest metrics here is the **Conversion Rate**, which tells you what percentage of visitors actually take a desired action, like signing up or making a purchase. You calculate it by dividing the number of people who take that action by the total number of visitors, then multiplying by 100 to get a percentage. For example, if 50 out of 500 visitors sign up, your conversion rate is 10%. The ideal conversion rate depends on several factors, including the business model, industry, pricing, and target audience. For example, in some fitness companies, converting 5% of website visitors into free trial users may be considered strong, while others may achieve rates of 10% or higher. Similarly, for deep tech companies selling to small and medium businesses (SMBs), converting more than 5% of free users to paid customers may be seen as a solid benchmark.[39]

---

[39] "SaaS Conversion Rate: How to Calculate, Benchmarks, Tips," startupvoyager.com, accessed June 21, 2025, https://startupvoyager.com/saas-conversion-rate/.

Figure 4-3 illustrates the flow of users through your specified conversion funnel, from the initial landing page visit to the final paid subscription. Each stage shows the absolute number of users and what percentage that number represents compared to the initial 1,000 visitors.

- **Landing Page Visit**: The funnel begins with 1,000 potential customers.

- **Checkout Page Visit**: 832 users proceed, retaining 83% of the initial traffic.

- **Free Trial Started**: 580 users start a free trial, which is 58% of the initial visitors.

- **Paid Subscription**: The process concludes with 180 users converting to a paid subscription, an 18% conversion rate from the top of the funnel.

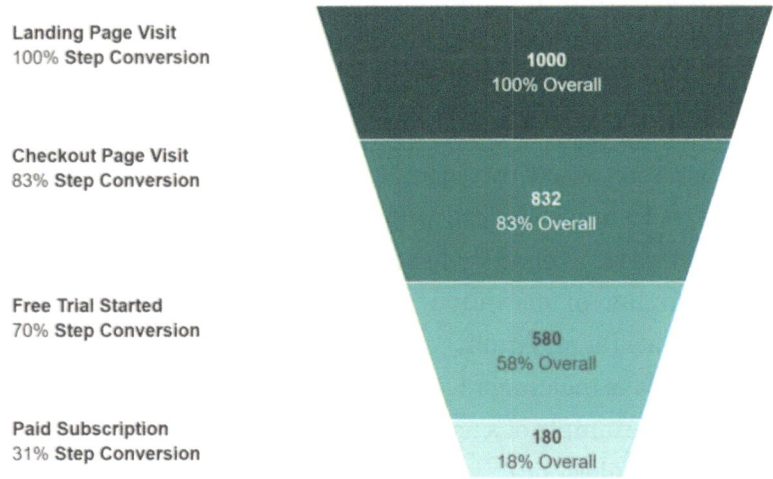

Customer Conversion Funnel

User Activity Over Last 30 Days

Landing Page Visit
100% **Step Conversion**

1000
100% Overall

Checkout Page Visit
83% **Step Conversion**

832
83% Overall

Free Trial Started
70% **Step Conversion**

580
58% Overall

Paid Subscription
31% **Step Conversion**

180
18% Overall

Figure 4-3. *Conversion Funnel Example*

Another important metric is the **Customer Acquisition Cost (CAC)**, which tells you how much money you spend to get one new customer. You find this by adding up all your sales and marketing expenses over a period and dividing that by the number of new customers acquired in the same period. For example, if you spend $10,000 on marketing and get 100 new customers, your CAC is $100. This cost should be less than the revenue you expect to earn from a customer (at minimum) and, ideally much less, to ensure profitability and growth.[40]

Click-Through Rate (CTR) is used when you run ads; it measures how many people click your ad compared to how many saw it. If 20 people click on an ad shown 1,000 times, your CTR is 2%. A higher CTR means your ads are more effective at grabbing attention.

Activation: Guide Users to Experience the Product's Value Proposition

Once you've acquired users, the next step is to ensure they take meaningful action and experience your product's core value. This is the transition from a casual visitor to an engaged user.

The **Activation Rate** is the percentage of new users who complete a key action that shows they've "activated" (the "aha moment"). This might be completing a tutorial, setting up a profile, or making a first purchase. The definition of "activation" is specific to each product; for Dropbox, it was a user uploading one file, while for early Facebook, it was a user making seven friends in ten days. You calculate it by dividing the number of users who reached this milestone by the total number of new users. For instance, if 200 out of 1,000 new users complete the activation step, your activation rate is 20%. There is no universal benchmark for a "good" activation rate because it varies widely by product, industry, and the definition of "activation." Some sources and growth experts mention that a 25% activation rate (at minimum) can be a positive indicator for early-stage products, suggesting that a meaningful portion of users are experiencing value.

Figure 4-4 illustrates an example of a user activation funnel where Activation Rate is 20% (for reaching activation milestone).[41]

---

[40] Matt Preuss, "Customer Acquisition Cost: A Critical Metrics for Founders," visible.vc blog, accessed July 02, 2025, `https://visible.vc/blog/customer-acquisition-cost/`.

[41] Lenny Rachitsky, "How To Determine Your Activation Metric," lennysnewsletter.com, accessed June 25, 2025, `https://www.lennysnewsletter.com/p/how-to-determine-your-activation`.

User Activation Funnel

Onboarding Activity Analysis

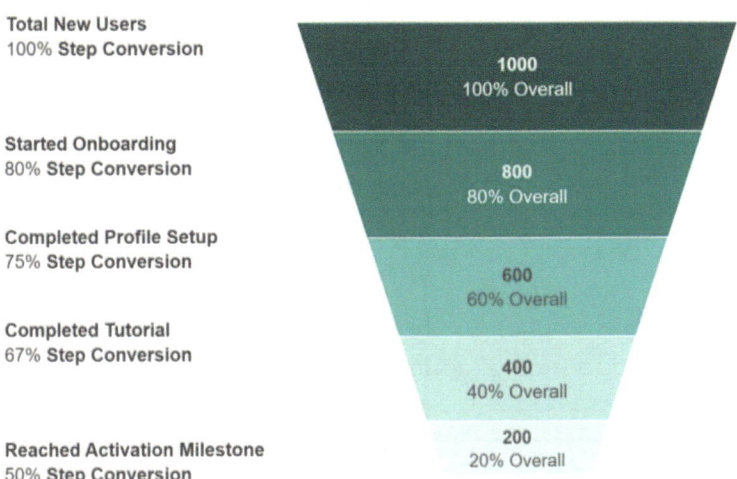

Total New Users
100% **Step Conversion**

1000
100% Overall

Started Onboarding
80% **Step Conversion**

800
80% Overall

Completed Profile Setup
75% **Step Conversion**

600
60% Overall

Completed Tutorial
67% **Step Conversion**

400
40% Overall

Reached Activation Milestone
50% **Step Conversion**

200
20% Overall

Figure 4-4. *Activation Funnel Example Showing User Activation Progression with Counts and Percentages*

Time to Activate tracks how long it takes a user to reach that activation milestone after signing up. The shorter this time, the better, because quick value realization reduces drop-offs.

For freemium or trial products, the **Free-to-Paid Conversion Rate** measures how many free users become paying customers (as illustrated in the last step of Figure 4-3). Based on data from over 1,000 products summarized by Lenny's Newsletter, general benchmarks help define "good" free-to-paid conversion rates across different product models (see Table 4-4).[42,43]

---

[42] Kyle Poyar and Lenny Rachitsky, "What Is A Good Free-To-Paid Conversion," lennysnewsletter. com, August 1, 2023, `https://www.lennysnewsletter.com/p/what-is-a-good-free-to-paid-conversion`.

[43] Kyle Poyar and Lenny Rachitsky, "What Is A Good Free-To-Paid Conversion," lennysnewsletter. com, August 1, 2023, `https://www.lennysnewsletter.com/p/what-is-a-good-free-to-paid-conversion`.

Table 4-4. Free-to-Paid Conversion Rate Benchmarks (Lenny's Newsletter)

Product Model	Good Conversion Rate	Why This Rate Is Typical
Freemium Self-Serve	3–5%	Freemium model attracts a high volume with low initial commitment. Only a small fraction of users needs to upgrade when convinced purely by product's value.
Freemium with Sales Assist	5–7%	The sales team's human touch helps achieve a better conversion rate than the self-serve model by proactively engaging with the most promising free users.
Free Trial	8–12%	By offering full product access for a limited time, a free trial creates urgency and attracts users who are more likely to buy.

While these benchmarks offer a valuable directional guide, they should not be treated as absolute truths. Because this data is rarely shared, any available figures come from a limited sample set and may not apply to your business. The "right" rate for your business is highly contextual and will be influenced by factors like your industry, price point, and product complexity. Therefore, use these figures as a starting point, but prioritize your own market-specific research for an accurate comparison.

Retention: Ensure Users Return to the Product on a Recurring Basis

Retention measures whether activated users continue to engage with your product over time. High retention is a strong signal that your product consistently delivers value, and it's far more cost-effective to retain existing customers than to acquire new ones.[44]

The Customer Retention Rate (CRR) tells you what percentage of your customers remain active after a certain period. To calculate it, you take the number of customers you have at the end of the period, subtract new customers gained during that time,

---

[44] Andrew Chen, Jeff Jordan, and Sonal Chokshi, "The Basics of Growth — Engagement & Retention," Andreessen Horowitz Podcast, accessed May 20, 2025, https://a16z.com/podcast/a16z-podcast-the-basics-of-growth-engagement-retention/.

and then divide by the number of customers you started with. Multiply by 100 to get a percentage. For example, if you start with 5,000 customers, gain 250 new ones, and end with 5,050, your retention rate is about 96%.

While numerous factors like product complexity and pricing influence retention, the most practical starting point for setting a benchmark is your business category. This metric, known as **Logo Retention** in the SaaS industry, measures the percentage of customers you keep over a given period. Its opposite, the **Churn Rate**, measures the customers you lose. If 10 out of 100 customers leave in a month, your churn rate is 10%.[45]

To provide concrete benchmarks, influential research from Lenny's Newsletter surveyed top growth experts and analyzed public data to establish guidelines for user retention. Table 4-5 summarizes what is considered "good" and "great" six-month user retention across different industries.[46,47]

Table 4-5. *Six-Month Retention Rate Benchmarks (Data from Lenny's Newsletter)*

Business Category	Good Retention	Great Retention
Enterprise SaaS	70%	90%
Mid-market/SMB SaaS	60%	80%
Consumer SaaS	40%	70%
Consumer Transactional	30%	50%
Consumer Social	25%	45%

As the table illustrates, what counts as a strong retention rate is highly contextual. The key is to understand the standard for your specific market and use it as a baseline to measure and improve upon.

---

[45] Lenny Rachitsky, "What Is Good Retention?," lennysnewsletter.com, June 9, 2020, `https://www.lennysnewsletter.com/p/what-is-good-retention-issue-29`.

[46] Lenny Rachitsky, "What Is Good Retention?," lennysnewsletter.com, June 9, 2020, `https://www.lennysnewsletter.com/p/what-is-good-retention-issue-29`.

[47] Lenny Rachitsky, "What Is Good Retention?," lennysnewsletter.com, June 9, 2020, `https://www.lennysnewsletter.com/p/what-is-good-retention-issue-29`.

Active Users metrics count how many unique users engage with your product daily (DAU), weekly (WAU), or monthly (MAU). A strong product often has a DAU to MAU ratio above 50%, meaning users come back frequently. The DAU to MAU ratio is also called the Stickiness Metric and was popularized by Facebook.[48,49]

Cohort Analysis is a powerful technique that groups users by a shared characteristic—typically their sign-up date—and tracks their behavior over time. This helps you see if product changes are improving retention for newer users compared to older ones and identify critical points where users tend to drop off.[50]

Figure 4-5 visualizes this concept using a heatmap. Each row represents a cohort (users who signed up in the same month), and each column tracks their engagement over the subsequent months since sign-up. The number in each cell shows the percentage of the original cohort that remained active.

The color gradient makes interpretation intuitive: green indicates strong retention, while red signals a drop-off. The blank cells represent future periods not yet reached by newer cohorts, creating the characteristic triangular pattern.

How to Read This Chart

- **Horizontally (Across a Row):** Tracks the life cycle of a single cohort to pinpoint when users lose interest

- **Vertically (Down a Column):** Compares newer cohorts to older ones at the same point in their life cycle—a crucial way to measure if your product is getting "stickier" over time

---

[48] Andrew Chen, "DAU/MAU is an important metric to measure engagement, but here's where it fails," Andrew Chen Blog, accessed June 22, 2025, https://andrewchen.com/dau-mau-is-an-important-metric-but-heres-where-it-fails/.

[49] Bennett Carroccio, "The Stickiest, Most Addictive, Most Engaging, and Fastest-Growing Social Apps—and How to Measure Them," Andreessen Horowitz, accessed June 20, 2025, https://a16z.com/the-stickiest-most-addictive-most-engaging-and-fastest-growing-social-apps-and-how-to-measure-them/.

[50] Andrew Chen and Brian Balfour, "Evaluate Retention Cohorts," reforge.com, accessed July 02, 2025, https://www.reforge.com/guides/evaluate-retention-cohorts.

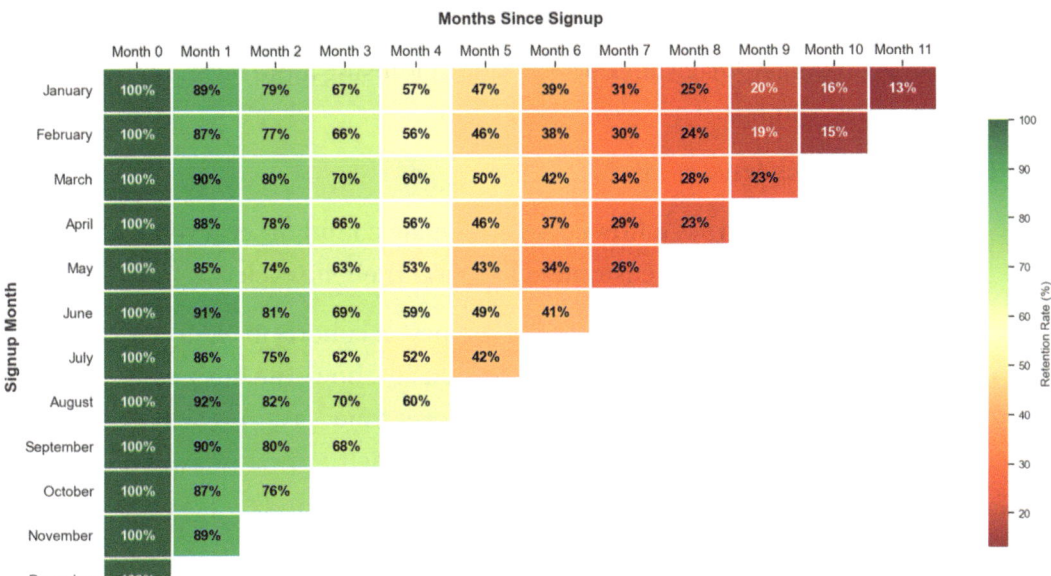

Figure 4-5. *Cohort Analysis—Engagement Retention Example*

Net Dollar Retention (NDR), also called **Dollar Retention**, measures how revenue from a customer cohort grows or shrinks over time. It accounts for all revenue dynamics after the initial sale: expansion (upsells), contraction (downgrades), and churn (lost customers). Unlike user retention, a healthy NDR for a SaaS business should exceed 100%, proving that the value delivered to customers is growing. As VC David Sacks states, the best SaaS companies achieve over 120% NDR each year, while less than 100% indicates a "leaky bucket."[51]

Figure 4-6 visualizes NDR using a cohort heatmap. Each row represents a cohort of customers acquired in a given month, and the columns track their revenue performance over the subsequent 12 months. The value in each cell represents the NDR for that cohort. For example, a value of 115% means that for every $100 of initial revenue, the cohort is now generating $115. The color scale is centered at 100%:

---

[51] David Sacks and Ethan Ruby, "The SaaS Metrics That Matter," sacks.substack.com, accessed June 30, 2025, https://sacks.substack.com/p/the-saas-metrics-that-matter.

- Green (>100%): Healthy growth. Expansion revenue is greater than revenue lost to churn and contraction.

- Red (<100%): A "leaky bucket." The cohort is losing more revenue than it is gaining.

How to read this chart:

- **Horizontally (Across a Row):** Follow a single cohort to see if you are successfully increasing customer value over their life cycle.

- **Vertically (Down a Column):** Compare newer cohorts to older ones. In this example, the increasing green intensity shows the business is improving its ability to expand accounts over time—a strong signal of product-market fit and a healthy business model.

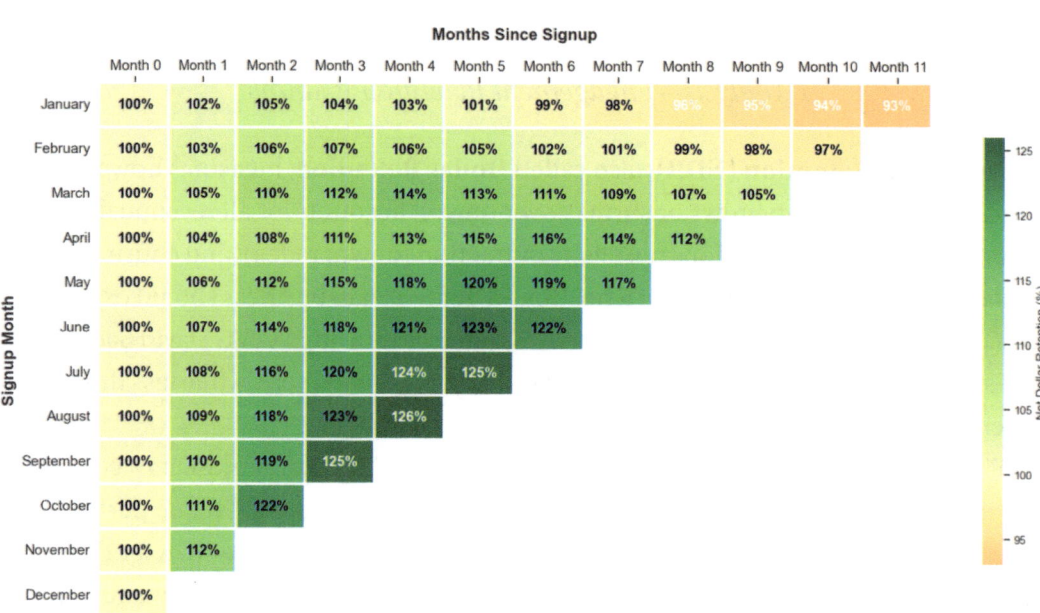

Figure 4-6. *Net Dollar Retention Example—Presented in Cohort Analysis*

While Net Dollar Retention (NDR) is typically discussed in percentages, visualizing the data in absolute dollars provides a more tangible understanding of its financial impact. This chart serves as a direct companion to the previous NDR heatmap, translating the percentages into concrete revenue figures.[52]

Figure 4-7 presents Revenue by Cohort by Month. This heatmap shows the actual dollar value of Monthly Recurring Revenue (MRR) generated by each customer cohort, assuming each started with an initial revenue of $1,000,000. Each cell displays the total MRR from that cohort, formatted in thousands for readability (e.g., "$1,250k" represents $1,250,000).

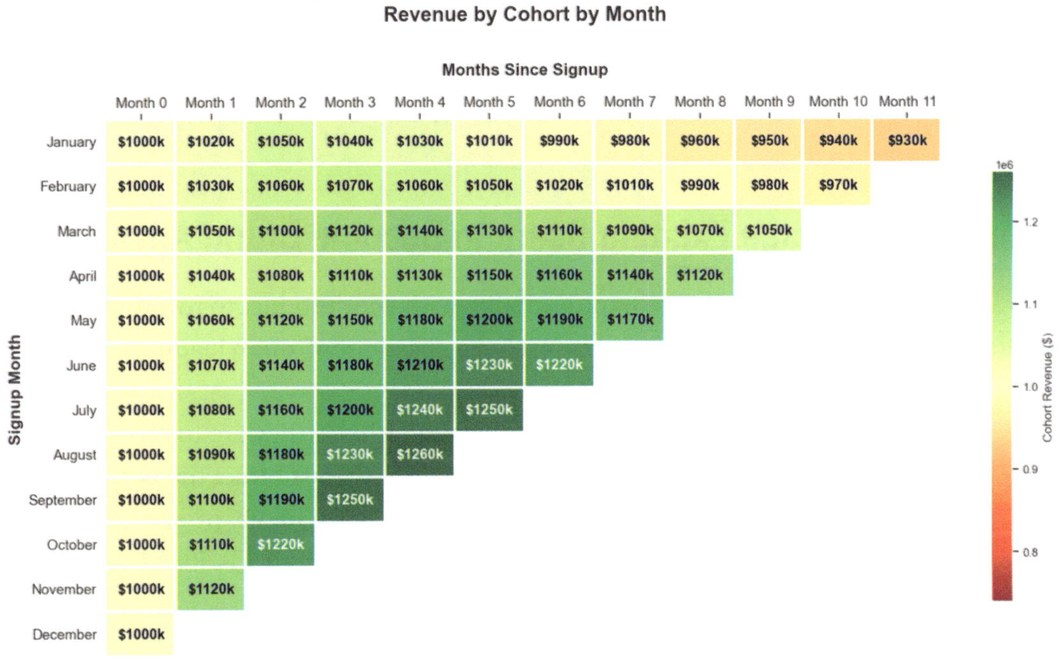

Figure 4-7. *Revenue by Cohort by Month Example—Presented in Cohort Analysis*

To ensure a direct and intuitive comparison, the color scale is identical to the one used in the percentage-based NDR chart.

---

[52] Bobby Pinero, "Cohorted Retention," equals.com, accessed June 23, 2025, `https://equals.com/guides/saas-metrics/cohorted-retention/`.

- **Green**: The cohort's revenue has grown beyond its initial $1,000,000. The September cohort, for example, grew to generate $1,250,000 by its third month.

- **Red**: The cohort's revenue has shrunk below its initial $1,000,000. The January cohort churned down to $930,000 by the end of its first year.

This view makes the concept of a "leaky bucket" tangible—you can see the actual dollars being lost. More importantly, it highlights just how powerfully a business with strong net retention can grow, with newer cohorts compounding real dollars on top of the initial sales.

We will dive deeper into the nuances of cohort analysis later in this book. There are numerous metrics that can be explored using cohort analysis that are widely used in startups.

Referral: Encourage Existing Users to Become Brand Advocates

This stage tracks whether users are happy enough with your product to recommend it to others. Referrals are powerful because they rely on social proof, often leading to higher conversion rates and lower acquisition costs.[53]

The Net Promoter Score (NPS) asks customers how likely they are to recommend your product on a scale from 0 to 10 (see Figure 4-8). You then subtract the percentage of unhappy customers (those scoring 0-6) from the percentage of happy promoters (9-10). For example, if 60% are promoters and 10% are detractors, your NPS is 50. A high NPS correlates with more referrals and better retention.[54]

---

[53] Andrew Chen, "How to Design a Referral Program," Andrew Chen Blog, accessed June 27, 2025, https://andrewchen.com/how-to-design-a-referral-program/.

[54] Andrew Chen, "A Practitioner's Guide to Net Promoter Score," Andrew Chen Blog, accessed June 28, 2025, https://andrewchen.com/a-practitioners-guide-to-net-promoter-score/.

Net Promoter Score (NPS) Scale

How likely are you to recommend our product to a friend or colleague?

Figure 4-8. *Net Promoter Score*

The **Viral Coefficient (K-factor)** measures how many new users each existing user brings in through referrals. It's calculated by multiplying the average number of invitations each user sends by the percentage of those invitations that convert into new users. For example, if each user sends 10 invites and 5% convert, your K-factor is 0.5. If your viral coefficient is 1 or higher, each user you bring in will generate at least one more new user through referrals (viral growth).[55]

Referral Rate is the percentage of new customers who come from referrals. If 400 out of 20,000 new customers were referred, your referral rate is 2%. Referred customers tend to stay longer and spend more.

Revenue: Convert the Value You Provide into a Profitable Income Stream

The final stage, Revenue, is the ultimate validation of your product's value. It focuses on monetizing your user base to create a sustainable business.

Monthly Recurring Revenue (MRR) is the total predictable income you get each month from subscriptions. To illustrate this concept, which is fundamental in books about data for startups, Figure 4-9 provides a detailed MRR analysis for a hypothetical startup during Q4 2024. This chart shows how a starting value is increased and decreased by various factors.[56]

---

[55] Mike Preuss, "K Factor: What is your SaaS Company's Viral Coefficient?," visible.vc, accessed June 26, 2025, `https://visible.vc/blog/k-factor-what-is-your-saas-companys-viral-coefficient/`.

[56] Matt Preuss, "Monthly Recurring Revenue (MRR) Explained: Definitions + Formulas," visible.vc, accessed June 27, 2025, `https://visible.vc/blog/mrr/`.

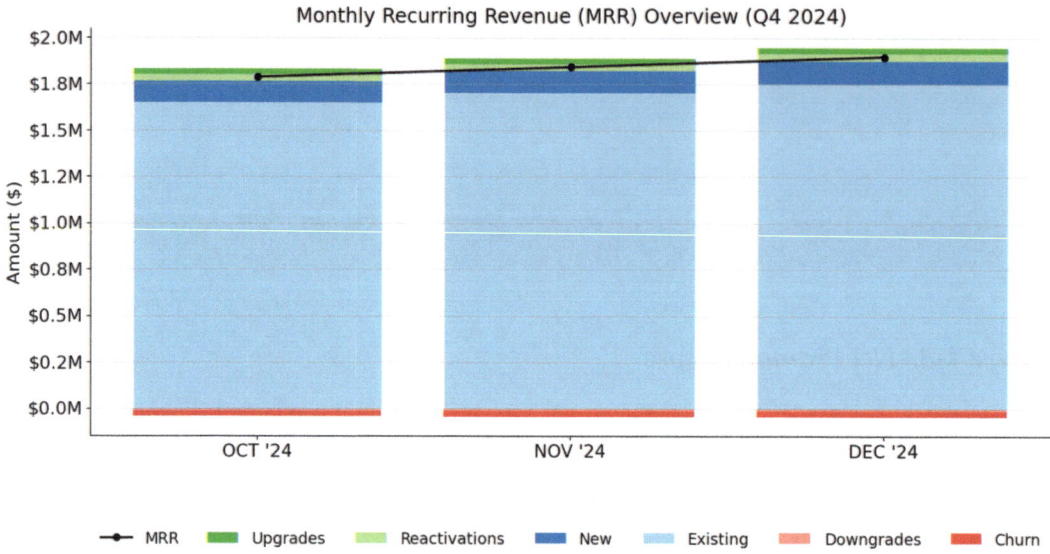

Figure 4-9. *Monthly Recurring Revenue (MRR)*

It is important to note how the bars and the line interact. The positive bars (blue and green) stack up to show the gross revenue additions for the month. The red bars represent revenue subtractions (Downgrades and Churn). The black line indicates the Net MRR, which is the final value after the subtractions have been applied to the additions. This is why the line sits below the top of the green bars—it represents the true, final revenue for the month.

The legend is ordered to match the chart's logic, starting with the final result (MRR) and then listing the bar components from top to bottom:

- **MRR**: The net Monthly Recurring Revenue after all additions and subtractions

- **Upgrades**: Additional income from customers moving to higher-tier plans

- **Reactivations**: Revenue from returning customers

- **New**: Income from new customer acquisitions

- **Existing**: The base revenue from customers at the start of the month

- **Downgrades**: Revenue lost from customers switching to lower-tier plans

- **Churn**: Revenue lost from customers who cancel their subscriptions

Annual Contract Value (ACV) is the average yearly revenue from a customer contract, especially important in B2B. If a customer pays $36,000 for a 3-year contract, the ACV is $12,000 per year.[57]

Average Revenue Per User (ARPU) shows how much revenue you make on average from each user. You calculate it by dividing total revenue by the number of users. For instance, $10,000 revenue from 1,000 users means an ARPU of $10.[58]

Figure 4-10 illustrates the Average Revenue Per User (ARPU) for Meta Platforms from 2011 to 2024, measured in US dollars. The data shows a significant and mostly consistent increase in ARPU over this period. It began at $5.00 in 2011 and rose to $49.63 by 2024. There was a slight dip in 2022, but the upward trend resumed in the following years.[59,60]

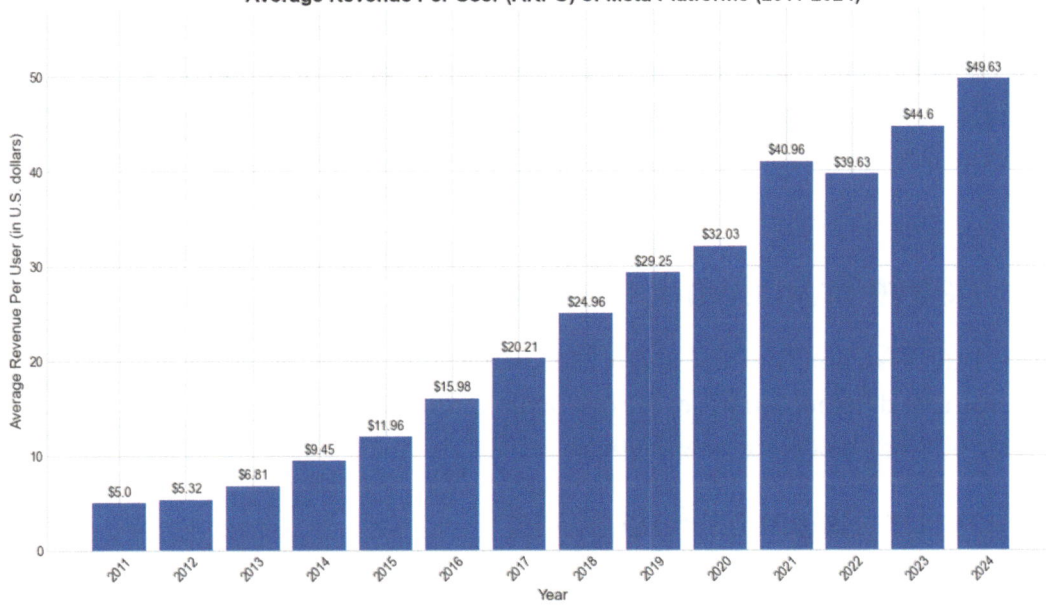

Figure 4-10. *Average Revenue Per User (ARPU) of Meta Platforms (2011-2024) from Source Data*

---

[57] Matt Preuss, "Breaking Down the Nuances of Annual Contract Value (ACV)," visible.vc, accessed June 28, 2025, `https://visible.vc/blog/annual-contract-value/`.

[58] Anu Hariharan, Frank Chen, and Jeff Jordan, "16 More Startup Metrics," Andreessen Horowitz, accessed June 29, 2025, `https://a16z.com/16-more-startup-metrics/`.

[59] Stacy Jo Dixon, "Average revenue per user (ARPU) of Meta Platforms from 2011 to 2024," Statista, January 2025, accessed July 4, 2025, `https://www.statista.com/statistics/234056/facebook-average-advertising-revenue-per-user/`.

[60] Stacy Jo Dixon, "Average revenue per user (ARPU) of Meta Platforms from 2011 to 2024," Statista, January 2025, accessed July 4, 2025, `https://www.statista.com/statistics/234056/facebook-average-advertising-revenue-per-user/`.

Customer Lifetime Value (LTV) estimates how much profit you expect to earn from a customer over their entire relationship with you. You calculate it by multiplying the profit you make per customer each month by the average number of months they stay with you. For example, if a customer generates $100 profit per month and stays for 20 months, the LTV is $2,000.

The **LTV to CAC Ratio** compares how much value a customer brings vs. how much it costs to acquire them. If your LTV is $3,000 and CAC is $1,000, the ratio is 3:1, which is considered healthy. Ideally, you want this ratio to be at least 3, meaning customers bring in three times what you spend to get them.[61]

While the AARRR metrics I have outlined serve as some of the most popular and practical tools for understanding startup growth, they represent only the beginning of a much broader conversation. The landscape of growth measurement is rich with nuance and complexity, offering countless additional metrics, frameworks, and perspectives. In the following sections of this chapter, we'll dive deeper into other frameworks that matter for startups, providing a more comprehensive view of how to track and drive progress.

Alternative Frameworks for Different Priorities

While AARRR is a powerful and widely used framework, other models have emerged that prioritize different aspects of the customer journey, making them better suited for specific business models or growth stages.

RARRA: The Retention-First Model

The RARRA framework flips AARRR on its head by starting with Retention, followed by Activation, Referral, Revenue, and finally Acquisition. Developed by Gabor Papp and Thomas Petit, this model is ideal for product-led growth companies, especially in competitive markets with high customer acquisition costs. The logic is that it's more efficient to build a product that users love and stick with, as a highly engaged user base will naturally drive referrals and revenue, making acquisition a more organic outcome.[62]

---

[61] Jamie Sullivan and Alex Immerman, "Why Do Investors Care So Much About LTV:CAC?" Andreessen Horowitz, accessed June 17, 2025, `https://a16z.com/why-do-investors-care-so-much-about-ltvcac/`.

[62] Gabor Papp, "Why Focusing Too Much on Acquisition Will Kill Your Mobile Startup," phiture.com, accessed June 24, 2025, `https://phiture.com/mobilegrowthstack/why-focusing-on-acquistion-will-kill-your-mobile-startup-e8b5fbd81724/`.

RACE: A Digital Marketing Perspective

The RACE framework (Reach, Act, Convert, Engage) was created by digital marketing expert Dr. Dave Chaffey to structure online marketing activities.[63]

- **Reach**: Build brand awareness and attract visitors.

- **Act**: Encourage interaction on your website or social media.

- **Convert**: Turn visitors into paying customers.

- **Engage**: Build long-term customer relationships and loyalty.

RACE is customer centric and data driven, providing a practical model for planning and optimizing a digital marketing strategy from initial awareness to long-term advocacy.

HEART: A Focus on User Experience

Developed by a UX research team at Google, the HEART framework measures user experience through five categories: Happiness, Engagement, Adoption, Retention, and Task success. Unlike AARRR's focus on business growth, HEART helps teams quantify user satisfaction to make objective, data-backed design decisions. It's particularly useful for software UX teams aiming to connect design choices directly to business goals and identify processes that might cause customer churn.[64]

Comparing the Frameworks

Each framework offers a different lens through which to view and measure your startup's growth. The best choice depends on your business model, growth stage, and strategic priorities. Table 4-6 compares different metrics frameworks.

---

[63] Dave Chaffey, "Race Marketing Model," davechaffey.com, accessed June 24, 2025, `https://www.davechaffey.com/digital-marketing-glossary/race-marketing-planning-model/`.

[64] "How to Choose the Right UX Metrics for Your Product," HEART Framework, accessed June 24, 2025, `https://www.heartframework.com/`.

Table 4-6. *Metrics Frameworks Comparison*

Framework	Primary Focus	Best For
AARRR	Acquisition, Activation, Retention, Referral, Revenue: customer acquisition and conversion funnel optimization	Startups and growth teams needing a simple, actionable model for the entire customer journey
RARRA	Retention, Activation, Referral, Revenue, Acquisition: customer retention and product-led growth	SaaS and mobile apps in competitive markets with high acquisition costs
RACE	Reach, Act, Convert, Engage: digital marketing planning and management	Marketers looking to structure and optimize their online activities across the customer life cycle
HEART	Happiness, Engagement, Adoption, Retention, Task success: user experience (UX) and customer satisfaction	Product and UX teams needing to make data-driven design decisions

While frameworks like AARRR provide a robust foundation for organizing your metrics, understanding the raw numbers is only the first step. To truly unlock the power of your data, you need to go beyond surface-level insights and employ more sophisticated analytical techniques. The next section delves into advanced methods that reveal the subtle patterns and underlying drivers of your startup's performance.

Summary

This chapter has provided a comprehensive guide to understanding and leveraging key metrics in the dynamic world of startups. We began by establishing why metrics are indispensable, drawing inspiration from companies like Airbnb and emphasizing that selecting the right metrics is crucial for survival and growth in resource-constrained environments. We distinguished between "good" metrics—those that are understandable, actionable, comparative, and presented as ratios—and misleading vanity metrics, which can create a false sense of security and lead to poor decisions.

A core theme explored was the concept of a **North Star Metric (NSM)**, the single most important measure that aligns an entire organization toward customer value and sustainable growth while acknowledging that this metric must evolve with the startup's stage. We delved into **Key Performance Indicators (KPIs)**, categorizing them into financial, customer, engagement, and growth metrics, and highlighting the critical distinction between leading and lagging indicators for proactive decision-making.

The chapter extensively covered prominent metrics frameworks, including the widely adopted **AARRR (Acquisition, Activation, Retention, Referral, Revenue)**, detailing how each stage measures customer journey progress and identifies optimization opportunities. We also examined alternative frameworks such as RARRA, RACE, and HEART, each offering a different lens based on specific business priorities.

In Chapter 5, we will elaborate on the techniques required to move beyond foundational metrics. We will explore methods like cohort analysis to understand user value over time, power user curves to identify your most engaged customers, and network effects measurement to determine if your product is building a defensible moat. While these powerful tools provide a more granular understanding of business dynamics, they should be seen as illustrative examples, not an exhaustive list. Every startup's context is unique, and the goal of the next chapter is to equip you with the thinking required to adapt these advanced methods and extract deeper, more actionable insights from your own data.

Appendix: The AARRR Metrics Toolkit

Table 4-7 summarizes the core metrics of the AARRR framework as discussed in this chapter. Use it as a quick guide to define, calculate, and set goals for the numbers that matter most at each stage of the customer life cycle. Understanding the metrics outlined there is fundamental for any startup founder or data professional. It's not just about crunching numbers. It's a diagnostic toolkit for your startup. Mastering these numbers allows you to pinpoint exactly where your growth engine is succeeding or failing.

Table 4-7. *Quick Reference: The AARRR Metrics Toolkit*

Stage	Metric	Description
Acquisition	Conversion Rate	**Purpose:** Measures the percentage of visitors who take a specific desired action (e.g., sign up). **Formula:** (Number of Conversions / Total Visitors) x 100 **Example:** If 50 out of 500 visitors sign up, your conversion rate is 10%. **Goal:** Varies widely based on industry, traffic source, and product complexity. Prioritize continuous improvement over a universal benchmark.
	Click-Through Rate (CTR)	**Purpose:** Measures the effectiveness of an ad or link in attracting a click. **Formula:** (Total Clicks / Total Impressions) x 100 **Example:** If an ad is shown 1,000 times and receives 20 clicks, the CTR is 2%. **Goal:** Varies by channel. A higher CTR indicates more compelling ad copy and targeting.
	Customer Acquisition Cost (CAC)	**Purpose:** The total cost to acquire a single new paying customer. **Formula:** Total Sales & Marketing Spend / # of New Customers Acquired **Example:** Spending $10,000 on marketing to get 100 new customers results in a $100 CAC. **Goal:** Keep it significantly lower than your LTV (a common target is LTV > 3x CAC).
Activation	Activation Rate	**Purpose:** The percentage of new users who experience the product's "aha moment" by completing a key action. **Formula:** (# of Users Who Hit Activation Milestone / # of New Users) x 100 **Example:** If 200 of 1,000 new sign-ups create their first project (the milestone), the activation rate is 20%. **Goal:** >25% is often considered a positive sign for early-stage startups. However, it's a context-dependent metric.

	Time to Activate	**Purpose:** Measures the average time it takes for a new user to reach the activation milestone. **Formula:** Average (Time of Activation - Time of Sign-up) **Example:** If users take, on average, 2 days to create their first project, the Time to Activate is 2 days. **Goal:** Shorten this time as much as possible through better onboarding.
	Free-to-Paid Conversion	**Purpose:** The percentage of free users who become paying customers. **Formula:** (# of New Paying Customers / # of New Free Users) x 100 **Example:** If 25 of 500 new free trial users convert, the rate is 5%. **Goal:** Varies by model (e.g., 3–8% for freemium, 15–25% for free trials). It's a context-dependent metric.
Retention	**Customer Retention Rate**	**Purpose:** The percentage of customers you keep over a specific period. **Formula:** ((End Customers - New Customers) / Start Customers) x 100 **Example:** If you start with 100 customers, gain 20, and end with 110, retention is 90%. **Goal:** Varies by industry and business model; >80% annually is strong for many B2B SaaS models.
	Churn Rate	**Purpose:** The inverse of retention. The percentage of customers you lose over a specific period. **Formula:** (# of Customers Lost / # of Customers at Start) x 100 **Example:** If you start with 100 customers and 5 leave, your churn is 5%. **Goal:** As close to 0% as possible. A "good" churn rate is highly dependent on the business model.

(continued)

165

Table 4-7. (*continued*)

Stage	Metric	Description
	Active Users (DAU, WAU, MAU)	**Purpose:** Measures the number of unique users who engage with your product daily (DAU), weekly (WAU), or monthly (MAU).
		Formula: A simple count of unique users over the period.
		Example: Your app has 10,000 DAU, 20,000 WAU, and 50,000 MAU.
		Goal: Grow this number consistently, as it reflects a healthy, engaged user base.
	DAU/MAU Stickiness	**Purpose:** Measures user engagement by showing how many of your monthly users return on a daily basis.
		Formula: (Daily Active Users / Monthly Active Users) x 100
		Example: With 2,000 DAU and 10,000 MAU, your stickiness is 20%.
		Goal: >50% is a strong target for products designed for daily use, showing high user engagement. Highly context dependent.
	Net Dollar Retention (NDR)	**Purpose:** Measures revenue change from an existing cohort of customers, including expansion and churn.
		Formula: ((Start MRR + Expansion - Downgrades - Churn) / Start MRR) x 100
		Example: Start with $100k MRR, add $20k in upsells, and lose $10k to churn. NDR is 110%.
		Goal: For SaaS >120% is excellent, indicating strong growth from the existing base.

Referral	Viral Coefficient (K-Factor)	**Purpose:** The average number of new users that each existing user generates.
		Formula: (# of Invites Sent per User) x (Conversion Rate of Invites)
		Example: If each user sends 4 invites and 25% convert, your K-Factor is 1.
		Goal: A sustained K-Factor greater than 1 indicates viral growth.
	Referral Rate	**Purpose:** The percentage of your total user base that was acquired through referrals.
		Formula: (# of Customers from Referrals / Total Customers) x 100
		Example: If 200 of your 1,000 customers came from referrals, the rate is 20%.
		Goal: Maximize this to fuel low-cost, organic growth.
	Net Promoter Score (NPS)	**Purpose:** Measures customer loyalty and the likelihood of word-of-mouth referrals.
		Formula: % Promoters (score 9–10) - % Detractors (score 0–6)
		Example: With 70% Promoters and 10% Detractors, your NPS is 60.
		Goal: Any score above 0 is considered good.
Revenue	MRR (Monthly Recurring Revenue) and ARR	**Purpose:** Tracks the predictable, recurring revenue from subscriptions.
		Formula: MRR = The net monthly recurring revenue after all additions and subtractions; ARR = Annual Recurring Revenue
		Example: With 100 customers paying on average $50/month (after all additions and subtractions), MRR is $5,000.
		Goal: To achieve consistent, predictable month-over-month growth.

(continued)

Table 4-7. (*continued*)

Stage	Metric	Description
	ACV (Annual Contract Value)	**Purpose:** The average annual revenue from a contract. **Example:** If a customer pays $36,000 for a 3-year contract, the ACV is $12,000 per year. **Goal:** The target ACV should align with your sales model (e.g., a low ACV needs a self-serve motion, while a high ACV can support an enterprise sales team). Growth indicates successful upselling.
	ARPU (Avg. Revenue Per User)	**Purpose:** The average revenue generated per user. **Formula:** Total Revenue in Period / # of Users **Example:** $10,000 revenue from 1,000 users means an ARPU of $10. **Goal:** Increase this over time through upselling and price optimization.
	LTV (Customer Lifetime Value, also known as CLTV or CLV)	**Purpose:** The total revenue you expect to generate from a single customer over their lifetime. **Formula:** The profit you make per customer each month multiplied by the average number of months they stay with you. **Example:** If a customer generates $100 profit per month and stays for 20 months, the LTV is $2,000. **Goal:** Should be significantly higher than CAC (e.g., LTV/CAC ratio of 3:1 or higher).
	LTV to CAC Ratio	**Purpose:** Compares a customer's lifetime value to their acquisition cost; a core measure of business model viability. **Formula:** LTV / CAC **Example:** If your LTV is $1,000 and your CAC is $250, your ratio is 4:1. **Goal:** A ratio of 3:1 or higher is considered a healthy and sustainable business model.

CHAPTER 5

Advanced Growth Analysis

Restaurants with great food seem to prosper no matter what. A restaurant with great food can be expensive, crowded, noisy, dingy, out of the way, and even have bad service, and people will keep coming. It's true that a restaurant with mediocre food can sometimes attract customers through gimmicks. But that approach is very risky. It's more straightforward just to make the food good. It's the same with technology. You hear all kinds of reasons why startups fail. But can you think of one that had a massively popular product and still failed?

—Paul Graham,

Programmer, writer, and investor[1]

Paul Graham's insight is the cornerstone of lasting success: a great product is non-negotiable. But in a competitive market, how do you prove your product is truly great? Answering this question is the difference between temporary traction and sustainable growth.

Let's frame this challenge through a familiar lens. Think of your startup as a restaurant where your goal is to serve an experience so compelling that customers don't just visit; they become regulars. But how can you be sure your product is hitting the mark? Are people coming back for more? Who are your most loyal patrons, and what keeps them returning? Answering these questions is the difference between running

---

[1] Paul Graham, "How to Start a Startup," Paul Graham, March, 2005, `https://paulgraham.com/start.html`.

© Piotr Sidoruk 2026

P. Sidoruk, *From Data to Dollars*, https://doi.org/10.1007/979-8-8688-1898-1_5

a flash-in-the-pan eatery and building an institution. These aren't just restaurant problems. They are the fundamental questions every startup should answer to achieve sustainable growth.

While foundational metrics provide an essential map of your business, they only show you what is happening. To truly navigate the path to product-market fit, you must dig deeper to understand the why. This chapter moves beyond simple snapshots to uncover the underlying dynamics of your business, equipping you with powerful tools for a more granular understanding of user behavior. In this chapter, you will learn how to

- **Utilize Cohort Analysis** to track user value over time, assess the health of your business model, and understand how customer behavior evolves.

- **Implement Power User Curves** to move beyond simple averages and identify your most engaged, high-frequency users who form the core of your product.

- **Measure Network Effects** to determine if your product is building a defensible moat and becoming more valuable as it grows.

It's crucial to remember that this is not an exhaustive list. Every startup possesses a unique business model and operates within a distinct industry. The applicability and utility of these techniques will vary based on your specific context, growth stage, and the nuances of your product. Consider these advanced methods as illustrative examples, designed to spark your thinking about how to extract deeper, more actionable insights from your own data.

Deeper Insights from Cohort Analysis: Tracking Your Business Model

Beyond showing if your product is getting "stickier," cohort analysis is the primary tool for assessing the health and trajectory of your entire business model. Investors love cohort analysis because it cuts through the noise of top-line growth to reveal the underlying quality of a business. As a16z emphasizes, strong retention is the foundation of growth, but true product-market fit is often seen when cohorts not only stick around but also expand their value over time. This is why metrics like Net Dollar Retention are so critical.[2]

But the analysis doesn't stop there. Cohort analysis can reveal seasonality (e.g., do users who sign up in the summer behave differently?) or the impact of specific marketing campaigns. By segmenting cohorts—for instance, by acquisition channel, geography, or initial plan—you can uncover which user segments are most valuable and where your product truly resonates. This allows you to focus your resources effectively.

Ultimately, cohort analysis tells a story. It shows if you are acquiring better customers, if your product is improving, and if your business is becoming more efficient over time. To see this in action, we can extend our cohort analysis to three of the most important metrics for a startup: Lifetime Value (LTV), Customer Acquisition Cost (CAC), and the all-important LTV/CAC ratio. While we've introduced LTV and CAC as key metrics, their true power is revealed when analyzed by cohort over time.[3]

Lifetime Value (LTV) by Cohort

Tracking the cumulative LTV of each cohort shows how much value you extract from customers over time. A healthy business should see LTV consistently increase as cohorts mature. Furthermore, by comparing LTV across different cohorts at the same point in their life cycle (e.g., Month 6), you can see if your ability to monetize users is improving.

Figure 5-1 shows Lifetime Value (LTV) by Cohort. This heatmap tracks the cumulative average LTV for each monthly cohort. The values show a clear progression as each cohort spends more time with the product. By reading vertically down a column, you can see that newer cohorts (e.g., September) are generating more value faster than older cohorts, signaling positive momentum in monetization.

---

[2] Andrew Chen, Jeff Jordan, and Sonal Chokshi, "Metrics and Mindsets for Retention & Engagement," Andreessen Horowitz Podcast, accessed June 28, 2025, `https://a16z.com/podcast/metrics-and-mindsets-for-retention-engagement/`.

[3] Christoph Janz, "Cohort Analysis," christophjanz.blogspot.com, accessed July 04, 2025, `https://christophjanz.blogspot.com/search/label/cohort%20analysis`.

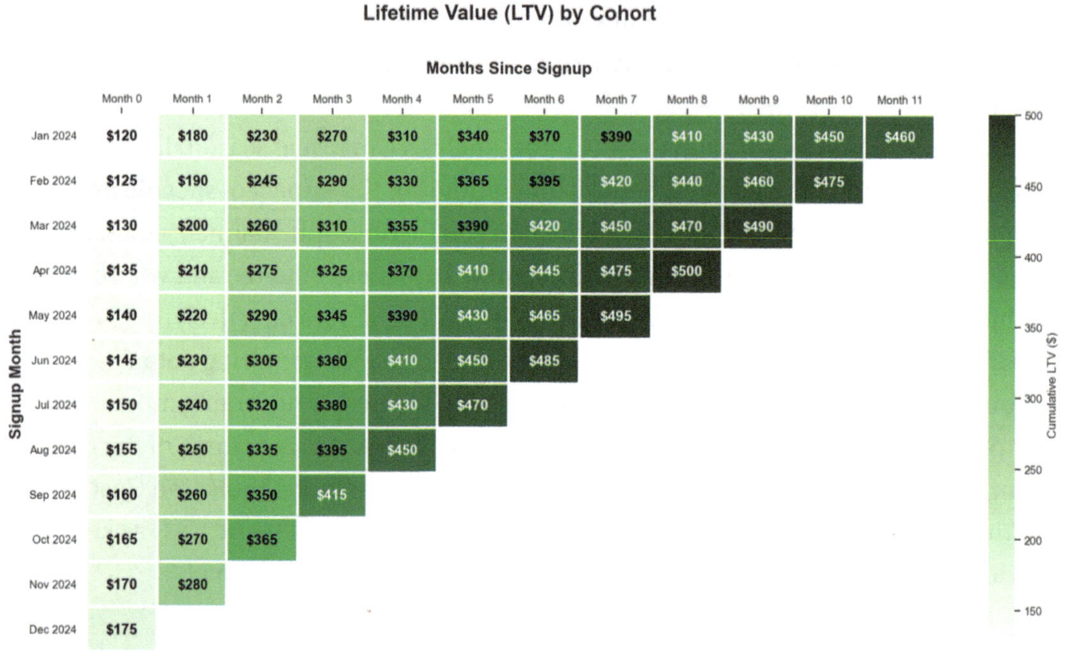

Figure 5-1. *LTV by Cohort*

Customer Acquisition Cost (CAC) by Cohort

Conversely, analyzing your Customer Acquisition Cost (CAC) by cohort shows how your acquisition efficiency is changing. Ideally, as your brand grows and you optimize marketing channels, your CAC should decrease. However, it can also increase as you scale and tap into more expensive channels. Tracking it by cohort is the only way to know for sure.

Figure 5-2 shows an example of the average CAC for customers acquired in each month. A desirable trend is a decrease in CAC over time, indicated by the lighter shades of red in more recent cohorts. This signals that the startup is becoming more efficient at acquiring new customers, a key indicator of a scalable growth engine.

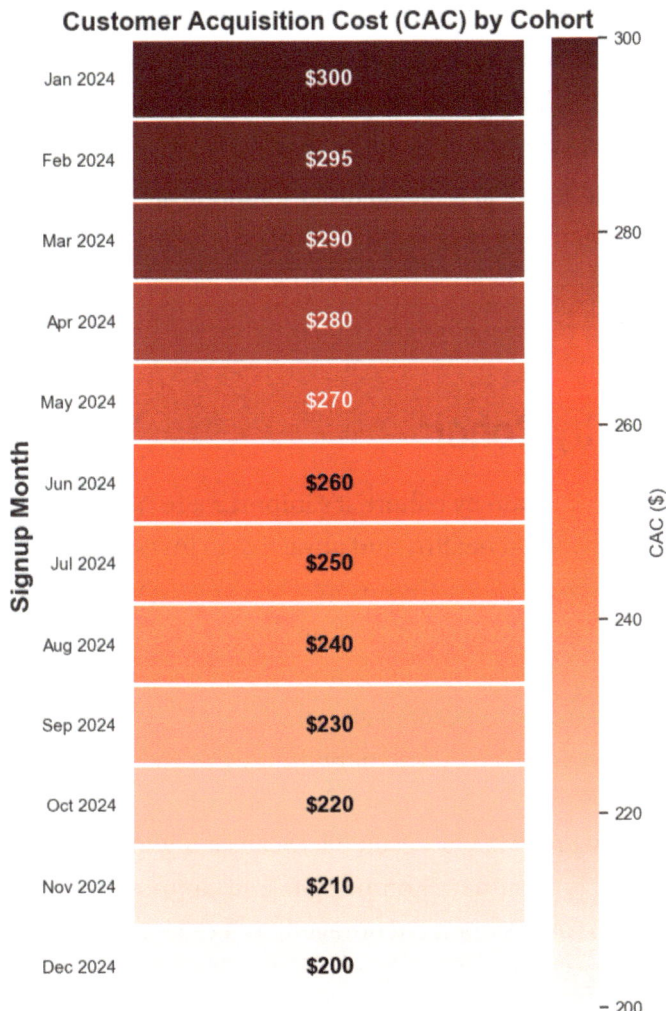

Figure 5-2. *CAC by Cohort*

Note   A blended Customer Acquisition Cost (CAC) that averages paid and organic sources is a dangerous fallacy. This common metric allows low-cost channels to mask the poor performance of expensive ones, leading to flawed investment decisions. The best practice is to calculate a Channel-Specific CAC to measure the true cost and effectiveness of each marketing channel independently from source data.[4]

LTV/CAC Ratio by Cohort

When analyzing LTV/CAC ratio by cohort, it's important to note that while a ratio above 3:1 is widely considered healthy and attractive to investors, I've had several conversations with investors where they expressed concern when the LTV/CAC ratio stayed very high (like 6x–8x) for multiple cohorts in a row. In my experience, this often signals to investors that the company might be playing it too safe with customer acquisition and missing out on potential growth.

Rather than maximizing short-term margins, they would often prefer to see a willingness to lower the LTV/CAC ratio and take on more risk, investing more aggressively in acquiring new users—even if it means a higher acquisition cost in the short term. If your goal is to outpace competitors and capture more of the market, a temporarily lower LTV/CAC is often encouraged, as it reflects a more aggressive and growth-oriented strategy.

Figure 5-3 synthesizes the previous two charts. It shows the evolving LTV/CAC ratio for each cohort, a critical indicator of business model viability. The goal is to see the ratio grow over time within each cohort and, more importantly, to see newer cohorts achieve a healthy ratio faster than older ones. The strong green colors in recent cohorts demonstrate a business that is not only acquiring customers more efficiently but also monetizing them more effectively.

---

[4] Christoph Janz, "9 Worst Practices in SaaS Metrics," SlideShare presentation, June 23, 2013, https://www.slideshare.net/slideshow/9-worst-practices-in-saas-metrics/23372152.

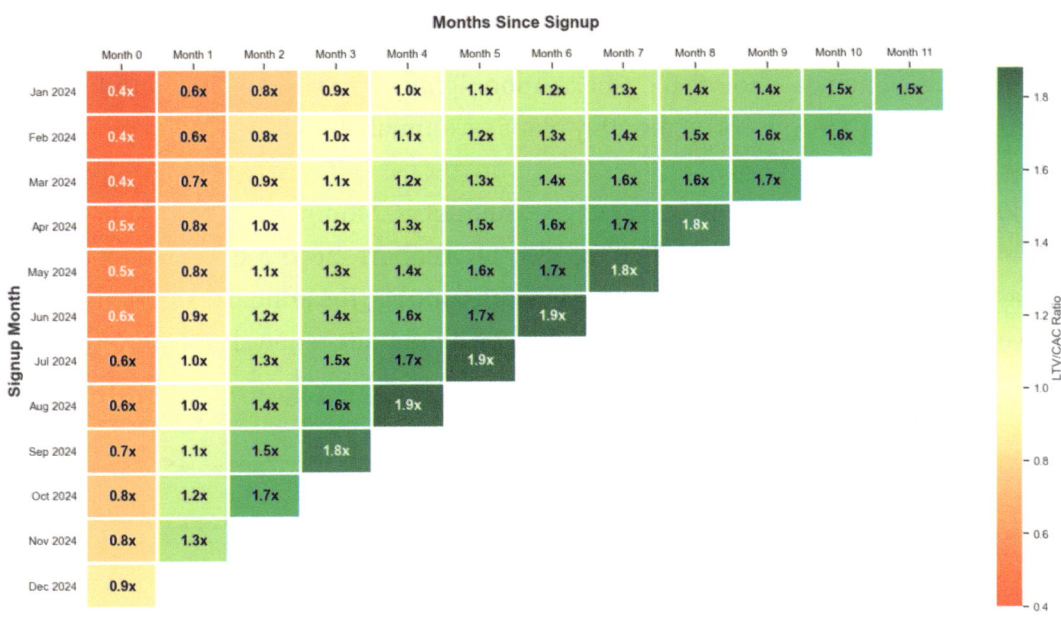

Figure 5-3. *LTV/CAC Ratio by Cohort*

Advanced Cohort Analysis: From Monitoring to Steering

So far, we have established cohort analysis as the definitive tool for viewing your startup's performance through a clean, unfiltered lens. It shows you if your product is getting stickier and if your unit economics are improving over time. Now, we move beyond monitoring the health of the entire user base to dissecting it. Advanced cohort analysis is about segmentation—the practice of slicing your cohorts into more granular groups to understand who your best customers are, why they are successful, and how to find more of them.

This is where cohort analysis becomes a steering wheel rather than just a dashboard. By filtering cohorts based on acquisition channels, user behaviors, or pricing plans, you can answer critical strategic questions that directly inform your product road map and marketing spend.

Segmentation by Financial Attributes

Not all revenue is created equal. A common and highly impactful analysis is to segment cohorts by the financial choices they make upon signing up. One of the powerful examples is comparing users on monthly plans vs. annual plans for a hypothetical subscription startup.

Figure 5-4 shows Retention by Plan Type (Annual vs. Monthly). This line chart tells a fascinating and cautionary tale. The monthly plan cohort shows a steady, predictable decay—the normal state of user attrition. In stark contrast, the annual plan cohort's curve is nearly flat for 11 months before falling off a dramatic renewal cliff at Month 12.

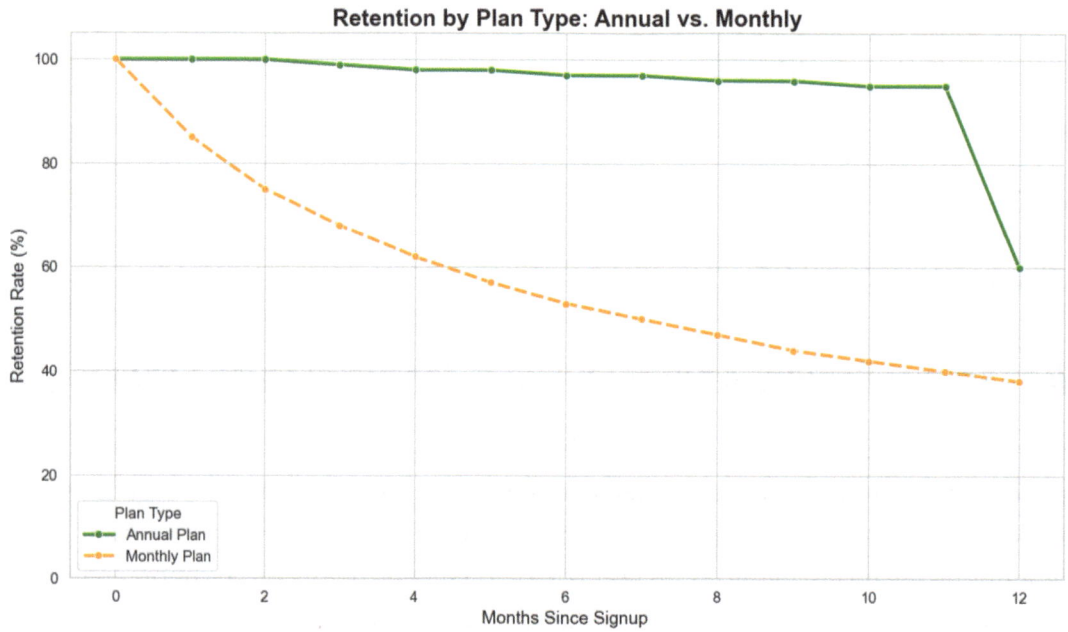

Figure 5-4. *Retention by Plan Type*

This pattern is a double-edged sword. On one hand, it demonstrates the immense power of the annual plan as a business model lever. It locks in customers, secures 12 months of cash flow upfront, and masks churn for an entire year. Even if a user becomes inactive in Month 2, you have already captured their full year's value.

On the other hand, the cliff reveals a critical underlying problem: engagement. The steep drop suggests many of these annual subscribers were not actively using the product and, when faced with the decision to renew, they churned. These were "zombie subscribers" whose inactivity was hidden by the long-term contract. For a startup, this is both a problem and an opportunity. The problem is that the core product isn't sticky enough to earn renewal. The opportunity is that the annual plan buys you a year to figure out how to fix that engagement issue while still making money. This single chart perfectly illustrates why segmented cohort analysis is essential for moving beyond surface-level metrics to truly understand your business.

Segmentation by User Characteristics and Behavior

Financial segmentation is just the beginning. The real power of cohort analysis is unlocked when you slice your user base by their attributes and behaviors. This helps you move beyond what is happening to why it's happening and to whom.

Geographical Segmentation

Let's continue our hypothetical startup analysis. Do users from the United States have a higher LTV than users from Europe? Answering this question is vital for prioritizing marketing spend and localization efforts.

Figure 5-5 directly compares the average LTV of users from the US cohort (solid line) against the Europe cohort (dashed line). The separation in the lines is stark: the US cohort is significantly more valuable over time. This single visual provides a clear mandate to either double down on the high-performing US market or investigate why the European market is underperforming. Is it a pricing, product, or language issue?

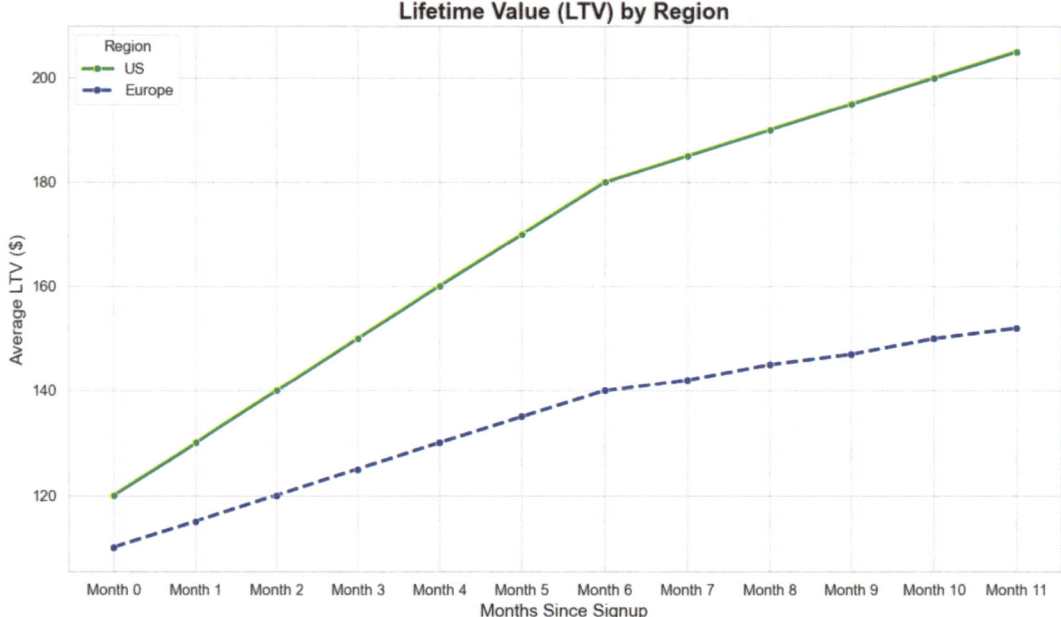

Figure 5-5. *LTV by Region*

Acquisition Channel Segmentation

This analysis helps you move beyond simply measuring Cost Per Acquisition (CPA) to understanding the long-term value of the customers each channel delivers. You might find that a channel with a higher CPA is worth it because it brings in customers with a much higher LTV.

Figure 5-6 reveals that customers acquired through Organic Search (solid line) have a significantly higher LTV than those from Paid Ads (dashed line). While paid ads might bring in more users quickly, this visual proves that your SEO and content marketing efforts are attracting your ideal, most profitable customers. This is crucial evidence for justifying long-term investment in content over short-term ad spend.

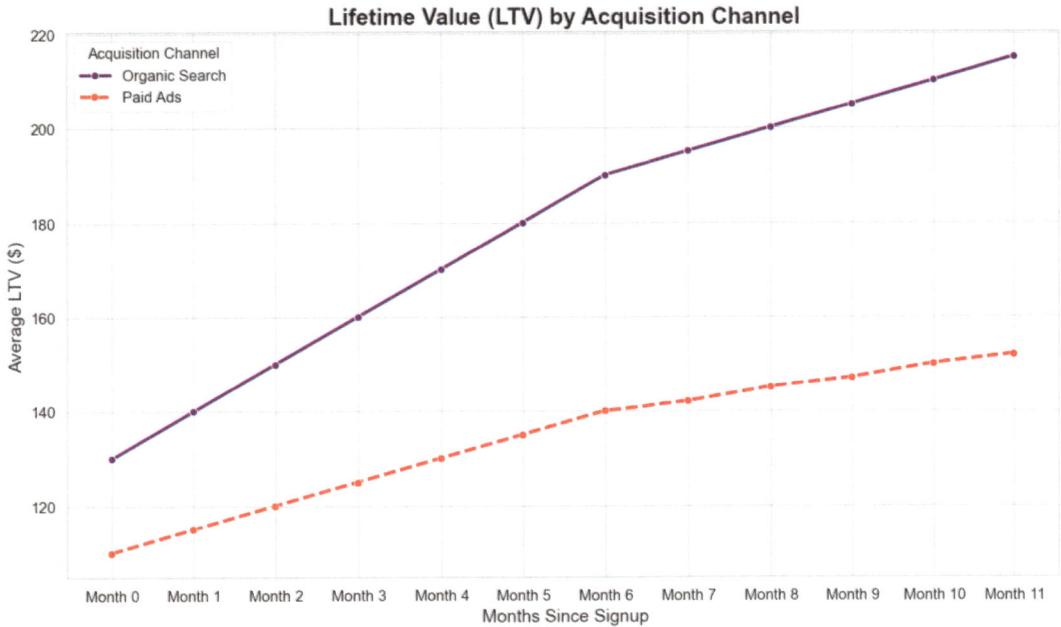

Figure 5-6. *LTV by Acquisition Channel*

Platform-Specific Segmentation (Creator Economy Example)

For marketplaces or platforms—especially in the creator economy—a user's experience is intrinsically tied to the supply side (the creators, instructors, or artists). Analyzing retention for cohorts grouped by their first subscribed creator can be incredibly insightful.

Figure 5-7 illustrates an example from a hypothetical fitness startup. We compare the retention of users who first subscribed to superstar trainer "Mason Jomoa" against the average retention of users subscribing to other trainers like "Ronald Schwarzenegger" and "Cristiano Donaldo." The result is clear: Mason Jomoa is an outlier, retaining his audience far better than the platform average. This is actionable intelligence. You can now study his methods, share those learnings with other trainers, or feature him more prominently to attract new users who are likely to stick around.

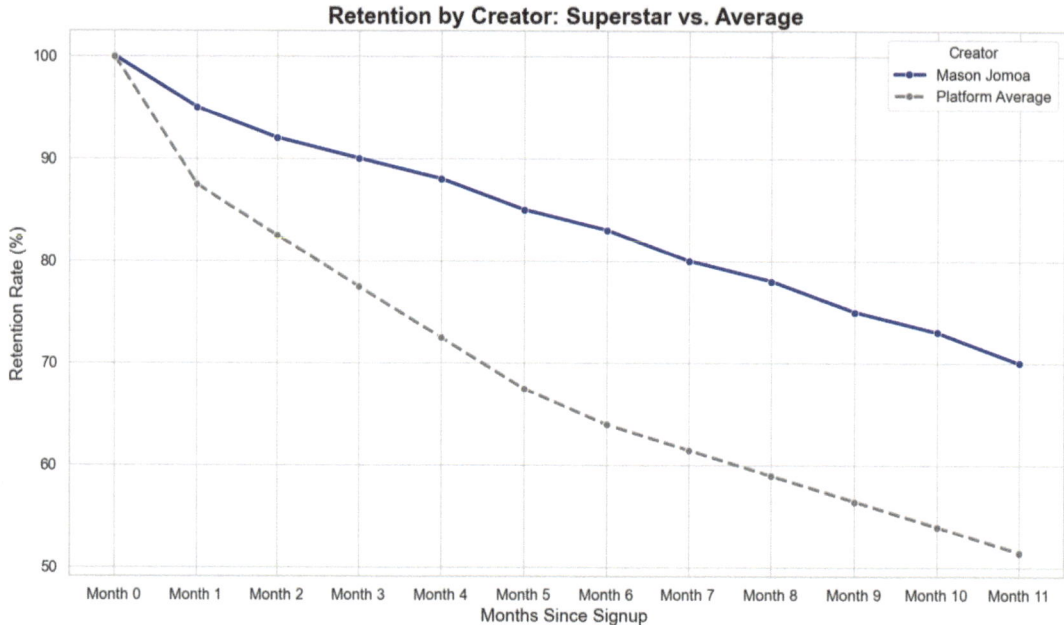

Figure 5-7. *Retention by Creator*

Engagement-Level Segmentation

One of the most powerful analyses is to segment users based on their intensity of engagement in their first week, for example, group users who performed a key action (like "listing an item" or "completing a project") vs. those who simply logged in. Tracking the LTV of these separate cohorts will almost certainly prove that early engagement is a powerful predictor of long-term value.

Figure 5-8 shows LTV by Engagement Level. This chart shows a dramatic difference in LTV between users who were highly engaged in their first week (solid line) vs. those with low engagement (dashed line). The gap between the two lines widens over time, confirming that the first-week experience is critical. This insight may give you a clear mandate: your top priority should be to redesign your user onboarding to drive every new user toward that key "aha!" moment of high engagement.

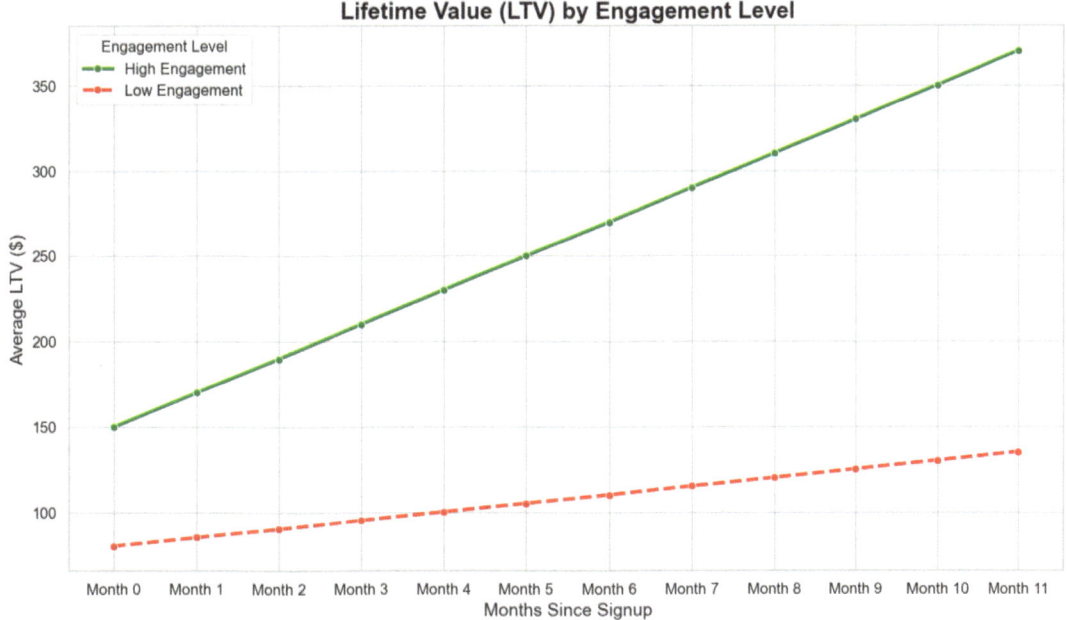

Figure 5-8. *LTV by Engagement Level*

Uncovering Context: Seasonality and Events

Your business does not operate in a vacuum. Cohort analysis is also brilliant for separating internal product improvements from external factors.

Seasonality is a common example. An e-learning platform might find that its September "back-to-school" cohort always has higher engagement than its December "holiday" cohort. Recognizing this pattern prevents you from mistakenly attributing the December dip to a product failure.

Figure 5-9 clearly shows a seasonal pattern. Retention peaks in the spring (March-May) and fall (September-October), with dips in the summer and winter. This allows you to set realistic expectations and evaluate product changes against a seasonal baseline rather than panicking during a predictable dip.

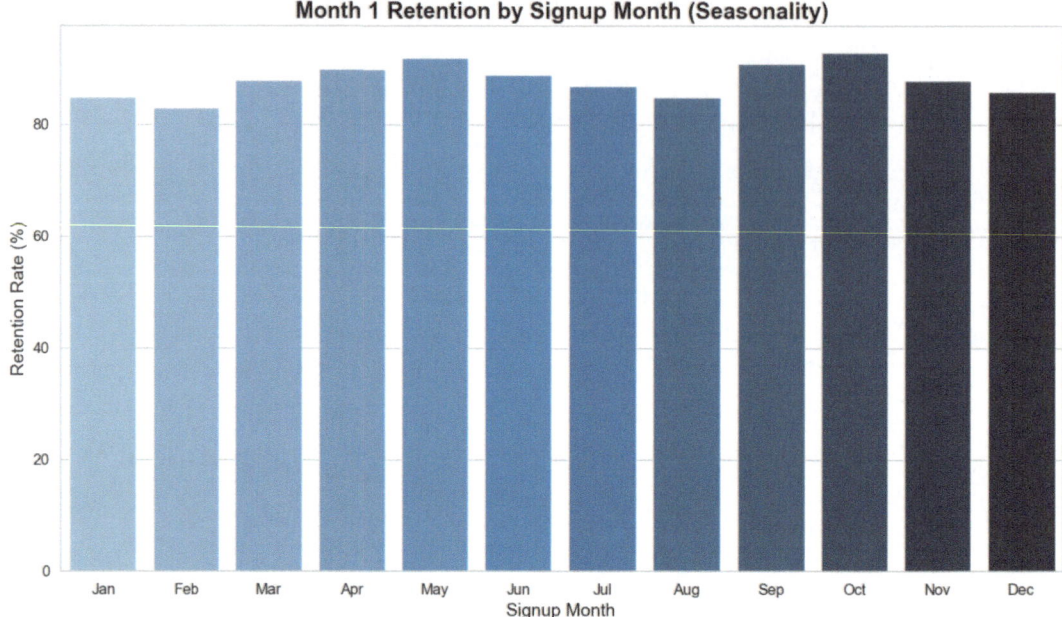

Figure 5-9. *Month 1 Retention by Sign-up Month*

Similarly, you can analyze event-driven cohorts. If your product was featured in a major news outlet, you can isolate the users who signed up during that period. By comparing their LTV and retention to a baseline cohort, you can measure the true impact and quality of that user influx.

Figure 5-10 compares the LTV of a cohort acquired during a major PR event (solid line) to a typical baseline cohort (dashed line). The event cohort's LTV is consistently higher, indicating that the event not only drove sign-ups but also attracted high-quality users. This provides a strong case for pursuing similar PR opportunities in the future.

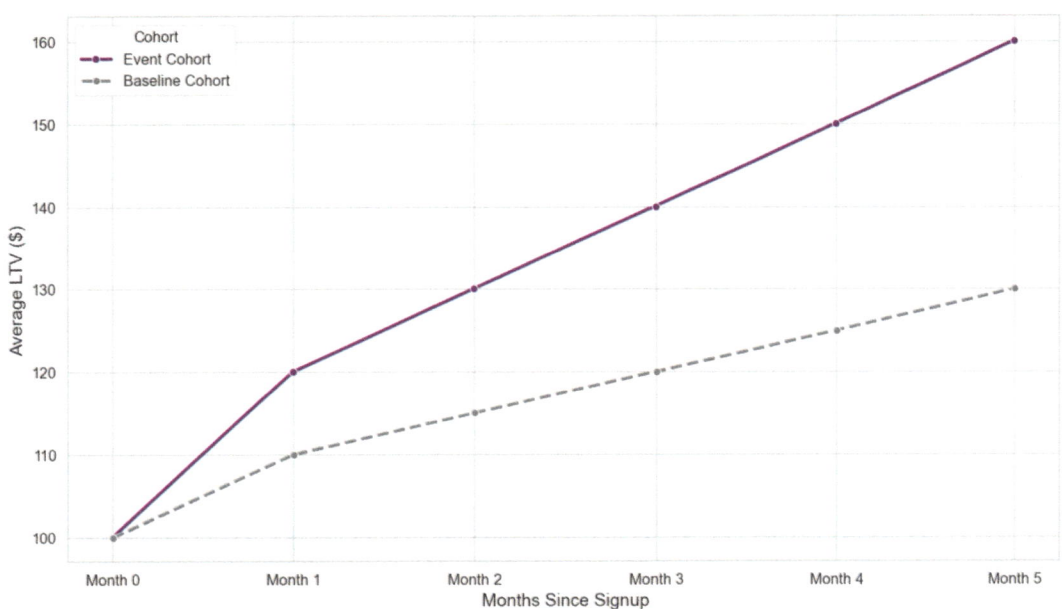

Figure 5-10. *LTV Comparison: Event Cohort vs. Baseline Cohort*

Alternative Visuals: Telling a Different Story

While heatmaps and line charts are workhorses, a stacked area chart can tell a powerful growth story. By stacking the revenue from each cohort on top of one another, this chart beautifully illustrates how new cohorts add layers of growth on top of a retained base, visualizing the compounding effect of strong retention.

Figure 5-11 provides a compelling visual of how a healthy business grows. Each colored layer represents the revenue from a single cohort. You can see how the total revenue (the top edge of the chart) is a sum of the retained revenue from old cohorts and the new revenue from recent ones. This is the visual definition of sustainable, compounding growth.

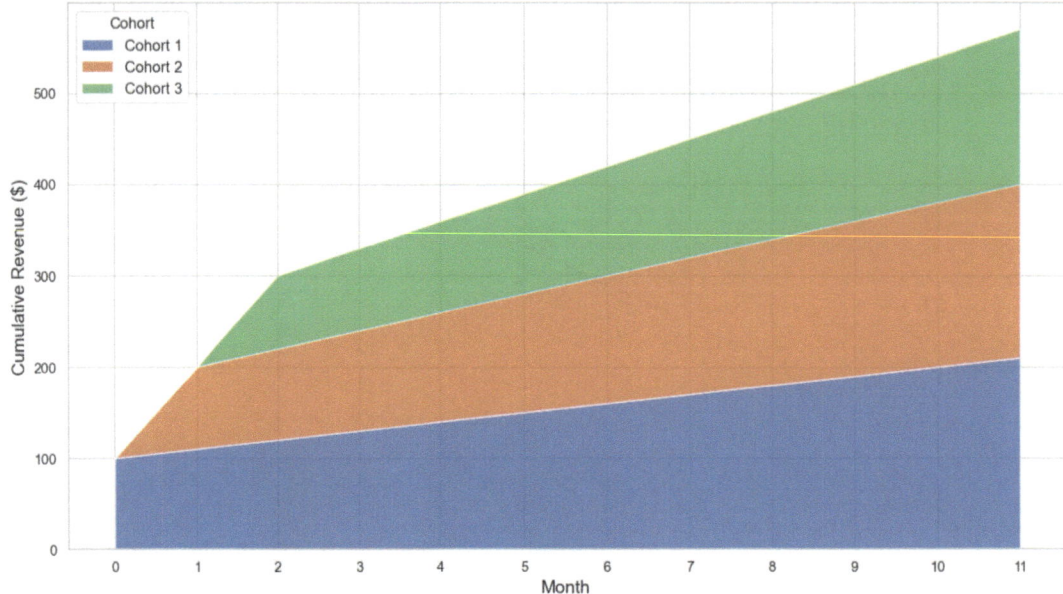

Figure 5-11. *Stacked Area Chart: Revenue Growth by Cohort*

The Infinite Possibilities of Cohort Analysis

The examples we've explored—from financial and behavioral segments to seasonality and event-driven analysis—are foundational. They are the essential toolkit for numerous startups. However, they represent only a fraction of what's possible with cohort analysis. The true power of this technique emerges when you move beyond these standard templates and tailor it to the unique context of your industry and business model.

The most insightful analyses come from segmenting users by dimensions that are specific to your product's value proposition. The possibilities here are genuinely vast. For instance, if you run an application for pregnant women, you could analyze cohorts by their sign-up month, but a far more powerful dimension could be the user's stage of pregnancy. A cohort of women in their first trimester will have vastly different needs, engagement patterns, and feature requests than a cohort in their third trimester. By segmenting this way, you can see if your product is successfully guiding users through their entire journey, from early pregnancy to postpartum care. Are you retaining them effectively as they transition from one stage to the next? This level of specific analysis is what turns data into a competitive advantage.

Ultimately, cohort analysis is a way of thinking. It's about asking the right questions and having the curiosity to slice your data in ways that reveal the underlying truth about your users. The key is to think creatively about the factors that truly define your customers' experience and segment them accordingly. This is how you move from simply measuring retention to actively shaping it, building a better product and a more durable business in the process.

As we've seen, cohort analysis provides an unparalleled lens for understanding user behavior and business health. However, a single analytical tool, no matter how powerful, offers only one perspective. To truly grasp the multifaceted nature of user engagement and product-market fit, it's essential to complement cohort analysis with other specialized metrics. One such powerful technique, particularly valuable for products seeking to understand the intensity and frequency of user interaction, is the power user curve.

Power User Curves

Andrew Chen, as a prominent venture capitalist and former growth executive, introduces the Power User Curve as a superior alternative to traditional engagement metrics like DAU/MAU (Daily Active Users divided by Monthly Active Users) for startups seeking to understand their most valuable users. This analytical framework, also known as the activity histogram or "L30," provides startups with a nuanced view of user engagement by plotting the distribution of users based on how many days they were active within a month.[5]

The Power User Curve offers several critical advantages over blended metrics that startups desperately need. Unlike DAU/MAU, which provides only a single aggregated number, the Power User Curve reveals the heterogeneity among user segments, showing whether a startup has developed a hardcore, engaged user base that returns daily. This granular insight proves invaluable for startups because it exposes the variance in user behavior that single metrics obscure, allowing founders to identify and nurture their most committed users. Figure 5-12 illustrates what Power User Curves may look like for social products.

---

[5] Andrew Chen, "The Power User Curve," Andrew Chen Blog, accessed March 24, 2025, `https://andrewchen.com/power-user-curve/`.

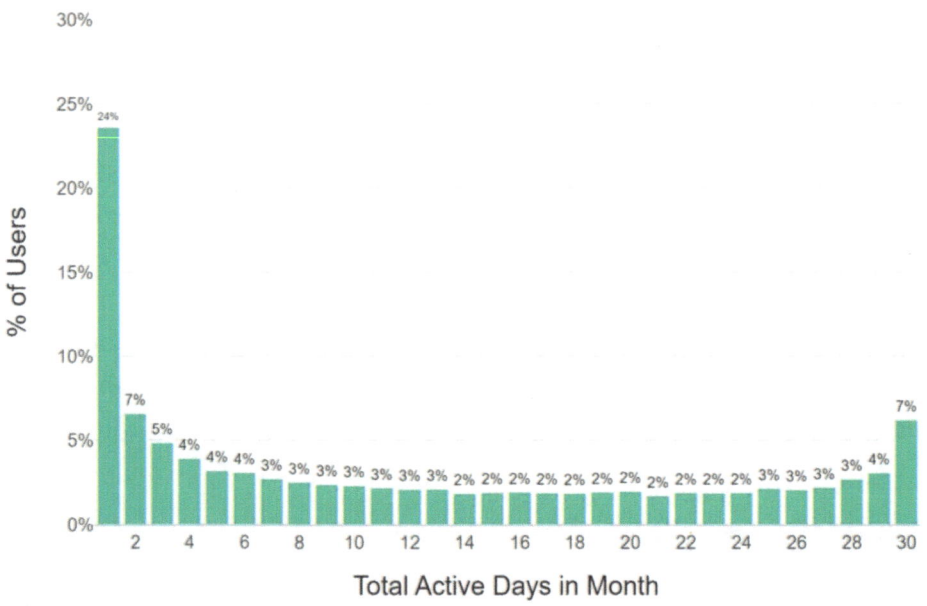

Figure 5-12. *Power User Curve for a Social Product*

Chen emphasizes that the shape of the Power User Curve tells a story about product-market fit and monetization potential. A "smile-shaped" curve, characterized by high numbers of light users on the left and a significant population of daily users on the right, indicates a healthy social or consumer product with strong engagement patterns. These smile-shaped curves typically correlate with products that can successfully monetize through advertising, as they generate sufficient daily impressions to support an ad-based business model.

Conversely, left-weighted curves (see Figure 5-13) reveal different user dynamics that require alternative strategic approaches. Professional networking products or investment platforms often exhibit this pattern, where most users engage only once or twice per month. Chen argues that such engagement patterns aren't necessarily problematic, but they demand business models capable of extracting significant value from infrequent user interactions.

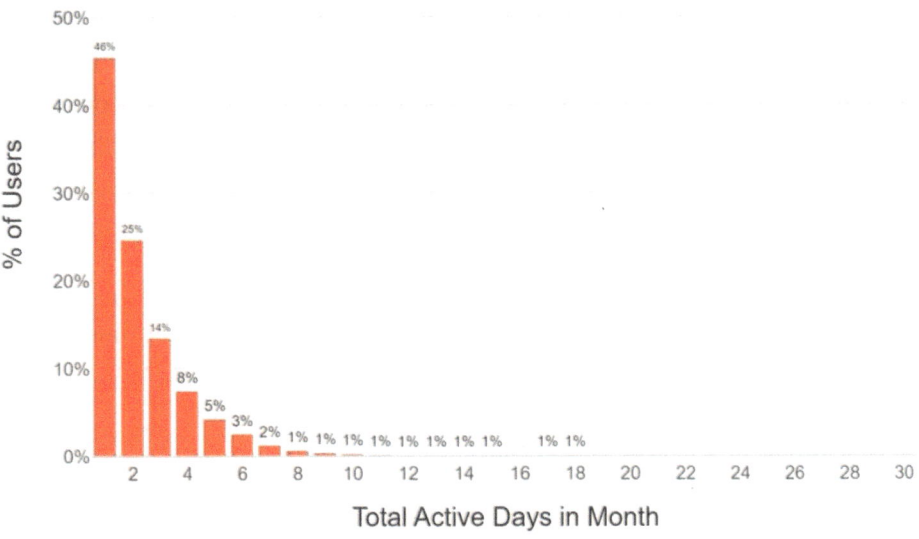

Figure 5-13. *Left-Weighted Power User Curve*

For B2B SaaS companies, Chen recommends analyzing 7-day Power User Curves (see Figure 5-14) rather than monthly distributions, as these products typically follow workweek usage patterns. This temporal adjustment allows SaaS startups to better understand their core user segments and identify opportunities to expand usage across different teams or functions within client organizations.

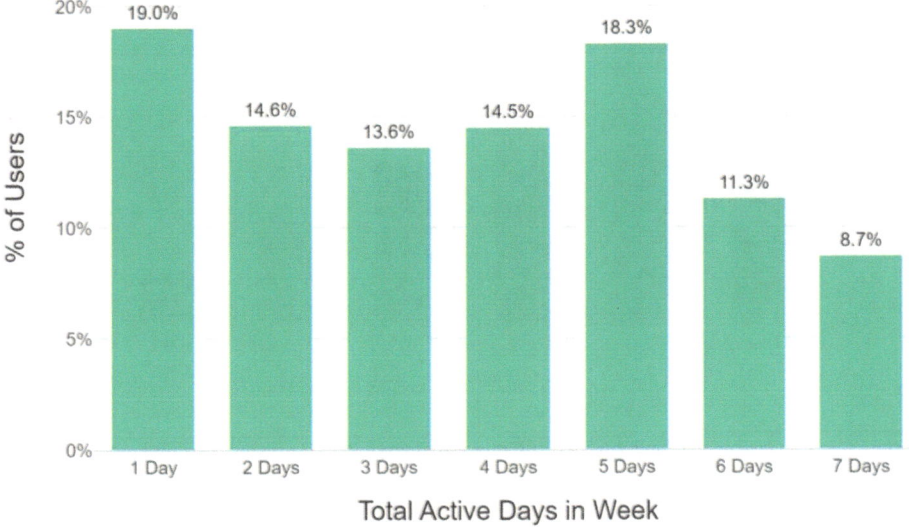

Figure 5-14. *Power User Curve (7-Day Period)*

Figure 5-15 presents an L7 Power User Curve for a hypothetical fitness app based on app opens. Analyzing a fitness product requires understanding its unique usage patterns, which may often be tied to weekly workout routines rather than daily social check-ins. An L7 Power User Curve is the perfect tool for this, as it aligns directly with how users plan their fitness schedules. The shape of this curve might suggest a clear story about the user base:

- **A Large Peak at 1-2 Days per Week:** This captures the majority of casual users, which aligns with findings that the average member works out about twice a week.

- **A Steady Decline for 3-5 Days:** This represents the smaller, more consistent group of users who are following a dedicated workout plan, often the recommended frequency for seeing results.

- **A Small Tail for 6-7 Days:** This is the most hardcore segment of your user base—the daily enthusiasts who are highly engaged and retained.

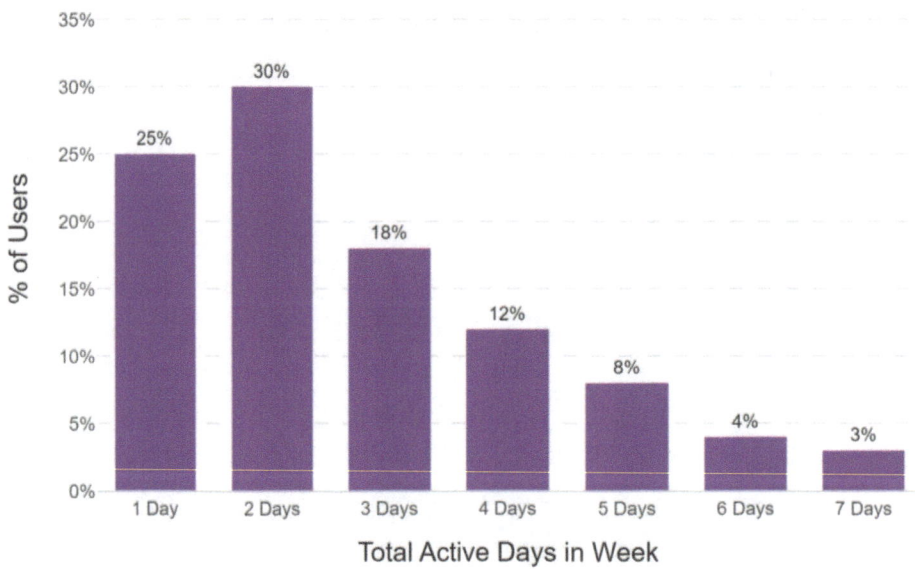

Figure 5-15. *L7 Power User Curve (Fitness App)*

Chen advocates for startups to move beyond measuring simple app opens or logins and instead focus on core activities that map closely to business value. Content platforms should analyze publishing frequency, marketplaces should examine transaction patterns, and social networks should track meaningful interactions rather than passive consumption. This customization ensures that Power User Curves reflect actual value creation rather than vanity metrics.

While the previous L7 curve for our fitness app showed promising engagement based on app opens, a deeper analysis is required to understand if users are truly adopting the product's core value proposition. The most insightful Power User Curves are not based on generic activity like logins but on the completion of a core user action. For our fitness app, this action is undoubtedly "completing a workout."

Figure 5-16 visualizes this. When we shift our focus from "days active" to "days with workouts completed," a different, more sobering picture emerges. The overall engagement is lower, and the distribution is more heavily weighted toward the left.

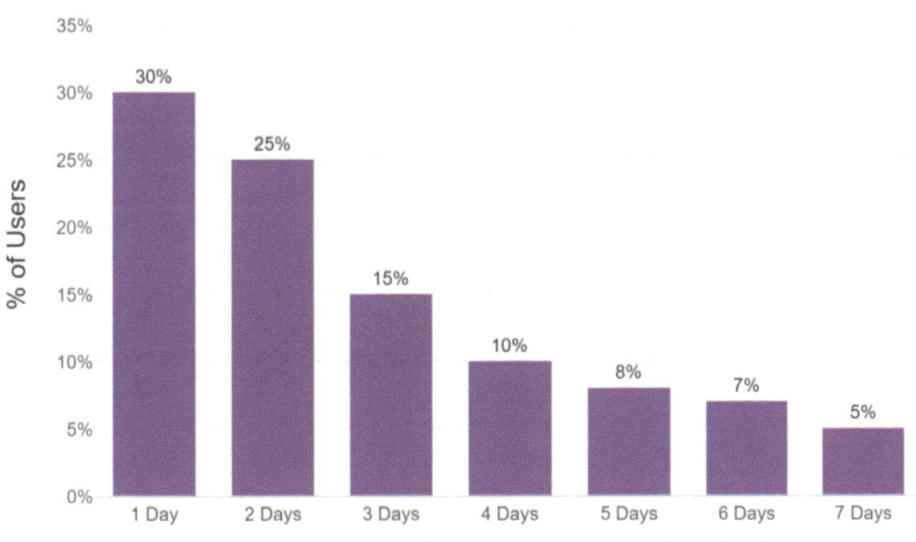

Figure 5-16. *L7 Power User Curve (Core Product Activity)*

This is not necessarily a sign of failure; rather, it is a critical diagnostic tool. This new curve prompts essential questions for the product team:

- Are users opening the app to explore its content but lack the motivation to start a workout?

- Is there friction in the user experience that prevents them from completing a session?

- Does the app suffer from bugs, a lack of new content, or performance issues that lead to drop-offs?

By comparing the "app open" curve with the "workout completed" curve, the startup can start further data exploration to identify the gap between casual interest and true engagement, providing a clear focus for future product improvements and experiments.

Tracking Product Evolution and Feature Impact

One of the most powerful applications of Power User Curves for startups involves cohort analysis over time. By comparing curves across different monthly or weekly cohorts, founders can determine whether their product improvements are successfully converting casual users into power users (see Figure 5-17). Chen notes that successful network effects products should show newer cohorts gradually shifting toward higher-frequency engagement as network density and liquidity improve.[6]

---

[6] D'Arcy Coolican and Li Jin, "The Dynamics of Network Effects," Andreessen Horowitz, accessed June 15, 2025, https://a16z.com/the-dynamics-of-network-effects/.

Power User Curve by Cohort

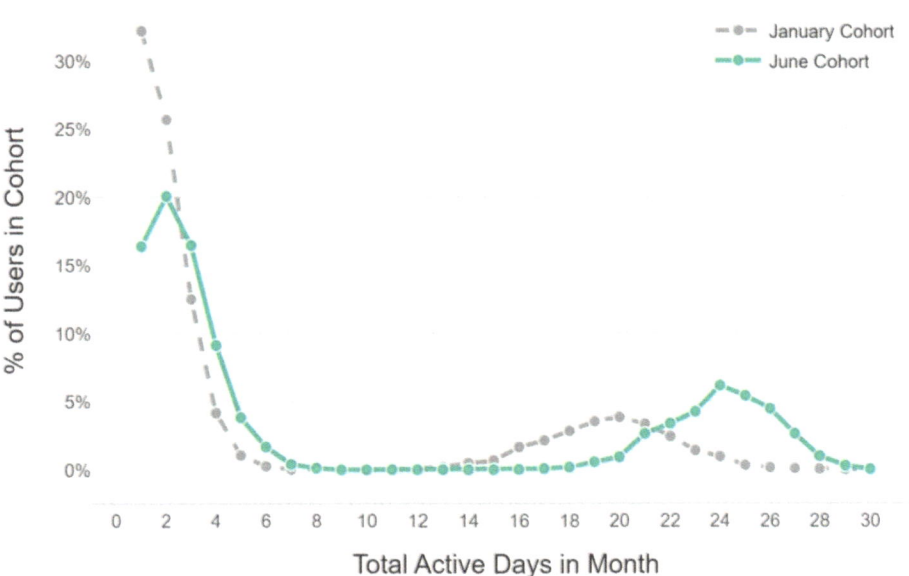

Figure 5-17. *Cohort Evolution: Power User Curve Improvement Over Time*

This temporal analysis proves particularly valuable for measuring the success of specific product releases or marketing initiatives. Startups can identify inflection points where engagement patterns changed and correlate these shifts with particular feature launches or growth experiments. For network effects businesses, this progression from left-leaning to smile-shaped curves often indicates that the product is achieving the critical mass necessary for sustainable growth.[7]

The framework becomes particularly insightful when applied to different user segments within the same product. For businesses with local network effects like ridesharing or local services, analyzing Power User Curves by geographic market can reveal which locations are developing the density necessary for strong network effects. Similarly, B2B products can segment curves by company size, industry, or user role to identify the most engaged customer personas.[8]

---

[7] D'Arcy Coolican and Li Jin, "16 Ways to Measure Network Effects," Andreessen Horowitz, accessed June 15, 2025, `https://a16z.com/16-ways-to-measure-network-effects/`.

[8] D'Arcy Coolican, "Hidden Networks: Network Effects That Don't Look Like Network Effects," Andreessen Horowitz, accessed June 15, 2025, `https://a16z.com/hidden-networks-network-effects-that-dont-look-like-network-effects/`.

Perhaps most importantly, Chen argues that Power User Curves should inform fundamental business model decisions for startups. Products with strong daily engagement patterns can pursue advertising or subscription models that depend on frequent user return. However, products with infrequent but valuable user interactions must develop monetization strategies that capture significant value during each engagement session.

The analysis also guides product development priorities by revealing which user segments generate the most business value. Startups can focus their limited resources on features that serve their power users while simultaneously working to move casual users toward higher engagement levels. This dual approach ensures both retention of the most valuable customers and expansion of the engaged user base over time.

Chen's Power User Curve framework provides a powerful microscope to examine the engagement of individual users and key segments. It helps identify who your most dedicated users are and whether product changes are successfully creating more of them. However, for many modern businesses, the story doesn't end with individual activity. The true magic—and the most durable competitive advantage—often emerges when one user's activity makes the product more valuable for everyone else. This phenomenon, known as network effects, requires its own distinct set of measurements to understand if your product is truly becoming more valuable as it grows.

Measuring Network Effects

Network effects represent one of the most powerful competitive advantages in modern software and marketplace businesses, yet measuring them effectively remains a challenge for most startups. As the team at Andreessen Horowitz notes in their comprehensive analysis, network effects exist on a spectrum rather than as a binary characteristic, and they evolve dynamically as products, users, and competition change. Understanding how to measure these effects is crucial for founders who want to build sustainable, defensible businesses that become more valuable as they grow.[9]

---

[9] D'Arcy Coolican and Li Jin, "16 Ways to Measure Network Effects," Andreessen Horowitz, accessed June 15, 2025, https://a16z.com/16-ways-to-measure-network-effects/.

The fundamental principle underlying all network effect measurement is deceptively simple: your product should become more valuable as more people use it. However, translating this concept into actionable metrics requires a nuanced approach across five key categories that a16z identifies: acquisition metrics, competitor analysis, engagement tracking, marketplace dynamics, and economic indicators. Each category provides different insights into whether your network is truly strengthening over time.

The most telling indicator of genuine network effects lies in your **organic user acquisition trends**. As your network grows and becomes more valuable, the share of users who join organically (without paid marketing) should increase relative to those acquired through paid channels (see Figure 5-18). This happens because users increasingly want to join on their own as they recognize the growing value of participation. For direct-side network effects like Facebook, this manifests as people inviting friends because their own experience improves with each new connection. In two-sided marketplaces like Airbnb or eBay, both suppliers and buyers are drawn to join organically due to the expanding revenue opportunities and variety of choices available.

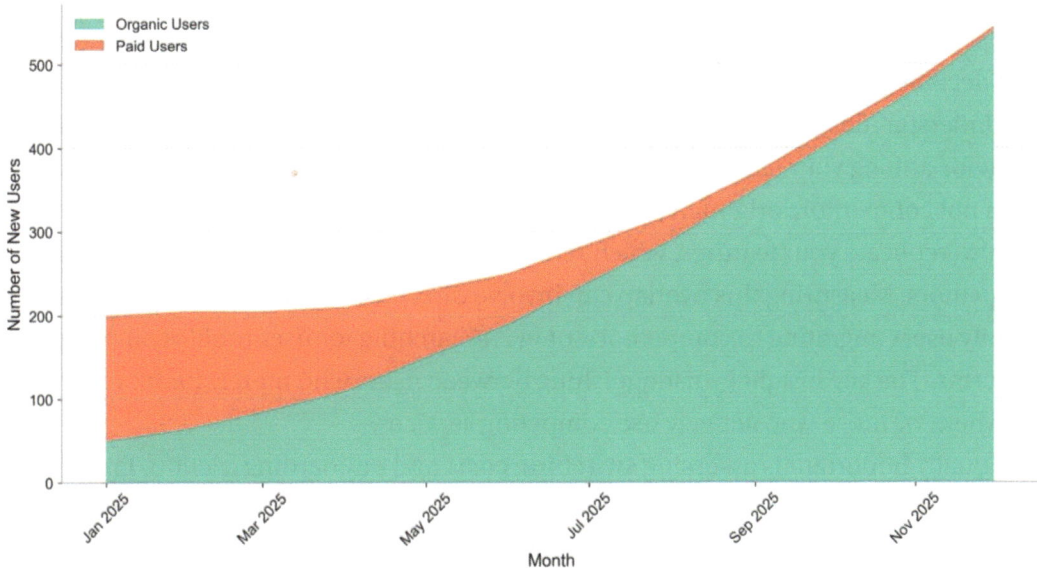

Figure 5-18. *Organic vs. Paid User Acquisition*

However, measuring organic acquisition requires careful consideration of your network's geographic scope. Companies with local network effects will see increasing organic adoption on a geography-by-geography basis rather than globally. A home services platform, for example, will need to track organic growth separately for each city or region since network value doesn't transfer across distant locations.

Beyond acquisition patterns, **traffic source analysis** reveals whether your platform is becoming a true destination for users. The a16z article illustrates this beautifully with OpenTable's evolution: initially, users would research restaurants elsewhere, visit the restaurant's website, and book through an OpenTable widget. As the network matured, users began starting their restaurant discovery process directly on OpenTable, skipping external research entirely. This shift toward direct traffic indicates that users find increasing value in your network as it grows, representing a classic "come for the tool, stay for the network" dynamic.

Customer acquisition cost (CAC) trends provide another lens into network effect strength, though the relationship is more complex than simple theory suggests. While CAC should theoretically decline as network effects accelerate, real-world factors like marketing channel competition and substitute availability can complicate this metric. The ridesharing industry exemplifies this complexity, where abundant driver alternatives have actually increased acquisition costs over time, while platforms like OpenTable have seen decreasing restaurant acquisition costs as they aggregated demand.

Understanding your competitive landscape requires measuring **multi-tenanting behavior** among your users. The a16z analysis emphasizes that if other companies (even not competitors originally) can easily replicate your network elsewhere, they will also replicate your features, which reduces usage and compresses margins for all competitors. Measuring this overlap can involve direct user surveys, churn analysis to identify users migrating to competitors, or even searching for user profiles on competing platforms. The key insight is distinguishing between users who merely maintain profiles elsewhere vs. those who actively use competing services.

Equally important is assessing **switching costs** and onboarding friction. Products requiring high upfront investment from users create natural moats against competition but may struggle with new user activation. Stitch Fix exemplifies high switching costs, as customers invest significant time explaining preferences and calibrating their styling experience. Conversely, Uber leveraged existing credit card information from ridesharing making it easy for people from other food delivery networks to easily join Uber Eats.

Retention cohort analysis offers perhaps the most direct measurement of network effects in action. In theory, newer user cohorts should demonstrate better retention than older cohorts because they experience your product when the network is larger and more valuable. However, the a16z article warns that early adopters often represent your most ideal customers, potentially skewing this analysis. Additionally, factors like competitive pressure, hyperlocal network effects that reset for new geographies, or even negative network effects from overcrowding can complicate retention trends.

Moving beyond basic retention, core action retention provides deeper insights into network effect strength. As already mentioned, rather than measuring app opens, focus on whether users increasingly take the actions that drive value from your network. As network density grows, retention around these core actions should improve more dramatically than surface-level engagement metrics.

For subscription or paid products, **dollar retention and paid user retention** become critical indicators. Users willing to pay demonstrate clear value perception, so products with strengthening network effects should see increasing financial retention among newer cohorts. Angie, a one-stop platform for nearly any home-related need, from plumbing and electrical work to landscaping, and cleaning, serves as an example where improved network coverage should translate to better subscriber retention as the service becomes more useful.[10]

Marketplace-specific metrics require special attention to match rates and market depth. **Match rate** measures how successfully your platform connects supply with demand (e.g., "What percentage of a rideshare driver's time is spent transporting passengers compared to driving without anyone in the car?"), while its inverse, measuring "zeros" or unsuccessful matches, identifies friction points that drive users to competitors.

Market depth, a concept adapted from financial markets, reflects how easily users can find what they need and access your service. In diverse marketplaces like Airbnb, each new listing enhances value by broadening the range of unique options available to users. In contrast, platforms with uniform supply, such as Lime, experience diminishing returns from additional inventory once there's enough density to reliably meet demand.

---

[10] "How It Works," angi.com, accessed June 21, 2025, `https://www.angi.com/landing/how-it-works`.

Inventory turnover and its inverse, **days to turn**, gauge how quickly your marketplace moves supply over time. These metrics are most effective in traditional marketplaces where users actively choose matches rather than in algorithm-driven on-demand platforms. For example, in a used car marketplace, if the cost of goods sold for a quarter is $750,000 and the average inventory value is $375,000, the inventory turnover would be 2.0—meaning the inventory is sold and replaced twice per quarter. Similarly, job marketplaces often track the time to receive first applications, while peer-to-peer platforms may focus on how rapidly transactions are completed to assess marketplace efficiency.

The **concentration of your marketplace participants** reveals network defensibility through fragmentation analysis. Greater fragmentation on both supply and demand sides creates more valuable and defensible networks, as no single participant can threaten the platform by leaving. This can be measured by tracking what percentage of Gross Merchandise Value (GMV) your top sellers or buyers represent.

Finally, **unit economics trends** provide the ultimate test of network effect strength. As participants derive greater value from your network, they should willingly pay more for access through subscriptions, listing fees, or transaction fees. Successful network effect businesses typically evolve from offering subsidies to turning on monetization to eventually increasing prices with minimal churn. For local network effect businesses, these improvements should manifest on a per-market basis as individual geographies mature and build network density.

The a16z framework emphasizes that no single metric tells the complete story of network effects. Two-sided marketplaces should focus heavily on marketplace and unit economics metrics, while social networks require deeper attention to engagement and activity patterns. The key is selecting the right combination of measurements that align with your specific product, audience, and market environment while maintaining focus on the core question: is your product becoming more valuable as more people use it?

Understanding these metrics isn't just about measurement—it's about building a sustainable competitive advantage. Network effects create powerful moats, but only when properly cultivated and measured. By implementing the comprehensive measurement framework outlined by a16z, startups can move beyond hoping they have network effects to systematically building and strengthening them over time.

The Challenge of Asymmetry: Power Laws in the Creator Economy

While traditional network effects often lead to broad, distributed value, the Creator Economy presents a more complex dynamic heavily influenced by what Andrew Chen calls the creator power law. This principle states that a small, highly concentrated number of top creators capture a vast majority of the audience and, consequently, the revenue. This creates a fragile ecosystem for startups built to serve creators, as their success becomes disproportionately dependent on a handful of top performers, often called "whales."[11]

In the Creator Economy, these are the superstar creators whose departure could cripple a platform's financials. This dynamic creates several challenges:

- **Revenue Concentration:** A platform's top-line revenue becomes fragile. If a single whale leaves the platform, the financial impact can be immediate and severe.

- **Acquisition Dilemma:** To achieve scale, startups must attract these large creators. However, the "long tail" of smaller creators they initially attract are often too small to generate meaningful revenue on their own.

- **Intense Competition:** Because every platform needs the same top creators, the competition to attract and retain them is fierce. This is compounded by the "battle for the bio link," a zero-sum game for the single most valuable piece of promotional real estate on a creator's social media profile.

The stark impact of this revenue skew is best understood visually. Figure 5-19 uses a treemap, a visualization designed to show part-to-whole relationships, to display a typical revenue distribution for a Creator Economy startup. The area of each colored rectangle is directly proportional to that creator segment's share of the total revenue, providing an immediate sense of scale. The chart offers a dramatic illustration of the "whale problem": the "Top 1% Whales" consume 70% of the visual space, representing their outsized control over the platform's income. In stark contrast, the "Bottom 90% Minnows" are relegated to a tiny part at the end of the chart.

---

[11] Andrew Chen, "Creator Economy 2.0: What we've learned, why it's hard, and what's next," Andrew Chen Blog, accessed June 14, 2025, `https://andrewchen.com/creator-economy-20/`.

Figure 5-19. *A Hypothetical Revenue Distribution for a Creator Economy Startup*

This dependency leads directly to what Chen identifies as the graduation problem. As creators become more successful on a platform, they begin to question the platform's take rate, especially since they are the ones bringing their own audience and doing the creative work. These whales are often tempted to "graduate" by building their own custom websites or apps, taking their revenue and audience with them. This differs significantly from marketplaces like Airbnb or Uber, which add value by aggregating both supply and demand independently.

To build a defensible business in this environment, creator-focused startups must provide value that goes far beyond commoditized tools for monetization. The most durable companies create their own network effects, for example, by helping creators acquire new audiences from within the platform or by developing proprietary technology, such as advanced AI tools, that is difficult for a creator to replicate on their own.

To proactively identify and manage this dependency, startups should monitor a specific set of metrics that go beyond top-line revenue. These metrics help quantify the extent of the "whale problem" and provide early warnings about platform risk.

- **Revenue Concentration Analysis:** This is the most direct measurement. By tracking what percentage of Gross Merchandise Value (GMV) or total revenue is generated by the top 1%, 5%, and 10% of creators, you can quantify your platform's dependency. An increasing concentration is a red flag that your reliance on whales is growing. This can also be expressed with a Gini coefficient, a statistical measure of distribution inequality.

- **Segmented Engagement and Retention:** A healthy platform should see engagement grow across all user segments, not just among the top performers. You should analyze engagement and retention metrics separately for each cohort ("Whales," "Dolphins," "Minnows"). For instance, plotting separate power user curves for each segment can reveal if only your whales are highly engaged, while the rest of your user base is passive. Similarly, tracking the churn rate of whales specifically is critical, as the loss of even one can have a major impact.

- **Multi-tenanting Analysis:** To assess the "graduation" risk, it is crucial to understand if your top creators are also active on competing platforms. This can be measured through user surveys or by monitoring their public activity on other services. High multi-tenanting among your whales suggests that your platform's switching costs are low and that these key creators could easily leave.

The phenomenon of power laws and the "whale problem" in the Creator Economy underscore a critical lesson: not all users or, in this case, not all creators are equal in their impact on your business. Understanding the distribution of value and engagement within your user base is paramount for building a resilient platform and effectively managing risk.

Understanding and measuring network effects is the final piece of our advanced analytical toolkit. When combined with cohort analysis and Power User Curves, these methods provide a multi-dimensional view of your business's health, moving far beyond surface-level metrics. Together, they form a powerful system for diagnosing issues and identifying opportunities, giving you the insights needed to steer your startup toward sustainable growth.

Summary

This chapter moved beyond foundational metrics to explore the advanced analytical techniques that reveal the "why" behind your business's performance. To gain deeper insights, we focused on three core methods:

- **Cohort Analysis:** This technique tracks user behavior over time, allowing you to assess the health of your business model by analyzing the value of specific customer groups as they mature.

- **Power User Curves:** By mapping the intensity of user engagement, these curves help you identify your most valuable, hardcore users and measure whether product changes are successfully creating more of them.

- **Network Effects Measurement:** We explored how to quantify this powerful competitive advantage through a variety of acquisition, engagement, and economic indicators to see if your product truly gets better as it grows.

While essential, these methods are fundamentally reactive. They provide an exceptionally clear view of what has already happened. But measurement alone does not equal progress. Knowing your LTV/CAC ratio is improving is one thing, but having a framework to drive that improvement across your entire team is another. The true power of data emerges when it is translated from a report into a shared strategy.

This raises the critical questions that separate insight from impact. How do you move from tracking performance indicators to setting actionable goals that align the whole company? How do you present your data to meet the shifting expectations of investors at each funding stage? And how do you ensure your metrics evolve with your strategy, all while maintaining the intellectual honesty required to build credibility?

In the next chapter, we will answer these questions, providing the tools to translate your data into decisive action, secure investor confidence, and drive sustainable growth.

CHAPTER 6

From Insight to Impact

Ideas are easy. Execution is everything.

—John Doerr,

Venture Capitalist, author of *Measure What Matters*[1]

John Doerr's words ring true for every startup: great ideas are common, but great execution is rare. Equipped with the measurement toolkit from previous chapters, you now have endless ideas for what to track. You can build dashboards for user engagement, funnels for conversion, and cohorts for retention until you are drowning in data. But this is where execution truly begins. It starts by asking the one critical question that separates successful startups from those that merely stall: So what? How do we use this data to align the entire team toward our North Star Metric? How do we focus on the numbers that investors actually care about? This chapter provides the answers, revealing how to translate your data into decisive action, secure investor confidence, and drive sustainable growth. In this chapter, you will learn to

- **Translate Metrics into Action with OKRs:** Go beyond tracking data by using the Objectives and Key Results (OKRs) framework to set clear, actionable goals. You'll learn the practical difference between OKRs and KPIs and how to use them together to align your team and execute your strategy.

---

[1] "John Doerr: Ideas are easy, execution is everything.," Kleiner Perkins YouTube channel, March 5, 2015, https://www.youtube.com/watch?v=4xWGSUZmkIc.

P. Sidoruk, *From Data to Dollars*, https://doi.org/10.1007/979-8-8688-1898-1_6

- **Navigate Real-World Investor Expectations:** Discover what venture capitalists truly look for in your metrics at the pre-seed, seed, and Series A stages. We will also explore how economic conditions shift the focus to capital efficiency, making metrics like the Burn Multiple critical for survival and success.

- **Adapt Your Metrics as You Grow:** Learn from case studies of iconic companies like Amazon, Facebook, and Uber. See how and why they evolved their North Star Metrics to adapt to new strategies and market realities, demonstrating that measurement must be dynamic.

- **Master the Art of Honest Data Storytelling:** Understand the dangers of "cheating with your charts"—from misleading cumulative graphs to manipulated axes. This section provides a guide to maintaining intellectual honesty, which is essential for building credibility with investors and your team.

Our road map from insight to impact begins here. To successfully navigate the path from data to growth, you need a compass that keeps your entire team oriented toward the same strategic goals. For many of the world's most successful companies, that compass is the Objectives and Key Results (OKRs) framework.

Objectives and Key Results (OKRs) in Startups

While metrics provide vital insights into your startup's performance, translating these insights into actionable progress requires clear goal setting. Objectives and Key Results (OKRs) represent a widely adopted goal setting framework that enables startups, as well as companies of all sizes, to establish clear, measurable goals and systematically track their progress toward achieving them. The framework is built upon two fundamental components: **Objectives**, which are defined as memorable, qualitative descriptions of what an organization or team aims to achieve. These objectives should be concise, inspirational, and engaging, designed to motivate and challenge the team.[2]

---

[2] John Doerr, *Measure What Matters: How Google, Bono, and the Gates Foundation Rock the World with OKRs* (Portfolio, 2018).

The second component, **Key Results**, consists of specific, measurable outcomes that indicate progress toward the objective. For each objective, John Doerr in *Measure What Matters* and Google's own OKR guide recommend setting three to five key results to maintain focus and clarity. Doerr emphasizes: "A limit of three to five OKRs per cycle leads companies, teams, and individuals to choose what matters most. In general, each objective should be tied to five or fewer key results." Google's re:Work guide echoes this: "Google often starts with the organizational OKRs and aligns priorities using three to five objectives with about three key results for each objective."[3,4]

The overarching purpose of OKRs is to foster alignment and engagement within an organization around a set of clearly defined and measurable goals. This methodology shifts the emphasis from merely listing tasks to focusing on achieving specific, measurable outcomes that directly contribute to the company's overarching objectives. This outcome-oriented approach encourages teams to strategically prioritize their work, ensuring that their efforts are directed toward activities that will have the most significant impact on the company's overall goals.[5]

How Startups Use OKRs for Goal Setting and Alignment

Startups typically adopt a **quarterly cadence** for setting OKRs, which aligns well with their often rapid pace of development and the need for frequent adjustments. The process generally begins with the company's leadership defining high-level objectives for the upcoming quarter or year. These broad company-level objectives then serve as the foundation for individual teams and even individual employees to develop their own OKRs, ensuring a clear line of sight and alignment from the top of the organization down to the specific tasks being undertaken. It is considered crucial to involve the team members in the process of setting OKRs.

This collaborative approach fosters a sense of ownership and encourages buy-in, as individuals feel more invested in achieving goals they have had a hand in defining. Furthermore, to promote transparency and ensure that everyone within the organization

---

[3] John Doerr, *Measure What Matters: How Google, Bono, and the Gates Foundation Rock the World with OKRs* (Portfolio, 2018).

[4] "Set goals with OKRs," rework.withgoogle.com, accessed June 29, 2025, `https://rework.withgoogle.com/en/guides/set-goals-with-okrs`.

[5] Ben Lamorte, *The OKRs Field Book: A Step-by-Step Guide for Objectives and Key Results Coaches* (Wiley, 2022).

understands the company's priorities and how individual efforts contribute to the bigger picture, OKRs are often made public and accessible to all employees. This transparency cultivates a culture of accountability, as individuals and teams are aware of their responsibilities in relation to the company's overarching goals and can also see the progress being made by their colleagues. By having shared visibility into goals, teams can more easily identify dependencies between their work and the work of others, allowing them to proactively address potential roadblocks and collaborate more effectively toward achieving common objectives.

Benefits of Using OKRs for Startups

The adoption of OKRs offers numerous benefits for startups, particularly in their pursuit of rapid and sustainable growth. One of the primary advantages is the focus and prioritization that OKRs bring. By limiting the number of objectives and key results, startups are compelled to concentrate their efforts on a few critical priorities, thereby avoiding the trap of spreading resources too thinly. This disciplined approach ensures that the team's energy is directed toward what truly matters for achieving the company's strategic goals.[6]

OKRs also ensure alignment, accountability, and transparency throughout the organization. The measurable nature of key results enables objective tracking of progress, while the public nature of OKRs fosters a culture of trust and collaboration. By setting ambitious objectives, OKRs encourage teams to think creatively and push beyond their usual limits, driving innovation and growth. The regular cadence of OKR reviews allows startups to quickly respond to change, maintaining agility and adaptability.

Ultimately, OKRs provide a valuable framework for startups to balance short-term execution with long-term vision, ensuring that immediate actions always contribute to the company's strategic direction.[7]

---

[6] Dima Raketa, "OKRs In Startups: Pathway To Opportunity Or Just A Utopian Dream?," forbes.com, accessed June 27, 2025, https://www.forbes.com/councils/forbesbusiness development council/2023/10/10/okrs-in-startups-pathway-to-opportunity-or-just-a-utopian-dream/.

[7] Bruce Gil, "Why Startups Should Use OKRs," whatmatters.com, accessed June 22, 2025, https://www.whatmatters.com/articles/why-startups-should-use-okrs.

Origin of OKRs

The concept of Objectives and Key Results can be traced back to Intel in the 1970s, where Andrew Grove, a key figure in the company's history, is credited with introducing the framework. Initially, this methodology was known as "Intel Management by Objectives" or "iMBOs." However, it was John Doerr, who later became a renowned venture capitalist at Kleiner Perkins, who played a pivotal role in popularizing OKRs on a much wider scale. In 1999, Doerr introduced the OKR framework to the founders of a young startup called Google. The methodology resonated strongly with Google's leadership, and it quickly became deeply ingrained in the company's culture and operational processes.

The significant impact that OKRs are widely credited with having on Google's exponential growth and overall success has led to its widespread adoption across the technology industry and in various other sectors around the world. The evolution of goal setting from the earlier "Management by Objectives" approach to the more collaborative and team-oriented OKR framework reflects a broader shift in management philosophies. There is now a greater emphasis on fostering autonomy and a sense of ownership at all levels within an organization. While the fundamental idea of setting objectives and tracking progress has existed for a long time, OKRs, as they have been popularized, represent a more dynamic and less bureaucratic approach to achieving ambitious goals.[8]

OKRs vs. KPIs

Some may confuse Objectives and Key Results (OKRs) with Key Performance Indicators (KPIs). While they leverage the same metrics, they serve different purposes:[9]

- **KPIs Tell You Where You Are:** They're constant signals of business health that you track continuously.

- **OKRs Tell You Where You Want to Be:** They're focused on specific goals with timeframes.

---

[8] Giulia Pines, "The Origin Story," whatmatters.com, accessed June 22, 2025, https://www.whatmatters.com/articles/the-origin-story.

[9] Carlos Gonzalez de Villaumbrosia, "OKRs vs. KPIs: What's the Difference?," productschool.com, accessed June 29, 2025, https://productschool.com/blog/skills/okr-vs-kpi-difference.

Think of KPIs as vital signs that your doctor tracks at every visit: weight, heart rate, blood pressure. They indicate overall health regardless of specific goals. OKRs, by contrast, are more like a fitness plan with specific targets: "Lose 8 pounds by summer's end" or "Reduce blood pressure to below 125/85 within 6 months."

For startups, it's best to use both: OKRs to set direction and KPIs to measure progress toward those goals. This combination provides clear direction, measurable targets, and data-driven insights to drive success. Table 6-1 presents the key differences between OKRs and KPIs.

Table 6-1. *OKRs vs. KPIs: The Simple Breakdown*

Aspect	OKRs	KPIs
Main Job	To drive change: achieve big and new goals.	To monitor business health and daily performance.
Focus	Where are we going? (Future).	How are we doing now? (Present).
What Success Is	Making significant progress (70% is often great).	Hitting a specific target (100% is the goal).
How Long It Lasts	Temporary; set for a specific time (like a quarter).	Ongoing; tracked continuously over time.
Simple Analogy	A Road Trip: The destination and key milestones.	The Car's Dashboard: Your speed, fuel, and engine health.

Case Study: The First Metrics Meeting

Picture this: You're the first data professional at a hypothetical fitness app startup focused on personalized workout tracking and community engagement. It's your first meeting with the leadership team to discuss key metrics, and you need to present a clear framework for measuring success using both KPIs and OKRs. The founders are eager to understand what metrics will drive their business forward and how to set ambitious yet achievable goals.

You walk the team through concrete examples covering user engagement, user acquisition, retention, monetization, product performance, feature adoption, and user satisfaction, showing how KPIs and OKRs work together in each area. These are just a range of example metrics and OKRs to take inspiration from; when defining your own, you should always tailor them to your company's unique context and ensure you follow best practices—remember, "a limit of three to five OKRs per cycle leads companies, teams, and individuals to choose what matters most."

Understanding the Context

Fitness apps represent a rapidly growing market. Success in this space requires tracking both operational performance through KPIs and strategic advancement through OKRs. The leadership team needs to understand how these two frameworks complement each other in driving startup growth. Table 6-2 presents hypothetical User Engagement KPIs and OKRs comparison that clearly presents these concepts.

Table 6-2. *Comparison: KPIs vs. OKRs—User Engagement*

KPIs	OKRs
KPI: Daily Active Users (DAU): 15,000+ users KPI: Average Session Duration: 12+ minutes KPI: Monthly Active Users (MAU): 45,000+ users	Objective: Build an engaged fitness community KR1: Increase DAU from 15,000 to 25,000 KR2: Achieve 75% weekly retention rate KR3: Generate 500+ user-generated workout posts monthly

Beyond general engagement, a startup's growth hinges on its ability to efficiently attract new users. Let's look at how KPIs and OKRs work together to optimize user acquisition efforts (see Table 6-3).

Table 6-3. *Comparison: KPIs vs. OKRs—User Acquisition*

KPIs	OKRs
KPI: Customer Acquisition Cost (CAC): <$25 KPI: Cost Per Install (CPI): <$3.50 KPI: Organic vs Paid Install Ratio: 60:40	Objective: Accelerate sustainable user growth KR1: Acquire 10,000 new users this quarter KR2: Reduce CAC by 20% through referral programs KR3: Achieve 40% of installs from organic channels

Acquiring users is only half the battle; retaining them is arguably more critical for long-term success. This is where a robust focus on user retention becomes indispensable. Here's how KPIs and OKRs can guide your efforts in minimizing churn and fostering loyalty (see Table 6-4).

Table 6-4. *Comparison: KPIs vs. OKRs—User Retention*

KPIs	OKRs
KPI: 7-day Retention Rate: >40%	Objective: Create habit-forming fitness experiences
KPI: 30-day Retention Rate: >25%	KR1: Improve 30-day retention from 25% to 35%
KPI: Churn Rate: <15% monthly	KR2: Reduce monthly churn to under 10%
	KR3: Achieve 60% of users completing onboarding

Ultimately, a sustainable startup must convert its value proposition into a profitable income stream. The following section illustrates how financial KPIs and strategic OKRs are employed to measure and drive revenue growth and monetization (see Table 6-5).

Table 6-5. *Comparison: KPIs vs. OKRs—User Revenue and Monetization*

KPIs	OKRs
KPI: Average Revenue Per User (ARPU): $8/month	Objective: Build a sustainable revenue model
KPI: Monthly Recurring Revenue (MRR): $180,000+	KR1: Increase ARPU from $8 to $12
KPI: Subscription Conversion Rate: 8–12%	KR2: Reach $250,000 MRR by quarter end
	KR3: Launch premium tier with 15% conversion rate

Beyond financial viability, the user experience is paramount, especially for product-led growth. Monitoring the technical performance and overall quality of your product is essential for user satisfaction and retention. Table 6-6 highlights key performance indicators and objectives related to ensuring a stellar product experience.

Table 6-6. *Comparison: KPIs vs. OKRs—Product Performance*

KPIs	OKRs
KPI: App Crash Rate: <0.8%	Objective: Deliver exceptional user experience
KPI: Average Load Time: <3 seconds	KR1: Achieve 99.5% crash-free sessions
KPI: App Store Rating: >4.2 stars	KR2: Reduce load time to under 2 seconds
	KR3: Maintain 4.5+ star rating with 1000+ reviews

Once a strong product foundation is in place, driving the adoption and consistent use of key features becomes a priority. Table 6-7 demonstrates how KPIs and OKRs can be leveraged to ensure users discover and engage with the core functionalities that deliver the most value.

Table 6-7. *Comparison: KPIs vs. OKRs—Feature Adoption*

KPIs	OKRs
KPI: Workout Completion Rate: >65%	Objective: Drive feature adoption and engagement
KPI: Social Feature Usage: 35% of MAU	KR1: Increase workout completion rate to 75%
KPI: Goal Setting Adoption: 45% of users	KR2: Get 50% of users to set fitness goals
	KR3: Launch AI coaching with 25% adoption rate

Ultimately, all efforts converge on satisfying your users and transforming them into advocates. This final set of metrics and objectives in the example focuses on understanding and enhancing overall user satisfaction, a leading indicator of long-term growth and brand loyalty (see Table 6-8).

Table 6-8. *Comparison: KPIs vs. OKRs—User Satisfaction*

KPIs	OKRs
KPI: Net Promoter Score (NPS): >40	Objective: Delight users and build advocacy
KPI: Customer Support Response Time: <4 hours	KR1: Achieve NPS score of 50+
KPI: In-app Survey Rating: >4.0	KR2: Reduce support tickets by 30%
	KR3: Generate 200+ App Store reviews monthly

Strategic Recommendations for Implementation

For KPI Tracking

- **Dashboard Setup:** Create dashboards tracking daily, weekly, and monthly KPIs.

- **Automated Reporting:** Implement automated alerts when KPIs fall below thresholds.

- **Cross-Functional Visibility:** Ensure all teams have access to relevant KPIs for their domain.

For OKR Implementation

- **Quarterly Cadence:** Set OKRs every quarter with mid-quarter check-ins.

- **Ambitious but Achievable:** Target 60-70% achievement rate for optimal motivation.

- **Company-wide Alignment:** Cascade company OKRs to team and individual levels.

Key Takeaways for the Leadership Meeting

- **Complementary Frameworks:** KPIs provide operational oversight, while OKRs drive strategic advancement.

- **Startup-Specific Focus:** Early-stage fitness apps should prioritize user engagement, retention, and product-market fit metrics.

- **Data-Driven Culture:** Both frameworks require consistent measurement and regular review cycles.

- **Scalable Implementation:** Start with core metrics and expand measurement capabilities as the team grows.

The combination of comprehensive KPI monitoring and ambitious OKR goal setting will provide the startup with both the operational insights needed for day-to-day optimization and the strategic direction required for breakthrough growth in the competitive fitness app market.

The hypothetical startup example demonstrates the practical application of KPIs and OKRs in a consumer tech startup. However, the power of this framework extends to highly specialized data projects within any industry. The next example will walk you through how to define concrete objectives and measurable key results for specific data initiatives, from churn prediction to AI chatbot development, illustrating how these principles apply to the technical core of many startups.

Playbook: Measuring Success of Data Projects with OKRs

Setting clear objectives and measurable key results is crucial for the success of data-driven projects. This playbook guides you through defining OKRs tailored to typical data initiatives such as churn prediction, AI chatbots, recommendation engines, and fraud detection. By practicing this framework, you will learn how to align data projects with business goals, track progress effectively, and demonstrate impact to stakeholders.

Customer Churn Prediction Model OKRs

To begin, let's address one of the most fundamental challenges for any startup: retaining customers. A customer churn prediction model is a classic data science project aimed directly at this goal. Table 6-9 outlines Objectives and Key Results (OKRs) for such a project.

Table 6-9. *Customer Churn Prediction Model OKRs*

Objective	Key Results
Reduce monthly churn rate by leveraging predictive insights	1. Deploy a churn prediction model with >85% AUC on hold-out data by end of Q3
	2. Ensure 60% of at-risk customers receive personalized offers within 48 hours of risk flagging
	3. Lower actual monthly churn rate from 8% to 6% within two quarters

AI-Powered Chatbot for Customer Support OKRs

Once a startup has a strategy to reduce churn, the focus often shifts to improving the efficiency of customer support. An AI-powered chatbot is a prime example of leveraging data to enhance user experience while managing operational costs. Table 6-10 presents OKRs for an AI-powered customer support chatbot project.

Table 6-10. *AI-Powered Chatbot for Customer Support OKRs*

Objective	Key Results
Improve customer support efficiency with AI	1. Handle ≥40% of incoming support tickets via chatbot by end of Q2 2. Maintain chatbot resolution accuracy ≥90% 3. Increase customer satisfaction (CSAT) score for chatbot interactions to ≥4.2/5

Personalized Recommendation Engine OKRs

Building on better support, the next step might be to drive deeper engagement and increase revenue. A personalized recommendation engine uses data to deliver more value to users, which in turn boosts key business metrics. Table 6-11 summarizes OKRs for a personalized recommendation engine project.

Table 6-11. *Personalized Recommendation Engine OKRs*

Objective	Key Results
Boost user engagement and conversions through personalized suggestions	1. Launch recommendation engine with mean reciprocal rank (MRR) ≥0.15 2. Increase click-through rate (CTR) on recommended items from 5% to 9% within three months 3. Achieve a 7% lift in average order value among users exposed to recommendations

Real-Time Fraud Detection Pipeline OKRs

Finally, all customer-facing initiatives must be built on a secure foundation. Protecting the business from financial loss is a critical, non-negotiable function. A real-time fraud detection pipeline is an essential data project for safeguarding revenue and maintaining user trust. Table 6-12 lists OKRs for this type of project.

Table 6-12. *Real-Time Fraud Detection Pipeline OKRs*

Objective	Key Results
Enhance transaction security with proactive fraud detection	1. Implement real-time scoring of 100% of transactions within 150 ms latency 2. Attain fraud recall $\geq$92% and precision $\geq$88% on live traffic 3. Reduce monthly fraud losses by 25% within one quarter

These examples illustrate how OKRs can transform abstract data initiatives—from predicting customer churn to deploying advanced AI—into quantifiable projects with clear business impact. By establishing precise objectives and measurable key results, startups can ensure their data science investments are strategically aligned and demonstrably contribute to core business goals.

As you move beyond internal goal setting, it's essential to understand how metrics function in the real world: how they are interpreted by external stakeholders, how their importance shifts as a startup grows, and how different industries and funding stages demand different approaches. The next section explores these practical realities, examining not only what investors look for but also how startups adapt their metrics and frameworks to drive credibility, secure funding, and build lasting, scalable businesses.

Metrics in the Real World

In the real world, investors and stakeholders demand more than just well-structured frameworks—they seek metrics that demonstrate tangible progress and business viability. This section examines how key startup metrics are prioritized and interpreted not only by venture capitalists but also by founders and other stakeholders across different industries and growth stages. By understanding these practical expectations and how they shift as your company matures, founders can tailor their metric reporting to build credibility, secure funding, and guide their startups toward scalable, sustainable success.

The Reality: VC's Key Metrics

While popular frameworks offer valuable structure and are widely recognized in the startup world, many investors—including David Sacks—emphasize that true business success comes down to tracking a small set of core metrics that matter most for a given industry and stage. Sacks, for example, highlights that one of the main advantages of SaaS businesses is their measurability: rather than monitoring dozens of indicators, only a handful of key metrics are truly essential for understanding performance and growth.[10]

While VCs utilize frameworks, their experience and specialization in certain industries (e.g., SaaS, fintech, AI) undoubtedly influence which metrics they prioritize and how they interpret them. Rather than focusing too much on beautifully designed frameworks with catchy acronyms, top-tier investors have converged on a practical set of metrics that have proven effective across thousands of portfolio companies and billions in deployed capital. It's important to keep this in mind when preparing for meetings: if you can, research the VCs you plan to meet and invest time in tailoring your narrative. Focus on the handful of metrics that matter most to that particular investor rather than presenting a generic set based on popular frameworks. Their time is valuable, and what matters most is answering their specific questions with clarity and relevance to their unique approach.

The Evolution of Startups' Metrics As the Startup Grows

In their comprehensive report, "The Ultimate Guide to Startup Metrics," Speedinvest Pirates—the growth unit of leading European venture capital fund Speedinvest—provides a qualitative road map for venture-backed startups navigating the complex landscape of metrics. Drawing on extensive experience with a diverse portfolio of startups, the guide offers not just examples of the most valuable metrics but also clear explanations of how to calculate them and insights into how venture capitalists interpret these numbers. By supplementing Speedinvest's framework with perspectives from

---

[10] David Sacks and Ethan Ruby, "The SaaS Metrics That Matter," sacks.substack.com, accessed June 30, 2025, https://sacks.substack.com/p/the-saas-metrics-that-matter.

other industry experts, founders can gain a nuanced understanding of which metrics matter most at each stage. In this section, we'll examine how these metrics play out in practice, highlighting both best practices and real-world challenges.[11]

Metrics are not universal. Their relevance shifts depending on a startup's industry, business model, and—most critically—its stage of development. As Speedinvest founder Dieter Rappold puts it, "a data-based decision making culture is the single most important competitive edge of startups against incumbents." The report organizes its guidance by funding stage—pre-seed, seed, and Series A—offering targeted benchmarks and focus areas for each phase of a startup's journey.

Pre-seed

At the pre-seed stage, the focus is less on hard revenue numbers and more on validating the core idea and showing early signs of market interest. According to the Speedinvest report, investors are primarily looking at the strength of the founding team and the power of their vision.

- **Product and Market Validation**: The main goal is to demonstrate that a problem exists and your solution is a compelling one. For B2B or deep tech startups, a key metric is the number of proofs of concept (PoCs) established with potential clients, with paid PoCs being an especially strong signal. For B2C fintech products, validation can come from proxies like the size of a waiting list, the number of beta customers, or an analysis of the probability of conversion based on early user actions.

- **Early Traction**: For startups with a freemium or open source model, early traction metrics are vital. These can include the number of downloads, sign-ups (which, in this context, are not considered vanity metrics), or—for open source projects—activity on platforms like GitHub, such as the number of contributors or stars.

---

[11] Julia Weinmayr et al., *The Ultimate Guide to Startup Metrics: What to Track, When and Why* (Vienna: Speedinvest Pirates GmbH, 2020).

- **Building the Foundation**: For marketplaces, the pre-seed stage is about solving the "chicken-and-egg" problem by building both the supply and demand sides. Key metrics involve tracking the outreach to suppliers and the conversion rate from a simple sign-up to an active user who regularly performs key actions on the platform.

Seed

The seed stage is where a startup must demonstrate it has found a "hook in the market," transitioning from a promising idea to a business with real traction and a tested product. As the report highlights, this is typically the launch phase where the focus shifts toward revenue, growth rates, and understanding the underlying unit economics.

- **Initial Revenue and Growth**: The emphasis is on demonstrating top-line growth and, more specifically, recurring revenue. As Markus Lang, Associate Partner at Speedinvest, emphasizes: "You need to show that your users are not only paying but also using the product." This makes user retention and engagement metrics critical, especially for businesses with annual subscription models where renewal data is not yet available.

- **Validating the Business Model**: For deep tech and industrial tech companies, the focus moves from securing PoCs to converting them into paying customers. The report suggests having 10 to 20 PoCs in the pipeline, with at least one successful conversion being an ideal signal. Metrics like Annual Contract Value (ACV) and the length of the sales cycle become crucial for investors to assess the business's viability.

- **Unit Economics**: For industries like health and consumer tech, startups must now have a firm grasp of their unit economics. This means tracking Customer Acquisition Cost (CAC), Lifetime Value (LTV), and the LTV/CAC ratio at all times. While the ideal ratio is often cited as 3x LTV > CAC, the report cautions that any calculations based on assumptions must paint a realistic picture.

Series A

By Series A, a startup is expected to have a proven business model and be ready to scale. The narrative shifts decisively toward monetization, sustainable growth, and strong unit economics. As Speedinvest Principal Chris Zemina states, "monetization becomes important, as often early-stage based companies raise on the product and the vision - however, over time monetization must follow."

- **Scalable Revenue**: Focus on Annual Recurring Revenue (ARR) and Monthly Recurring Revenue (MRR). The report highlights a generalized rule of thumb for growth: aiming for 3x year-over-year growth for the first two years, followed by 2x for the subsequent three years. Investors will analyze the history of this growth to project future potential.

- **Profitability and Efficiency**: The LTV/CAC ratio comes under intense scrutiny, with a "gold standard" of 5x being desirable, though 3x to 4x is also considered strong.

- **Retention and Customer Health**: Proving customer retention is non-negotiable at this stage. For enterprise-focused businesses, a retention rate of around 80% is a rough benchmark. For consumer products, tracking retention cohorts (e.g., daily, weekly, or monthly active users) shows that the user base is not just growing but is also engaged and loyal over time. In specific industries like insurance, metrics such as the loss ratio (how much of the total premium gets claimed) become vital indicators of the business's long-term health.

While the Speedinvest report provides an excellent foundation, the world of venture capital is not a monolith; different investors and firms emphasize different metrics based on their unique experiences and investment theses. Examining these alternative viewpoints can offer a more nuanced and complete picture of what it takes to secure funding and build a scalable business. These perspectives often add critical details missing from a high-level overview or even present slightly different benchmarks.

One of the most interesting data-driven views comes from an analysis by Aumni, a venture capital analytics firm, which studied over 10,000 data points from portfolio companies. Their findings reveal a dynamic shift in priorities as a startup matures. For example, while early-stage fundraising focuses heavily on survival and potential, the

Aumni report highlights that metrics like cash on hand and cash runway dramatically "plummet in importance" from the seed stage to Series D and beyond. In their place, metrics signaling operational efficiency become critical as the company is expected to function as a mature, sustainable enterprise. This data suggests that the narrative must evolve from "can we survive?" to "how efficiently can we operate?"[12]

Further granularity can be found in guides like The VC Corner, which provides more specific benchmarks that vary by business model, adding a layer of detail to the standards presented by Speedinvest. For instance, where a general LTV/CAC ratio of 3x is often cited as a healthy benchmark, The VC Corner specifies a target of 3–5x for SaaS companies but a higher 4–6x for B2C businesses, reflecting the different economic models at play. The guide also offers more concrete growth targets, suggesting that a seed-stage company should aim for 15–20% month-over-month growth, whereas a Series A company might target 8–12%. These more precise figures can give founders a sharper, more actionable set of goals tailored to their specific market and stage.[13]

Startup Metrics by Industry and Stage According to Speedinvest[14]

These different perspectives underscore a crucial point: while the principles of good metrics are universal, their application is highly contextual. The most important metrics for a B2B deep tech company at the seed stage will differ significantly from those for a consumer fintech app targeting a Series A. To make this practical, the following tables consolidate the key metrics that investors typically prioritize across different industries and funding stages, drawing from the insights discussed. Table 6-13 illustrates Deep Tech startup metrics by stage.

---

[12] "What KPIs do venture firms care about across stages?," aumni.fund, accessed June 25, 2025, https://www.aumni.fund/blog/kpis-across-stages.

[13] Ruben Dominguez Ibar, "Key Startup Metrics VCs Care About: How to Track, Improve & Use Them to Raise Capital," thevccorner.com, accessed June 20, 2025, https://www.thevccorner.com/p/key-startup-metrics-vcs.

[14] Julia Weinmayr et al., *The Ultimate Guide to Startup Metrics: What to Track, When and Why* (Vienna: Speedinvest Pirates GmbH, 2020).

Table 6-13. *Deep Tech Startup Metrics by Stage*

Stage	Key Metrics
Pre-seed	Number of Proof of Concepts (PoCs), Paid PoCs, GitHub activity (for open source), Downloads and sign-ups (for freemium)
Seed	Number of PoCs, Conversion of pilots to paying customers, Annual Contract Value (ACV)
Series A	Recurring Revenue, Booked Revenue vs. Annual Recurring Revenue (ARR), Customer Acquisition Cost (CAC), Churn Rate, Conversion Rate (CR), Retention Rate

Just as deep tech startups have unique metric priorities, the financial technology (fintech) sector also presents its own set of critical indicators that evolve with the company's growth. Let's explore the key metrics that matter for fintech ventures at each stage (see Table 6-14).

Table 6-14. *Fintech Startup Metrics by Stage*

Stage	Key Metrics
Pre-seed	Number of Beta Customers, Waiting Lists, Probability of conversion, POCs
Seed	Customers Count, MRR, Engagement Metrics, NPS, ACV, CAC, LTV, LTV/CAC ratio, Converted POCs
Series A	Growth rate, Retention & Engagement Metrics (DAU, WAU, MAU), Monetization metrics

Beyond the specialized demands of deep tech and fintech, the broader health and consumer tech space also possesses distinct metric requirements. These sectors often balance user engagement with monetization, leading to a unique set of focuses at different startup stages (see Table 6-15).

Table 6-15. *Health & Consumer Tech Startup Metrics by Stage*

Stage	Key Metrics
Pre-seed	Convince the investors you have two things: the vision and a good team.
Seed	Growth, MRR and Revenue Metrics.
Series A	Growth, MRR, Revenue and Retention Metrics, CAC, LTV, LTV/CAC

Finally, for startups operating in the industrial tech sector, the metrics shift to reflect longer sales cycles, complex integrations, and the emphasis on pilot conversions. Understanding these nuances is crucial for demonstrating progress and securing investment (see Table 6-16).

Table 6-16. *Industrial Tech Startup Metrics by Stage*

Stage	Key Metrics
Pre-seed and Seed	Number of pilots, Conversion of pilots to paying customers, ACV, Sales cycle length, NPS
Series A	Conversion and Retention Rates, MRR, ARR, CAC, LTV, ACV

As we've observed, the landscape of critical metrics shifts significantly as a startup matures and its industry evolves. But beyond industry specifics and growth stages, external economic conditions also play a profound role in how investors evaluate a startup's health. The current environment, characterized by tighter capital markets, places a renewed emphasis on efficiency and sustainable growth. The next section delves into how leading venture capitalists interpret metrics like the Burn Multiple, which become paramount indicators of capital efficiency during economic downturns.

Conserving Fuel: Capital Efficiency As the New Growth Imperative

If you're a SaaS company you really just want to prioritize survival during this period...growth is always nice, don't get me wrong, but I would rather be a little more conservative on growth and guarantee survival than trying to juice growth an extra 10, 20, 30 percent but risk the fate of your company...Instead of having tailwinds, you've got very strong headwinds over the next couple of years and if you try to maintain your speed with intense headwinds...you're going to burn an incredible amount of fuel—or you can just accept the fact that you're going to have intense headwinds, slow down your speed, and just burn a lot less fuel and conserve.

—David Sacks,

Co-founder of Craft Ventures[15]

---

[15] "David Sacks - SaaS Metrics, Tech Cycles & Fundraising in a Down Market," World of DaaS With Auren Hoffman YouTube channel, accessed June 22, 2025, https://www.youtube.com/watch?v=jyo1rF4iEZ4.

The economic environment fundamentally shifts metric interpretation and importance. David Sacks emphasizes the **Burn Multiple** as a critical efficiency measure during economic uncertainty. The Burn Multiple, calculated as Net Burn divided by Net New ARR, answers the crucial question: "how much is the startup burning in order to generate each incremental dollar of ARR?"[16,17,18]

Sacks recommends "a Burn Multiple of less than one is amazing, but anything less than two is still quite good." He urges SaaS companies to plan more conservatively during downturns. He notes that many startups set targets based on a burn multiple of two, but this leaves them vulnerable—if they fall short on new ARR or experience higher churn, their burn multiple can quickly escalate to unsustainable levels. Instead, he recommends companies adopt a much leaner approach, aiming for a burn multiple closer to one, to ensure they have a safety margin and can weather unexpected challenges.[19]

To make this concept tangible, we can visualize it using a Burn Multiple Matrix for a hypothetical startup, as illustrated in Figure 6-1. It provides a quick diagnostic tool for assessing the financial health and growth efficiency of a company. To use it, you need two key figures from your financials:

- **Annual Net Burn:** Locate your company's total net cash burned over the last 12 months on the vertical (Y) axis.

- **Net New ARR:** Find the amount of new annual recurring revenue you added during that same period on the horizontal (X) axis.

---

[16] Janelle van Deventer, "What is Burn Multiple, and how do you calculate it?," Capchase Blog, accessed June 22, 2025, `https://www.capchase.com/blog/what-is-burn-multiple-and-how-do-you-calculate-it`.

[17] Justin Kahl and David George, "A Framework for Navigating Down Markets," Andreessen Horowitz, accessed June 25, 2025, `https://a16z.com/a-framework-for-navigating-down-markets/`.

[18] David Sacks, "The Burn Multiple: How Startups Should Think About Capital Efficiency," sacks.substack.com, accessed June 26, 2025, `https://sacks.substack.com/p/the-burn-multiple-51a7e43cb200`.

[19] David Sacks and Ethan Ruby, "The SaaS Metrics That Matter," sacks.substack.com, accessed June 30, 2025, `https://sacks.substack.com/p/the-saas-metrics-that-matter`.

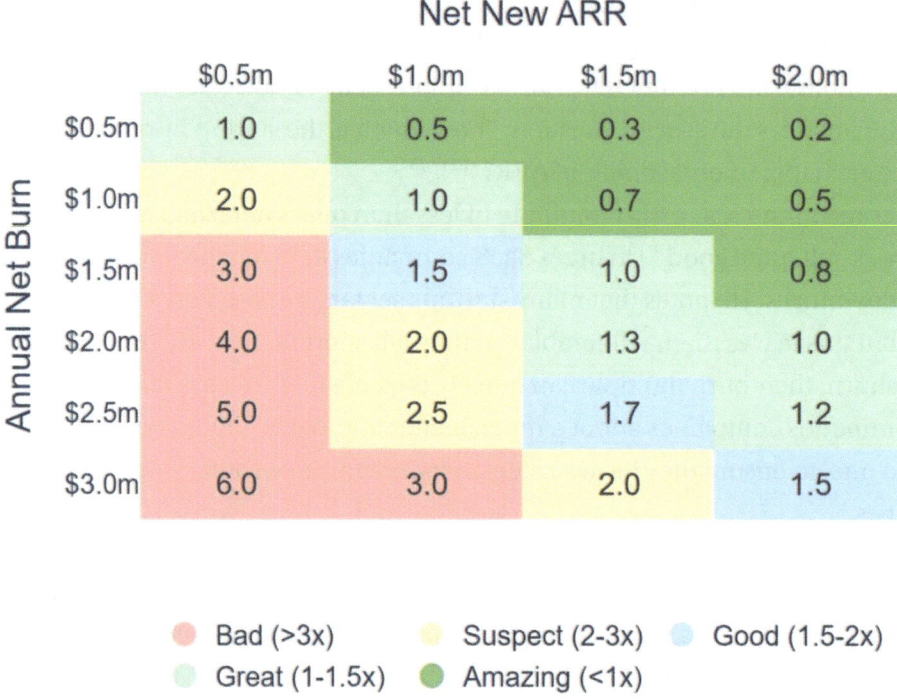

Figure 6-1. *The Burn Multiple Matrix—Example*

The cell where your burn and new ARR intersect reveals your Burn Multiple. The number inside the cell is the multiple itself, and the color provides an immediate, qualitative assessment of your performance.

Let's walk through a few scenarios using Figure 6-1 to see how this works. Imagine a startup has an Annual Net Burn of $2.0m (find "$2.0m" on the vertical axis) and has generated $1.0m in Net New ARR (find "$1.0m" on the top axis). By tracing these values to their intersection, we find a cell with the number 2.0. This means the startup spent $2 to generate every $1 of new annual revenue. The cell is shaded blue, categorizing this performance as "Good"—a respectable, though not exceptional, level of efficiency for a company investing in its growth.

Now consider a more concerning case. Another company has an Annual Net Burn of $3.0m but only managed to add $0.5m in Net New ARR. Finding this intersection on the matrix reveals a Burn Multiple of 6.0. This cell is shaded red, flagging the performance as "Bad." Such a high multiple is unsustainable and signals deep-seated issues in the company's go-to-market strategy or product-market fit.

Conversely, a highly efficient company might burn just $0.5m to generate $1.0m in Net New ARR. This yields an "Amazing" Burn Multiple of 0.5, shaded in dark green. This startup is effectively printing money, a clear sign of a powerful business model and exceptional execution.

The true power of Figure 6-1 lies in this immediate visual feedback. The color gradient, from a healthy green to a cautionary yellow and a dangerous red, allows founders, executives, and investors to bypass the mental math and instantly gauge the efficiency of a company's growth engine. It transforms a simple calculation into a strategic compass, pointing toward either sustainable growth or a need for urgent course correction.

David Sacks acknowledges that while another popular metric **Hype Ratio**— calculated as capital raised (or burned) divided by ARR—is a common way to assess capital efficiency, he believes the Burn Multiple is a more useful metric because it centers on recent performance rather than cumulative capital history. The Burn Multiple, by focusing on how much net burn is required to generate new ARR, gives a clearer picture of a startup's current efficiency and operational health.

Runway Calculation becomes critically important during uncertain times. It represents the amount of time—typically measured in months—that a startup can continue operating before its available funds are exhausted. While this measure may not be as critical for businesses that are already profitable, it's essential for early-stage startups that haven't yet started earning revenue from their product or service. Monitoring runway closely helps these companies ensure they have enough resources to reach their next milestone before running out of money.[20]

As Tessa Clarke, co-founder of OLIO, observes: "It's no longer about growth at all costs; business basics are now back in vogue." This shift requires startups to reconsider their North Star metrics, potentially moving from growth-focused metrics like user acquisition to revenue-focused metrics like Annual Recurring Revenue in recognition that "revenue trumps growth in this new environment."[21]

---

[20] Matt Preuss, "How to Calculate Runway & Burn Rate," visible.vc, accessed June 18, 2025, `https://visible.vc/blog/how-to-calculate-runway/`.

[21] Tessa Clarke, "Startups need new metrics for a tough new era," sifted.eu, accessed June 20, 2025, `https://sifted.eu/articles/startup-metric-new-era-olio`.

Practical Implementation Framework

David Sacks and industry sources agree that frameworks are useful for providing structure, but it's the right, practical metrics that truly inform decision-making. The essential point is to focus on metrics that accurately reflect the company's underlying health and progress rather than relying on vanity metrics that look impressive but offer little real value for guiding business choices.

Effective metrics serve as essential tools for startups, much like reliable instruments in an aircraft, enabling founders to continuously adjust, iterate, and maintain strong control over their business. However, it is important to avoid overwhelming complexity; tracking an excessive number of metrics, especially before product launch, can be counterproductive and distracting.[22]

The emphasis on selecting meaningful metrics reflects the evolving landscape of venture capital evaluation, where practical and verifiable measurements take precedence over purely theoretical models. Cultivating a culture centered on data-based decision-making has become a key competitive advantage for startups striving to outperform established incumbents.

Metrics Evolution in Startup Growth: Case Studies of Leading Companies

Successful startups rarely rely on a static set of metrics. As companies evolve, so do their key performance indicators, reflecting shifts in business models, market conditions, and strategic priorities. This section presents case studies from leading startups like Amplitude, Facebook, Amazon, and Uber, illustrating how their North Star Metrics and growth measurements have adapted over time. These examples provide valuable lessons on the dynamic nature of metrics in scaling startups.

---

[22] "Consumer Startup Metrics | Startup School," Y Combinator YouTube Channel, accessed June 14, 2025, https://www.youtube.com/watch?v=fdD4y4Civp4.

Amplitude: From Weekly Querying Users to Weekly Learning Users

Amplitude's evolution provides a compelling example of how product analytics companies must refine their North Star Metrics as they mature. According to Justin Bauer, Amplitude's Chief Product Officer, the company initially focused on **"weekly querying users" (WQUs)** as their primary metric when they were a small 20-person product team.[23]

The original WQU metric made sense for an analytics platform because their main value proposition was helping product managers and analysts answer questions about user behavior. This metric captured their core differentiator—that Amplitude users didn't just view dashboards but explored data deeply to understand what drives user behavior. WQUs proved to be both a reflection of customer value delivery and a leading indicator of account retention and expansion.

However, as Amplitude evolved, they recognized that WQUs didn't truly reflect the type of impact they wanted to have with customers. Their mission wasn't simply to help users build better analyses—it was to help companies build better products, which is inherently a team sport. By focusing solely on users who queried, they were limiting the potential of their solution.

In 2018, Amplitude made a strategic shift to **"weekly learning users" (WLUs)**, defined as users who are active and share a learning that is consumed by at least two other people within seven days. This new metric better captured their mission of helping teams complete the Build > Measure > Learn loop. The transition involved extensive data analysis, examining everything from Amplitude engagement and feature adoption to customer product velocity and usage of instrumentation. They also conducted network analysis to understand relationships between users and how connectedness differed for their most successful customers.

This metric evolution resulted in three supporting indicators: Activated Accounts (organizations reaching at least 5 WLUs), Broadcasted Learnings (content consumed by 2+ people within 7 days), and Consumption of Learnings (total reach of all broadcasted learnings). The new framework led to a 50% increase in WLUs per active organization in the six months following implementation.

---

[23] Justin Bauer, "We're Evolving Our Product's North Star Metric. Here's Why.," amplitude.com blog, accessed June 23, 2025, `https://amplitude.com/blog/evolving-the-product-north-star-metric`.

Facebook: The Power of Meaningful Engagement over Vanity Metrics

Facebook's approach to metrics stands in stark contrast to competitors like MySpace, demonstrating how meaningful engagement metrics outperform vanity metrics. One of Facebook's most famous early insights was identifying an "Aha moment": new users who added at least seven friends within their first ten days were significantly more likely to become long-term, active users. This discovery emerged from rigorous data analysis by the growth team, led by figures like Chamath Palihapitiya, and shaped Facebook's onboarding and product development strategies during its crucial growth phase from millions to billions of users.[24]

While this **"7 friends in 10 days"** metric was a powerful leading indicator of user retention, it was not Facebook's North Star Metric (NSM). Instead, Facebook's true NSMs were **Monthly Active Users (MAU)** in its earlier years and, as the company matured, **Daily Active Users (DAU)**. These metrics directly reflected the platform's core value proposition—connecting people and fostering meaningful engagement at scale.

As Sequoia Capital points out, Facebook's choice of North Star Metric evolved alongside user behavior and technology:

> *Change the metric as your business evolves. Your top-line metric may need to change over time. For example, before the mobile age, users checked products such as Facebook less frequently because of lack of access and connectivity. As mobile use increased, these companies revised their primary metric from MAU to DAU.*

> —Data Science Team,
>
> Sequoia Capital[25]

In contrast, growth advisor and investor Buckley Barlow argues that a key reason for MySpace's downfall and Facebook's success lies in the North Star Metric (NSM) each company prioritized. MySpace chose to focus on Registered Users—a metric that looks

---

[24] "Chamath Palihapitiya On The Growth Principles That Got Facebook To Billions Of Users," startuparchive.org, accessed June 25, 2025, `https://www.startuparchive.org/p/chamath-palihapitiya-on-the-growth-principles-that-got-facebook-to-billions-of-users`.

[25] Jamie Cuffe, Avanika Narayan, Chandra Narayanan, Hem Wadhar, and Jenny Wang, "Defining Product Success: Metrics and Goals," Sequoia Capital Articles, accessed July 02, 2025, `https://articles.sequoiacap.com/defining-product-success-metrics-and-goals`.

impressive but doesn't necessarily reflect real engagement—as their NSM. In contrast, Facebook made Monthly Active Users their primary metric. This difference in focus helped Facebook thrive while MySpace struggled.

While it's important for social media platforms to keep track of how many users sign up, MySpace's problem was that Registered Users only indicated how many people had created accounts, not how many were actually using the service regularly. If users don't find ongoing value in a product, they'll eventually stop using it. By measuring Monthly Active Users, Facebook was able to track real engagement and understand which users were consistently finding value in the platform.

Amazon: The Constant Evolution

Amazon's journey from a small startup to one of the world's largest companies is a story of relentless evolution. In the early days, Amazon's survival depended on rapid experimentation and a willingness to adapt. It's believed that the company's first North Star Metric was simple: **the number of new detail pages** created on the website, each representing a new book or product added to the catalog. This metric reflected Amazon's initial obsession with offering the broadest selection possible, believing that more products would naturally attract more customers and drive growth. The approach evolved to focus also on other metrics like **in-stock availability** and **page views**.[26]

However, as Amazon grew, it became clear that not all growth metrics translated into real business value. The leadership team noticed that increasing the catalog size didn't always lead to higher sales. This realization marked a shift from input metrics (like catalog breadth) to output metrics such as actual sales, customer retention, and engagement. The company began to focus on metrics that directly reflected **customer satisfaction** and **business performance** rather than just operational activity.

As Amazon survived the dot-com bubble and expanded into new categories, its metrics culture evolved further. The launch of initiatives like Amazon Marketplace and Amazon Prime required new ways to measure success. For Marketplace, metrics shifted to include third-party sales and platform engagement. With Prime, the focus expanded to subscription growth and delivery speed, both of which were tightly linked to customer loyalty and repeat purchases.[27]

---

[26] Colin Bryar and Bill Carr, *Working Backwards: Insights, Stories, and Secrets from Inside Amazon* (St. Martin's Press, 2021).

[27] Brad Stone, *The Everything Store: Jeff Bezos and the Age of Amazon* (Little, Brown and Company, 2013).

Today, Amazon's North Star Metric is widely recognized as the "**number of purchases per month**." This metric encapsulates Amazon's core mission to be "Earth's most customer-centric company." By measuring successful transactions, Amazon tracks both customer satisfaction and business growth in a single, actionable number. This unified metric aligns teams across the company, from logistics to software development, ensuring that everyone is working toward the same customer-focused goal.[28]

Amazon's approach to metrics is a reflection of its broader evolution as a company. From a startup obsessed with catalog size to a global leader driven by customer outcomes, Amazon's willingness to adapt its North Star Metric has been key to its long-term success. This evolution demonstrates the company's commitment to continuous improvement and its ability to pivot as both technology and customer expectations change.

Uber: The Discovery of Weekly Trips

Uber's rise from a disruptive San Francisco startup to a global transportation juggernaut was guided by a continual search for a single metric that could align every team around marketplace health. In its early days, Uber's founders were obsessed with **Estimated Time of Arrival (ETA)**, aiming to minimize every rider's wait time. This focus on ETA reflected a belief that speed and availability were Uber's true value proposition—faster pickups would drive demand, which in turn would attract more drivers, creating a self-reinforcing cycle of growth and reliability.[29,30]

As Uber expanded beyond its hometown, raw ETA data alone proved insufficient for measuring the nuanced health of a two-sided marketplace. Simply knowing how quickly cars arrived did not reveal whether riders were retained or drivers remained engaged. The company briefly augmented its dashboards with completed rides, monthly active users, and driver utilization metrics—but these fragmented views failed to provide a unified compass for decision-making.

---

[28] Tope Longe, "North Star Metric Examples from Tech Giants," uxcam.com blog, accessed June 22, 2025, `https://uxcam.com/blog/north-star-metric-examples/`.

[29] Andrew Chen, "Uber's virtuous cycle. Geographic density, hyperlocal marketplaces, and why drivers are key," Andrew Chen Blog, accessed July 02, 2025, `https://andrewchen.com/ubers-virtuous-cycle-5-important-reads-about-uber/`.

[30] Thi Nguyen, "ETA Phone Home: How Uber Engineers an Efficient Route," uber.com blog, accessed June 30, 2025, `https://www.uber.com/en-FR/blog/engineering-routing-engine/`.

The pivotal breakthrough came when Ed Baker—fresh from scaling Facebook's Daily Active Users framework—joined Uber as VP of Growth and led the search for a new North Star Metric. Baker and CEO Travis Kalanick recognized that Uber needed a metric capturing both sides of the marketplace simultaneously. They landed on **Weekly Trips**, reasoning that every trip required a rider and a driver: rising weekly trip counts signaled growing demand and healthy supply in one simple number.[31]

This new metric transformed Uber's growth philosophy. Uber teams began optimizing for trip growth, using weekly trip counts as a diagnostic tool. When trips dipped in a city, teams could drill into surge pricing, driver incentives, or local competition to diagnose and remedy the issue. The number of weekly trips thus became both a barometer of network effects and a trigger for rapid operational responses.

Andrew Chen, who witnessed this transformation firsthand on Uber's growth team, later described how Weekly Trips forged a truly metric-driven product organization. Every experiment—from pricing tweaks to new product features—was measured by its impact on weekly trips, creating a culture of continuous iteration and data-backed alignment across engineering, operations, and marketing.[32]

Today, Weekly Trips stands as Uber's enduring North Star. It distills the complex dynamics of thousands of local markets into one actionable figure, ensuring that every software release, driver incentive, and marketing campaign is mobilized toward expanding the core of Uber's marketplace: the seamless orchestration of riders and drivers worldwide. This evolution—from ETA to Weekly Trips—demonstrates how the right guiding metric can transform a company's strategy and sustain growth on a global scale.[33]

Lessons Learned

These case studies demonstrate that successful startups don't simply choose a North Star Metric and stick with it forever. Instead, they evolve their metrics as their business models mature, market conditions change, and their understanding of customer value deepens.

---

[31] "Scaling Growth | Gustaf Alstromer (YC Partner + Airbnb) & Ed Baker (Uber)," ycombinator.com, accessed June 21, 2025, https://www.ycombinator.com/blog/scaling-growth-panel/.

[32] Andrew Chen, The Cold Start Problem: How to Start and Scale Network Effects (Random House UK Ltd, 2023).

[33] Patrick Gentry, "Growing your company: Why the North Star metric works," venturebeat.com, accessed June 13, 2025, https://venturebeat.com/entrepreneur/growing-your-company-why-the-north-star-metric-works/.

The key insight is that the best North Star Metrics directly capture customer value while serving as leading indicators of business success, requiring continuous refinement as companies grow and markets evolve. Table 6-17 presents the summary of North Star Metric case studies discussed above.

Table 6-17. *Case Study Summary Table*

Company	North Star Metric	Description
Amplitude	Weekly Learning Users (WLUs)	Users who are active and share insights consumed by at least 2 others within 7 days
Facebook	Daily Active Users (DAUs)	Number of users actively engaging with the platform daily, evolved from MAUs
Amazon	Number of Purchases per Month	Monthly transaction volume reflecting successful customer value delivery
Uber	Weekly Trips	Total completed rides reflecting successful marketplace matching between riders and drivers

Cheating with Your Charts

If you can't prove what you want to prove, demonstrate something else and pretend that they are the same thing. In the daze that follows the collision of statistics with the human mind, hardly anybody will notice the difference.

—Darrell Huff,

How to Lie with Statistics[34]

Darrell Huff's words still sting decades later for those who take data seriously. In a startup, every metric, every graph, and every slide in your pitch deck carries weight. Investors pore over charts daily and have seen every trick in the book. Yet when misleading charts creep into your narrative, the danger isn't just fooling your audience— it's fooling yourself.

---

[34] Darrell Huff, *How to Lie with Statistics* (New York: W. W. Norton & Company, 1954), 74.

A manipulated chart can mask a faltering growth engine, exaggerate minor gains, or offer false comfort while real problems go unaddressed. Worse, once you start believing your own propaganda, course corrections come too late. Remember that investors will ask to see the raw data, and being caught bending the truth can cost you not only a deal but your reputation. To stay honest (and persuasive), recognize these common traps and learn how to tell a genuine story with your numbers.[35]

Using Cumulative Charts to Hide a Growth Slowdown

A cumulative line chart shows total users, revenue, or downloads over time. While it feels impressive to show a steady, upward spline, it can obscure whether growth is accelerating, flattening, or outright reversing. A startup that adds users at a slowing pace still produces a rising cumulative curve, making it impossible to spot an emerging plateau at a glance. Investors expect to see month-over-month or week-over-week deltas, not just the sum total. Hiding a slowdown in cumulative form may sneak past a cursory glance, but once charts switch to period-specific bars or difference lines, the real trend appears.

Manipulating the Y-Axis to Make Small Gains Look Dramatic

A graph that doesn't start its vertical axis at zero instantly exaggerates any uptick. For example, compressing the Y-axis to span a narrow band around your latest figures makes a two-percent bump look like a hockey-stick surge. Unless you explicitly note the truncated scale, readers assume the dramatic rise reflects actual magnitude. Always anchor your axis at zero or clearly call out the scale break to avoid misleading impressions of your traction or efficiency gains.

Cherry-Picking the Best-Performing Cohort and Presenting It As Typical

Cohort analysis can reveal deep insights—but only if cohorts are representative. Selecting a single cohort (for instance, the one launched after a special promotion) and holding it up as your "average" user journey distorts reality. That top-performing batch will almost always outperform your broader user base. Meaningful cohort analysis groups users by consistent criteria and shows multiple cohorts side by side so investors can see whether the star cohort is an outlier or the rule.

---

[35] Becca Cudmore, "Five Ways To Lie With Charts," nautil.us, accessed July 04, 2025, `https://nautil.us/five-ways-to-lie-with-charts-235151/`.

Focusing on a Vanity Metric While Ignoring a More Important One

It's tempting to lead with "downloads," "pageviews," or "invite requests," metrics that swell quickly and look good in a pitch. But if your true north is Daily Active Users (DAU), Monthly Recurring Revenue (MRR), or customer retention, glossing over these core KPIs is a red flag. Vanity metrics can rise even when engagement, churn rates, or monetization flatline. Calling out DAU as just "one more number" undermines your credibility—investors know that downloads don't pay the bills nor signal sustainable growth.

Forcing Power User Curves to Smile

The "smile" shape in a Power User Curve has become an iconic signal of a healthy, habit-forming product. It's the hallmark of elite social products and daily-use apps, indicating a strong base of both casual and deeply engaged users. However, this specific shape is only the ideal for a certain class of startup. For many others—such as SaaS tools, ecommerce sites, or specialized products like the fitness app we analyzed—a left-weighted curve is not only normal but expected.

Despite this, I have seen many startups become fixated on the "smile" and try to force their data to fit this narrative at all costs. The pressure to present a chart that looks like a top-tier social product can lead teams to manipulate their data. This can involve cherry-picking a favorable week, changing the definition of "active," or, as the chart below illustrates, massaging the data (combining 6 and 7 days into one bar) to create an artificially perfect "smile" (see Figure 6-2).

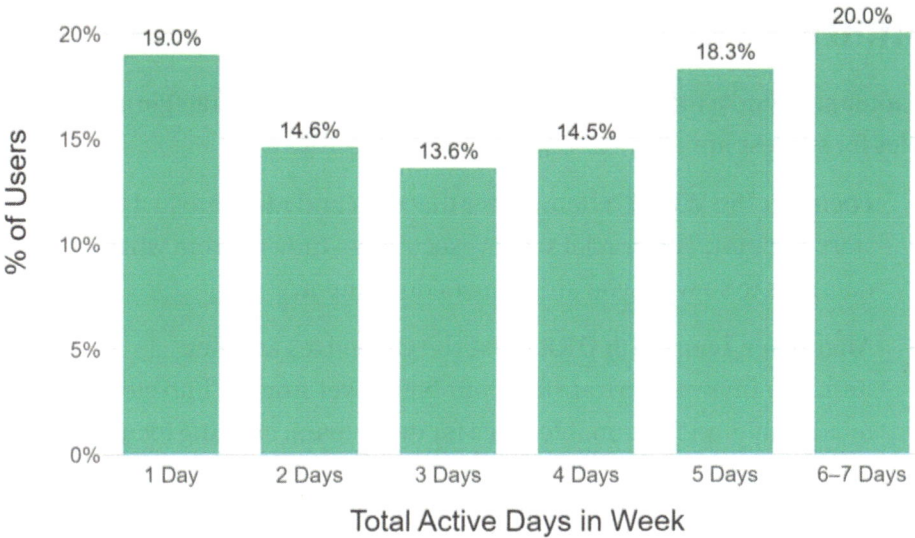

Figure 6-2. *Power User Curve with 6-7 Days Combined*

This is more than just a cosmetic tweak; it's a dangerous self-deception. An honest Power User Curve, in whatever shape it takes, is an invaluable diagnostic tool. It tells you the truth about how users interact with your product. A lumpy or left-weighted curve isn't a failure; it's a road map pointing directly to where your product or engagement strategy needs work. Investors fund teams that face reality squarely, not those that present a polished fiction.

The temptation to manipulate a Power User Curve is a microcosm of the central challenge this chapter addresses: the difference between using metrics to impress and using them to improve. Throughout our exploration—from the guiding light of a North Star Metric, through the structured funnels of AARRR, to the deep diagnostic power of cohort and power user analysis—the underlying principle has remained the same. The most valuable metrics are those that deliver unvarnished truths, even when those truths are uncomfortable. They are not endpoints but starting points for the critical conversations that build great companies.

As we wrap up, the following summary will bring together the key themes from this chapter. It will reinforce the foundational frameworks and advanced techniques discussed, providing a final, cohesive guide to help you build a culture of data-driven decision-making grounded in intellectual honesty and a relentless focus on creating genuine value.

Summary

This chapter provided a road map for converting raw data into strategic action. The core principles for success are

- **Focus on "So What?":** Remember that data and ideas are just the starting point. The crucial step is execution—moving from what the data says to so what you are going to do about it.

- **Align Your Team with OKRs:** Use the Objectives and Key Results framework to translate your high-level strategy into clear, measurable, and actionable goals for every team, creating focus and accountability.

- **Speak the Investor's Language:** Understand that the metrics that matter most to investors will change based on your startup's stage and the current economic climate. Prioritize capital efficiency and demonstrate a clear path to sustainable growth.

- **Evolve Your Metrics as You Grow:** The North Star Metric that guides you at the seed stage may not be the right one for Series A. Be prepared to adapt your measurement strategy as your company evolves.

- **Build Trust Through Honesty:** Your credibility with your team and investors is your most valuable asset. Maintain intellectual honesty in all your data storytelling by avoiding misleading charts and vanity metrics.

In Chapter 7, we will elaborate on the specific skillset required in startups from data professionals, equipping you with the technical expertise needed to effectively collect, analyze, and leverage the metrics discussed herein.

PART III

Tools and Skills

CHAPTER 7

Programming Skills for Data Professionals

In the beginning it'll be really expensive to run, then you can get it to be more efficient, and then over time we'll get to the point where a lot of the code in our apps and including the AI that we generate is actually going to be built by AI engineers instead of people engineers.

—Mark Zuckerberg,[1]

Founder of Meta

If you're reading this chapter, you might be wondering about the future of programming in the age of AI agents. The quote from Zuckerberg may not sound particularly optimistic for anyone looking to develop their programming skills. However, when you're the first data professional in a startup, AI isn't a threat—it's your greatest ally.

The goal of this chapter isn't to teach you all the details of SQL, Python, or R. Numerous excellent resources already cover those, and this chapter includes several resources that were instrumental in my career as a data professional. Instead, we'll focus on why these tools matter specifically in startup environments, when to choose one over another, and how AI is changing the equation.

This chapter will guide you through the essential programming skills needed to thrive as a startup's first data hire. We'll explore why programming remains crucial in the AI era and how to leverage these tools strategically in resource-constrained environments.

---

[1] "Joe Rogan Experience #2255 - Mark Zuckerberg," PowerfulJRE YouTube Channel, January 10, 2025, https://www.youtube.com/watch?v=7k1ehaEObdU.

P. Sidoruk, *From Data to Dollars*, https://doi.org/10.1007/979-8-8688-1898-1_7

Here's what we'll cover:

- The evolving role of programming in data work, examining past predictions about automation and why human skills remain essential

- The unique demands of startup data roles, where you'll function as a versatile generalist rather than a narrow specialist

- Core programming tools including SQL as the foundation of data work, and the strategic choice between Python and R for analysis

- Data storytelling and visualization techniques that transform raw numbers into compelling business narratives

- Practical application through building investor data rooms—a real-world scenario that showcases the integration of technical skills, business acumen, and communication

- AI collaboration strategies that treat artificial intelligence as a powerful assistant rather than a replacement

- Curated learning resources to build and advance your technical foundation

By the end of this chapter, you'll understand not just what programming tools to use but when and how to use them effectively in a startup context. You'll see how technical skills, business understanding, and storytelling combine to create unprecedented value for early-stage companies and how AI amplifies rather than replaces these capabilities.

Looking Back at Past Predictions: Were They Right?

The conversation about AI replacing developers isn't new; it's an evolution of a long-running theme in computing. Technology publisher Tim O'Reilly argues that this isn't the end of programming but "the end of programming as we know it." He points out that programming has always been evolving toward higher levels of abstraction, from physical circuits to machine code, assembly language, and modern high-level languages like Python. Each step made it easier to instruct computers, bringing more people into the field and sparking new waves of creativity. O'Reilly believes AI will not replace programmers but will transform their jobs, making some current tasks obsolete while revealing what truly requires human thinking.[2,3,4]

Partners at Andreessen Horowitz echo this sentiment, suggesting generative AI and large language models are reshaping software development into a new kind of high-level programming abstraction. They note that while AI can boost productivity, formal languages like Python and Java are here to stay because understanding foundational concepts remains essential for managing the non-deterministic nature of AI applications.[5]

Adding to this perspective, GitHub CEO Thomas Dohmke argues that the future of programming lies in a hybrid workflow, where AI tools complement rather than replace human expertise. Dohmke cautions against a complete reliance on AI, pointing out the inefficiency of spending several minutes trying to describe a simple fix in natural language to an AI agent when a skilled developer could manually code the same change in just a few seconds. For example, correcting a typo in a variable name might take a developer a moment, while articulating the same request for an AI could involve a lengthy back-and-forth conversation to get the context right. He believes the key to leveraging AI successfully is for developers to maintain their hands-on coding skills, enabling them to flexibly switch between AI assistance for large tasks and direct manual coding for quick, precise edits. While AI can automate repetitive work and allow developers to focus on higher-level system design, Dohmke asserts that you cannot scale on AI-generated code alone.[6]

[2] Cole Stryker, "Generative AI for Developers," IBM Think, accessed July 9, 2025, `https://www.ibm.com/think/topics/generative-ai-for-developers`.

[3] Anthony I. Wasserman and Steven Gutz, "The Future of Programming," Communications of the ACM 25, no. 3 (March 1982): 196–203, `https://doi.org/10.1145/358453.358459`.

[4] Tim O'Reilly, "The End of Programming As We Know It," O'Reilly Radar, February 4, 2025, `https://www.oreilly.com/radar/the-end-of-programming-as-we-know-it/`.

[5] Guido Appenzeller, Matt Bornstein, Yoko Li, and Derrick Harris, "Who's Coding Now? AI and the Future of Software Development," a16z podcast, May 16, 2025, `https://a16z.com/podcast/whos-coding-now-ai-and-the-future-of-software-development/`.

[6] Thomas Dohmke, "GitHub CEO: The AI Coding Gold Rush, Vibe Coding & Cursor," The MAD Podcast with Matt Turck YouTube Channel, June 12, 2025, `https://www.youtube.com/watch?v=Gp-oPGYrQDs`.

Startups still require skilled coders who can debug complex issues, manage system architecture, and provide the crucial human oversight needed to build robust products. For early-stage startups in particular, AI is a gift to the "one-man data army" who must single-handedly manage all data work. It allows you to boost your productivity and test more ideas, automate work, and focus on more creative, high-impact tasks.

Why Data Professionals Need Programming Skills

In the startup environment, programming skills are essential for several interconnected reasons. First, data analysis and transformation require the ability to manipulate data programmatically—something spreadsheets simply cannot handle at scale. Second, data science capabilities including machine learning and statistics are increasingly expected as core competencies. Third, the typical startup first data hire must be able to build and maintain data infrastructure, create data pipelines, and automate routine tasks.

The balance between programming, statistics, and business knowledge is what separates great startup data professionals from merely competent ones. While technical skills get you in the door, it's the combination of all three that drives real business impact.[7]

The First Data Person: A One-Man Show

Working as the first data professional in a startup means embracing ambiguity and wearing multiple hats. Unlike large corporations where specialized roles proliferate, startups thrive on data generalists—professionals who can seamlessly move across domains. When resources are limited, having someone who can build data infrastructure from scratch while also understanding and helping shape business strategy creates tremendous value. This has led to the "rise of the data generalist," a shift where adaptability and a broad skill set are prized over deep specialization as AI and advanced data platforms allow data teams to become more compact and efficient.[8]

---

[7] Mengying Li, "The Startup Founder's Guide to Hiring a Data Scientist," First Round Review, accessed July 8, 2025, https://review.firstround.com/the-startup-founders-guide-to-hiring-a-data-scientist/.

[8] Tarush Aggarwal, "Rise of the Data Generalist: Smaller Teams, Bigger Impact," 5X, April 30, 2024, https://www.5x.co/blogs/rise-of-the-data-generalist.

In larger companies, this specialization has given rise to a sprawling ecosystem of distinct roles, each with a narrow focus. You'll find job titles like BI Engineer, Data Scientist, Machine Learning Engineer, Analytics Engineer, Data Architect, Product Analyst, and dozens of others, from MLOps Engineer to Quantitative Analyst. Each role operates within a well-defined lane, contributing a specific skill set to a larger, more structured data organization.

In the startup world, however, this entire spectrum of responsibilities is often consolidated into a single person. This role goes by many names—Data Generalist, Startup Data Scientist, or Data Lead. I've heard many people calling me by just the informal "data guy"—but the title is less important than the function. The key differentiator is the sheer versatility required and an unrelenting focus on delivering **immediate business value**. It's a role defined not by a specific technical skill, but by the ability to wear all of these hats at once to solve whatever problem is most critical for the business's survival and growth.

The debate between hiring a generalist vs. a specialist is crucial for early-stage companies. Generalists are adaptable and can handle diverse functions, which is invaluable when roles are fluid and resources are scarce. Specialists become more critical as a company grows and requires deep expertise in specific areas.[9]

For the first data hire, however, a generalist approach is usually preferred. In my experience, you might find yourself:

- **Playing Marketer:** Designing email campaigns and A/B tests after segmenting the user base

- **Acting As Salesperson:** Building data narratives that convince potential investors of your startup's growth potential

- **Functioning As Compliance Officer:** Ensuring data practices meet regulatory requirements

- **Managing Upward:** Helping founders understand task complexity and timeframes

- **Serving As Architect:** Making strategic decisions about data infrastructure that will scale with the company

---

[9] Erik Bernhardsson, "The Data Team: A Short Story," Erik Bernhardsson's Blog, July 7, 2021, https://erikbern.com/2021/07/07/the-data-team-a-short-story.html.

This reality requires a mindset shift. Success depends less on perfection and more on pragmatism—finding the 80% solution that allows you to move on to the next priority. It's about substance over titles; the focus is on delivering business value, regardless of the specific role. Founders should look for a "hybrid-type data scientist" as their first hire—someone with strong analytical skills who also has a solid background in data engineering and ETL processes.

Startups typically hire data professionals who are adaptable generalists rather than siloed specialists. The ideal candidate thrives in ambiguous, fast-paced environments and is a proactive collaborator, regularly partnering with product, engineering, and leadership teams. The focus is on delivering practical, results-driven solutions, making the role unsuitable for those who prefer to work in isolation, require rigid structures, or are interested in purely theoretical or academic research.[10]

The Startup Data Professional's Toolkit

To succeed in this environment, you need a blend of key competencies. Fundamental technical skills are a must, including proficiency in SQL and a programming language like Python or R. Core to the role are strong analytical skills in mathematics, statistics, and predictive modeling. A successful first hire must also possess the business acumen to define a data problem based on business needs, establish a framework for collecting and measuring data, and understand the domain of the business. Finally, what often separates a good data professional from a great one is the ability for communication and storytelling—presenting complex findings clearly to both technical and non-technical stakeholders.

As the startup grows, your role will likely evolve. You may choose to become a manager and lead a team, or you might specialize in a particular area. Larger companies tend to have more specialized roles, such as Data Analyst, Data Scientist, or Data Engineer.

In a startup, your value is defined by more than just your technical proficiency. The environment demands a unique blend of skills that differ from those required in larger corporations.

---

[10] Yanir Seroussi, "Substance Over Titles: Your First Data Hire May Be a Data Scientist," Yanir Seroussi's Blog, February 5, 2024, https://yanirseroussi.com/2024/02/05/substance-over-titles-your-first-data-hire-may-be-a-data-scientist/.

– **Version Control with Git:** Mastery of Git is critical, even if you are the only data person at a startup and no one else views your code for ad hoc tasks or exploratory data analysis. While Git is well known as a tool for large, collaborative teams, it is just as essential for the solo data professional.[11]

  – **A Personal Safety Net:** First and foremost, you should use Git for your own benefit. It acts as a personal backup system and a detailed log of your work. It gives you the confidence to experiment and make changes, knowing you can easily roll back to a previously working version if you make a mistake. This revision history is invaluable for tracking how you calculated certain metrics, which is crucial when stakeholders revisit old requests and you need to trace your steps or compare past and present results.

  – **Future-Proofing for Growth:** Your startup's success will lead to team expansion. Having a version control system in place from the beginning is a form of future-proofing that makes it significantly easier to onboard new data professionals. When you need to hire help, an established Git workflow becomes critical for enabling effective collaboration, allowing the growing team to work on codebases efficiently, track changes, and seamlessly integrate their work.

– **The "Full-Stack" Mindset:** Startups require versatility. You need to be comfortable working across the entire data stack—from building data pipelines to creating dashboards and presenting findings to stakeholders.

– **Business Acumen and Storytelling:** You must be able to translate business problems into data questions and then narrate the story behind the numbers to non-technical stakeholders. In a startup, narratives often matter more than the raw facts.

---

[11] Jon Loeliger and Matthew McCullough, *Version Control with Git: Powerful Tools and Techniques for Collaborative Software Development*, 2nd ed. (O'Reilly Media, 2012).

- **Scrappiness and Pragmatism:** With limited resources, delivering simple, functional solutions quickly is key. Focus on the "80% solution" that solves the immediate problem. This "fail fast" and iterative mindset fosters quicker feedback loops and helps you identify where data can provide the most value.

- **Strategic Stakeholder Management:** When you're the primary data person, you must expertly manage the flow of requests from multiple stakeholders. This means knowing when a quick, ad hoc query is sufficient and when to invest in designing a complex new dashboard. It requires the shrewdness to negotiate priorities, explaining to stakeholders when a request isn't worth the effort or should be postponed. You can often satisfy their core needs with an alternative chart or a simpler analysis that answers the same underlying question. Ultimately, this demands strong people skills and the intelligence to guide your colleagues toward the most impactful data-driven decisions.

Figure 7-1 presents a comprehensive radar chart visualizing the diverse skill set beneficial for a startup's first data hire. The skills are assessed on a hypothetical and subjective scale from zero to 10. This chart serves as an illustrative example of how such a skill set might be depicted. It's important to recognize that this is a simplified representation; the specific skills and their relative importance may vary significantly depending on the particular startup, its team structure, industry, and current stage of development.

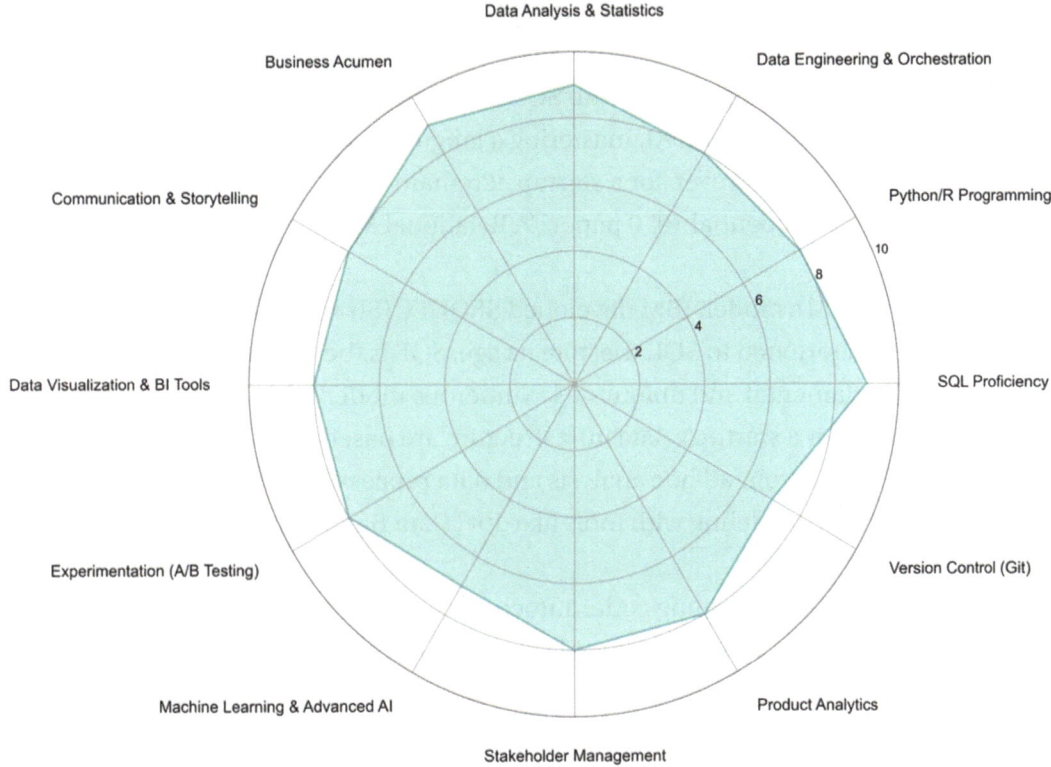

Figure 7-1. *Comprehensive Skills of a Startup's First Data Hire*

Furthermore, some startups may prioritize skills not featured on this chart. For example, a company might hire a more business-oriented data analyst who possesses deep domain knowledge but fewer technical skills. Conversely, another startup might require a highly technical professional proficient in multiple programming languages like Julia, Java, or C++, who focuses primarily on complex engineering tasks with less emphasis on business-related topics or storytelling.

With a clear picture of the diverse skill set a startup data generalist needs, it's time to open the toolkit. In the following sections, we will explore the essential programming languages and tools that bring these skills to life. We'll begin with the bedrock of any data operation: the language that allows you to talk to your data directly.

SQL: The Timeless Foundation

We begin our programming journey with SQL (Structured Query Language). In a landscape rapidly evolving with AI, mastering a language with roots in the 1970s may seem counterintuitive. However, for a startup, it remains a necessity. SQL's story begins with Dr. E. F. Codd's influential 1970 paper, "A Relational Model of Data for Large Shared Data Banks."[12]

Inspired by Codd's model, IBM developed SEQUEL (Structured English Query Language), later shortened to SQL. Despite its age, SQL is the "wheel" of data interaction—fundamental and ubiquitous. Numerous modern data warehouses and databases that power a startup's data infrastructure are based on SQL. In a startup, SQL is used for everything from ad hoc analysis and data processing to testing data pipelines and performing data modeling with tools like dbt (Data Build Tool). We will elaborate more on this in Chapter 8.

While AI assistants can debug code, autocomplete and draft queries, a foundational understanding of SQL is still indispensable for validating the logic and results. Modern SQL dialects are evolving to incorporate AI, and platforms like Google BigQuery are at the forefront of this transformation. BigQuery has enhanced SQL by integrating powerful features directly into the language, most notably through BigQuery ML (BQML).

BigQuery ML democratizes machine learning by allowing data analysts and SQL practitioners to create, train, and deploy ML models using SQL queries (BigQuery supports the GoogleSQL dialect). This approach removes significant barriers, as it eliminates the need for extensive programming knowledge in languages like Python or Java and allows professionals to leverage their existing SQL skills.[13]

One of the key advantages of BQML is that it brings machine learning to the data rather than the other way around. This increases the speed of model development by removing the need to export and format large amounts of data from the warehouse, reducing complexity and accelerating the path to production.

---

[12] Edgar F. Codd, "A Relational Model of Data for Large Shared Data Banks," Communications of the ACM 13, no. 6 (June 1970): 377–387, https://dl.acm.org/doi/10.1145/362384.362685.

[13] Google Cloud, "Introduction to AI and ML in BigQuery," Google Cloud Documentation, accessed July 9, 2025, https://cloud.google.com/bigquery/docs/bqml-introduction.

Using familiar SQL commands, data professionals can build a variety of models to address common business needs:

- **Predictive Models:** Use linear and logistic regression for tasks like price prediction or customer churn analysis.

- **Customer Segmentation:** Apply K-means clustering to identify distinct user groups.

- **Forecasting:** Build time series models to predict future trends, such as sales or demand.

Furthermore, BigQuery ML integrates with other Google Cloud services, allowing users to access Vertex AI models and Cloud AI APIs for tasks like text generation, translation, and natural language processing. It also includes powerful features for the entire machine learning life cycle, such as advanced feature engineering to preprocess data during model creation. While this SQL-based approach is powerful for many standard use cases, it lacks the flexibility for more advanced solutions that require significant customization. In such cases, the granular control offered by a language like Python remains essential.

Beyond machine learning, BigQuery enhances SQL with other advanced functions for complex analysis. These include powerful operators for reshaping data, similar to spreadsheet functionality, and the clause for filtering the results of window functions. BigQuery also supports recursive queries for analyzing hierarchical data and a rich set of functions for time series and range analysis. For a startup, the combination of these advanced SQL functions with the broader Google Cloud ecosystem provides a powerful, accessible starting point to manage the entire analytics workflow—from ingesting data and scheduling transformations to calculating key metrics and visualizing them in tools like Looker Studio or Google Sheets. For instance, an early-stage company already using Google Analytics for website tracking has a natural pathway to export that raw event data directly into BigQuery, where they can then use SQL to calculate core KPIs and build their first performance dashboards.[14]

---

[14] Google Cloud, "Query Syntax," Google Cloud Documentation, accessed July 9, 2025, `https://cloud.google.com/bigquery/docs/reference/standard-sql/query-syntax`.

For startups, SQL has evolved beyond simple database queries. Tools like dbt have revolutionized how data teams use SQL for data transformation. It allows you to break down transformations into modular, reusable models, validate data through built-in testing, document code, and visualize the pipeline's flow with clear lineage graphs.

This makes SQL not just a query language but a complete data transformation framework that addresses the four key challenges of traditional SQL work: code modularity, unit testing, documentation, and data flow visualization.

Python and R: The Workhorses of Analysis

While SQL is the essential tool for accessing and retrieving data, Python and R provide the advanced capabilities required for in-depth statistical analysis, predictive modeling, and sophisticated data visualization. Both are powerful, open source languages with passionate advocates, but the reality is that most everyday work performed by startup data professionals can be accomplished effectively with either language. This fundamental interchangeability represents one of the most important insights for aspiring data professionals navigating the startup ecosystem.

Many startups explicitly acknowledge this flexibility in their job postings, seeking candidates who are fluent in "either Python or R." However, despite this functional equivalence, market dynamics have created distinct preferences. Python has emerged as the more popular choice for startup environments, while R maintains strong positioning in specialized domains requiring deep statistical expertise.

The choice between these languages often reflects the professional background of the data team. The "programmer turned data professional" archetype typically gravitates toward Python, while "statisticians turned startup data professionals" who began their careers with R, STATA, or SAS often continue leveraging their statistical foundation while gradually expanding into other areas.

The Python Advantage in Startup Environments

The technical infrastructure compatibility represents one of Python's most significant advantages in startup environments. Cloud products like AWS Lambda provide native support for Python across multiple runtime versions, enabling data professionals to deploy analytical functions directly into production environments. This seamless integration allows startup data teams to move from analysis to production deployment without switching languages or requiring extensive infrastructure support.

In contrast, cloud platform support for R typically requires more complex setup. While Microsoft Azure offers some R support, the implementation typically involves containerized environments and additional configuration steps. This complexity can represent significant obstacles for resource-constrained startups, even when R's analytical capabilities might be superior for specific use cases.[15]

The R Stronghold: Specialized Statistical Excellence

Despite Python's numerical dominance, R maintains significant relevance in industries where statistical rigor is paramount. The pharmaceutical industry exemplifies this trend, with major companies like Roche, Biogen, GSK, and Janssen dedicating substantial resources to R-based projects. R's application extends from internal analyses to regulatory submissions with the FDA and other regulatory bodies.[16,17]

Table 7-1 demonstrates the breadth of companies utilizing R for specialized applications, ranging from pharmaceutical giants conducting clinical trials to financial institutions performing risk analysis.

---

[15] Marck Vaisman, "Using R with Azure Machine Learning," Microsoft Developer YouTube Channel, May 1, 2023, https://www.youtube.com/watch?v=ZjsTg2v5aSQ.

[16] Dario Radečić, "R in FDA Submissions: Lessons Learned from 5 FDA Pilots," Appsilon Blog, May 22, 2025, https://www.appsilon.com/post/r-in-fda-submissions.

[17] Appsilon, "Case Studies," Appsilon, accessed July 9, 2025, https://www.appsilon.com/case-studies/collection.

Table 7-1. *Examples of R Use Cases in Various Industries*

Company	Industry	R Use Case
Google	Tech	Various Data Science applications. Some of Google's job ads point out knowledge of both R and Python as minimum qualifications.[18]
Roche, Biogen, GSK, Janssen	Pharma	Clinical data analysis, drug discovery, and regulatory submissions.[19]
Perts	Education	Students experience analysis.[20]
Banks and fintech startups	Finance	Credit scoring, risk analysis.[21]
Suffolk	Construction	Safety management (using both R and Python).[22]
NASA	Space	Space missions planning.[23,24]

R's strength lies in its specialized statistical capabilities. While Python has a robust scientific ecosystem, R's packages are often developed in a unique synergy between academic researchers and corporate R&D teams. This is especially true in the pharmaceutical industry, where leaders like Roche—often in collaboration with firms such as Posit (formerly RStudio)—have built sophisticated ecosystems to produce validated analysis for regulatory submissions. This collaboration produces highly

---

[18] Google, "Data Scientist III, Product," Google Careers, accessed July 9, 2025, https://www.google.com/about/careers/applications/jobs/results/134060691198747334-data-scientist-iii/.

[19] Posit, "Roche's end-to-end R submission for a new drug application," Posit Blog, accessed July 7, 2025, https://posit.co/blog/roches-first-end-to-end-r-journey-to-submission/.

[20] Startup Jobs, "Data Analyst/R Programmer," Startup Jobs, accessed July 7, 2025, https://startup.jobs/data-analyst-r-programmer-perts-6554595.

[21] Obaid Pervaiz Gill, "Credit Risk Modelling Using Machine Learning: A Gentle Introduction," R-bloggers, August 2020, https://www.r-bloggers.com/2020/08/credit-risk-modelling-using-machine-learning-a-gentle-introduction/.

[22] Posit, "Transforming Safety Management at Suffolk with Predictive Analytics," Posit, May 28, 2025, https://posit.co/about/customer-stories/suffolk/.

[23] Posit, "Preparing for future missions to the Moon & Mars with dynamic workforce scenario planning," Posit, accessed July 8, 2025, https://posit.co/about/customer-stories/nasa/.

[24] Josh Blumenfeld, "Easier Access to NASA Earth Science Data in the Cloud," NASA Earthdata, June 5, 2024, https://www.earthdata.nasa.gov/news/easier-access-nasa-earth-science-data-cloud.

specialized and validated tools for specific statistical tasks, from bioinformatics and the rigorous analysis required for clinical trials to financial modeling. Ultimately, this deep integration makes R a compelling strategic choice, giving these companies access to a mature ecosystem of validated statistical packages, a specialized talent pool of fluent researchers, and the significant cost benefits of an open source platform.

The pharmaceutical industry's adoption of R represents a significant shift from traditional tools like SAS. Companies are making this transition for several reasons: cost-effectiveness, access to advanced statistical techniques, and the flexibility to implement cutting-edge methodologies developed by the open source community.

Python's Versatility in Startup Applications

Python's versatility gives it a decisive edge in startup environments where data professionals must function as "full-stack" generalists. Table 7-2 illustrates how major companies that began as startups have leveraged Python's capabilities across diverse applications.

Table 7-2. *Examples of Python Use Cases in Various Industries*

Company	Industry	Python Use Case
Netflix	Streaming	Recommendation systems and other data science applications[25]
Uber	Transportation	ETA Calculation, Demand Forecasting & Supply Chain Optimization[26,27]
Spotify	Music Streaming	Recommendation systems, data engineering and other[28]
Airbnb	Travel	Backend development, price optimization, search algorithms, data analysis, and other[29]

---

[25] "Python at Netflix," Netflix Technology Blog, April 29, 2019, `https://netflixtechblog.com/python-at-netflix-bba45dae649e`.

[26] Emily Reinhold, "Rewriting Uber Engineering: The Opportunities Microservices Provide," Uber Engineering Blog, April 20, 2016, `https://www.uber.com/en-PL/blog/building-tincup-microservice-implementation/`.

[27] Jonathan Pepin, "How Uber Engineering Massively Scaled Global Driver Onboarding," Uber Engineering Blog, September 2, 2016, `https://www.uber.com/en-PL/blog/driver-onboarding-funnel/`.

[28] Geoff van der Meer, "How We Use Python at Spotify," Spotify Engineering Blog, March 20, 2013, `https://engineering.atspotify.com/2013/03/how-we-use-python-at-spotify`.

[29] "Airbnb's Tech Stack: The Programming Language Behind the World's Largest Vacation Rental Marketplace," Themewaves, August 7, 2023, `https://themewaves.com/airbnbs-tech-stack-the-programming-language-behind-the-worlds-largest-vacation-rental-marketplace/`.

Python's design philosophy emphasizes readability and simplicity, making it an ideal choice for startup environments where development speed and team collaboration are crucial. The pandas library serves as the industry standard for data cleaning, transforming, and analyzing data, while scikit-learn provides accessible tools for building predictive models.

Web development capabilities represent another significant advantage for startups. Frameworks like Flask and Django allow data professionals to build and deploy data-driven web applications using the same language they use for analysis. This capability is invaluable in startup environments where team members may need to transition from analysis to production deployment rapidly.

Career Pathways and Professional Development

The pathway from software engineering to data science has become increasingly common in startup environments. For professionals with existing Python experience, the transition requires learning statistical concepts and specialized libraries rather than mastering an entirely new programming language. This creates a natural advantage for Python in startup hiring, where speed of onboarding and versatility are crucial factors.

Conversely, professionals with traditional statistical backgrounds often begin their careers using R, STATA, or SAS. When these statisticians transition into startup environments, they frequently encounter pressure to learn Python for its broader applicability and integration capabilities. However, their deep statistical knowledge combined with R expertise can provide significant value in startups requiring sophisticated analytical approaches.

The challenge for R-skilled professionals entering startup environments often lies not in their analytical capabilities but in adapting to the infrastructure and deployment requirements of modern data teams. Learning to integrate R workflows with cloud platforms and production systems requires additional technical skills that Python professionals develop more naturally.

Strategic Decision Framework

Table 7-3 provides a simple comparison of key factors startup founders should consider when choosing between R and Python for their data science initiatives.

Table 7-3. *Python vs. R Comparison*

Factor	R	Python
Learning Curve	Easier for statisticians	Easier for programmers
Primary Strength	Statistical analysis (especially for industries like pharma)	General-purpose programming and versatility
Cloud integration	Limited	Good
Number of professionals	Low	High
Industry preference	Pharma, healthcare, finance, academia	Tech, web development, AI/ML

The decision framework should be based on specific organizational needs rather than perceived language superiority. For startups operating in specialized statistical domains like healthcare, pharmaceutical research, or financial modeling, R's depth of statistical capabilities may provide competitive advantages that justify additional infrastructure complexity.

However, for general-purpose data analysis and machine learning applications, Python's ecosystem advantages make it the more practical choice for most startup environments. The language's superior integration with cloud platforms, broader job market, and full-stack development capabilities align well with the resource constraints and rapid development cycles typical of startup operations.

Industry-Specific Considerations

Different industries show distinct preferences based on their analytical requirements and regulatory environments. The pharmaceutical industry continues to embrace R due to its specialized statistical packages and regulatory compliance capabilities. Financial services firms utilize both languages, with R favored for complex statistical modeling and Python preferred for algorithmic trading and system integration.

Technology startups tend to favor Python for its versatility and development speed. The language's extensive machine learning libraries, including TensorFlow and PyTorch, make it particularly attractive for startups developing AI-powered applications. Python's role in cloud computing and serverless architectures further reinforces its position as the preferred choice for modern startup infrastructure.

Manufacturing and traditional industries are increasingly adopting both languages, with the choice often dependent on the specific analytical requirements and existing technical infrastructure.

The Future of Programming Languages for Data

For the modern data professional, being fluent in both Python and R is a definitive career advantage. This bilingual approach allows practitioners to move beyond the limitations of a single ecosystem and leverage the unique strengths of each language as needed. With the growing availability of tools that support interoperability, combining Python and R in integrated workflows is more practical than ever, allowing organizations to benefit from both worlds without imposing artificial constraints on their analytical teams.

In a startup environment, where limited resources and the need for rapid development are paramount, Python often emerges as the more pragmatic choice. Its large talent pool, seamless cloud integration, and full-stack capabilities give it a clear edge. However, this practicality comes with a crucial caveat for hiring managers. A narrow focus on Python proficiency can lead to overlooking critical skill sets. For instance, a programmer with a strong technical background—a "programming nerd"—may be exceptionally skilled in Python but lack the deep statistical and analytical knowledge commonly found in professionals with a strong background in R.

Furthermore, technical prowess alone is never enough to ensure success, especially for the first data hire in a startup. The dynamic and often ambiguous nature of a new venture demands more than just coding ability. It requires strong soft skills for collaboration and communication, as well as the business acumen to translate data-driven insights into strategic, actionable plans. For this reason, the argument that a candidate is "excellent at Python" should never be the sole justification for a hiring decision.

Ultimately, the most successful data professionals are those who understand both languages and can select the appropriate tool based on the specific analytical requirements and organizational context. For startup founders and hiring managers, this means recognizing that both languages can effectively meet most analytical needs. This understanding allows for more flexible hiring decisions, shifting the focus toward a candidate's broader profile, including their collaborative skills, strategic thinking, and alignment with the company's long-term goals.

Data Visualization and Storytelling

My urgency to write this book increased when I realized how poorly under-stood the concept of data storytelling was and how the term was in danger of becoming just another empty buzzword. Despite its immense potential, it was frequently positioned as just an extension of data visualization.

—Brent Dykes,

Effective Data Storytelling[30]

A compelling chart can often communicate more than a dense table of numbers. For a data professional, particularly in a startup environment, a critical skill is not just creating visuals but choosing the right tool for the specific task at hand. This decision involves weighing important trade-offs between flexibility, interactivity, and audience needs. In a startup, however, the most critical factor is often speed. The best choice is frequently the one that accomplishes the task in the least amount of time, allowing the team to iterate and make decisions quickly. It's up to you which tools are best for particular tasks:

- **Spreadsheets (Excel, Google Sheets):** For quick, small-scale analysis or one-off reports, spreadsheets are often the fastest and most accessible tool. Their universal familiarity makes them invaluable for ad hoc tasks, though they can become difficult to manage when dealing with larger, more complex datasets.

- **Python/R Libraries (Matplotlib, Seaborn, ggplot2, Plotly):** For deep, custom analysis and static visualizations that need to be tightly integrated into analytical workflows, these libraries offer unparalleled flexibility. A data professional might choose Python's matplotlib for initial data exploration or R's ggplot2 for creating publication-quality statistical graphics. While powerful, this approach can be more time-intensive than using other specialized tools.

---

[30] Brent Dykes, *Effective Data Storytelling: How to Drive Change with Data, Narrative and Visuals* (Wiley, 2019).

- **Business Intelligence (BI) Tools:** When the task is to create interactive, shareable dashboards for non-technical stakeholders, tools like Tableau, Power BI, and Looker are often the better choice. They are designed for speed and efficiency, making them ideal when time is of the essence for generating standardized reports. Their strength lies in empowering business users to explore data within a guided environment. We will explore how to choose the right BI tool in Chapter 9.

- **Product Analytics Tools:** For specialized use cases focused on user behavior within a product, these tools provide out-of-the-box solutions that are significantly faster to implement than building similar analytics from scratch. We will elaborate more on them in Chapter 10.

The onus is on the data professional to assess the project's goals, the audience's data literacy, and the required outcome to make a strategic choice from this diverse toolkit. In a fast-paced startup, this often means prioritizing the tool that delivers actionable insights in the least amount of time.

Brent Dykes argues that this limited view of focusing only on visuals misses the core of what makes data storytelling effective. He defines a data story as a structured approach that skillfully combines three essential pillars: data, narrative, and visuals. While visuals are used to enlighten an audience by revealing patterns that might be hidden in raw numbers, it is the narrative that provides context and structure. The narrative explains what is happening, guides the audience through the findings, and builds toward a main insight.

When these elements are properly combined, they work together to fully engage the audience, bridging logic and emotion to influence decisions and drive change. A data visualization on its own can present information, but without a supporting narrative, it lacks the persuasive power to transform that information into a compelling story. Therefore, while choosing the right visualization tool is an important decision for a data professional, it is only part of the process. The ultimate goal is to weave the outputs of these tools into a story that is memorable and leads to action.

Effective data storytelling combines data, narrative, and visuals. You must know your audience and tailor the story, connecting metrics to business outcomes and providing context that explains not just what happened but why.

Building an Investor Data Room

Perhaps no single task better illustrates the fusion of technical dexterity, business acumen, and narrative skill required of a startup data professional than preparing an investor data room. An Investor Data Room is a secure virtual repository where startups store and share critical documents with potential investors during the fundraising process. This digital space serves as the backbone of due diligence, enabling investors to verify claims made in pitch decks and assess the company's viability, growth potential, and associated risks.[31]

The term "data room" originated in the 1900s when companies used physical rooms filled with printed documents for investor review. Today's virtual data rooms maintain the same fundamental purpose: providing investors with comprehensive access to essential business information in a controlled, secure environment. As every startup is different, there is no single universal playbook for what exactly you should include in your data room. However, as a16z highlights, a well-organized data room typically includes core documents that tell your startup's story:[32]

- **Pitch Deck:** The narrative backbone outlining company vision, product strategy, market opportunity, traction, and team

- **Capitalization Table:** Current ownership structure, investor stakes, and any convertible instruments

- **Historical Financials:** Monthly path from gross revenue through net income and cash flow, breaking out different revenue streams and major costs

- **Usage Data and Key Metrics:** Business-specific metrics demonstrating growth, engagement, and market traction

- **LTV/CAC Analysis:** Demonstrating unit economics and the path to profitability

---

[31] Visible.vc, "Startup Data Room," Visible.vc Blog, accessed July 8, 2025, `https://visible.vc/blog/startup-data-room-b/`.

[32] Justine Moore, "The Insider's Guide to Data Rooms: What to Know Before You Raise," a16z, August 25, 2022, `https://a16z.com/the-insiders-guide-to-data-rooms-what-to-know-before-you-raise/`.

The Data Professional's Crucial Role

While startup founders are typically responsible for putting together the data room, the data professional's impact on its success can be substantial. Preparing an investor data room represents perhaps the most comprehensive showcase of a startup data professional's diverse skill set. This process requires the unique combination of technical abilities, business acumen, and storytelling expertise that defines the first data hire in a startup environment.

The data room creation process demands

1. **Data Extraction:** The first step is extracting data from its source. For structured data residing in production databases and data warehouses, SQL is the most efficient tool for pulling foundational datasets for key metrics. This includes extracting historical financial performance, user engagement metrics, and operational data that investors need to evaluate the business. However, for unstructured or semi-structured sources like JSON APIs, web pages, or text documents, a programming language like Python is essential for parsing and extracting the required information.

2. **Programming for the Full Analytical Workflow (Python/R):** Whether using Python or R, these general-purpose languages are essential for the full analytical workflow. This process begins with data preparation (cleaning raw data to ensure accuracy) and transitions to data transformation. During this second stage, data is aggregated, reshaped, and combined to calculate critical metrics like Monthly Recurring Revenue (MRR), Customer Lifetime Value (LTV), Customer Acquisition Cost (CAC), and churn rates. Then creating clear, persuasive visualizations that tell the company's growth story. This step may include Business Intelligence and Product Analytics tools as well as data science and machine learning techniques for data segmentation that supports the next step: shaping the narrative in a creative way.

3. **Data Storytelling Skills:** Weaving metrics and visuals into a coherent narrative that demonstrates deep understanding of the business and its future prospects. This involves contextualizing data within the broader market opportunity and competitive landscape.

4. **Business Understanding:** Knowing which metrics matter most to investors and how to present them in the most favorable yet honest light. This includes understanding investor preferences and tailoring the narrative accordingly.

Driving the Narrative Around Your Stats

The data room is where data storytelling becomes most critical. Investors don't just want to see numbers—they want to understand the story behind the metrics. Building context around your data means explaining the circumstances and influences that contributed to trends, highlighting big wins and acknowledging challenges.

When preparing your data room, accuracy is paramount. Investors have limited time and will scrutinize every detail for consistency. Be prepared to share raw data when investors request it—they often want to verify the underlying calculations behind your key metrics.

If any definitions or tracking methodologies for your metrics have changed over time, be completely transparent about these changes. Document when changes occurred, why they were necessary, and how they impact metric comparisons over time. Never leave investors questioning your data quality or meaning. This transparency demonstrates maturity and builds trust—qualities investors highly value.

Adapting to Investor Preferences

Not all metrics carry equal weight with different investors. Research your potential investors to understand their preferences and focus areas. If investors ask for specific metrics you haven't been tracking, start collecting that data immediately and communicate your timeline for providing comprehensive analysis. Remember, you can always ask investors what specific metrics they expect to see in your data room.

When your data reveals challenges or failed experiments, frame these as valuable learnings rather than weaknesses. If certain features didn't perform as expected, present this as insight gained about your users, product, and industry. This data might indicate you should focus on different user segments or pivot certain aspects of your strategy.

Sometimes, seemingly negative metrics can tell a positive story when properly contextualized. If overall conversion rates look concerning, dig deeper to show how your core user segments perform significantly better, or demonstrate how recent product improvements are trending in the right direction. This is where creativity, business understanding, and deep data analysis skills become invaluable.

The Million-Dollar Moment

In these situations, the value of a business-oriented data generalist becomes most apparent. Sometimes one creative insight from the startup's data professional—one well-crafted narrative that reframes challenging metrics or reveals hidden opportunities—can determine whether the startup secures investment or not. The ability to see patterns, understand context, and communicate insights clearly can literally be worth millions of dollars in funding decisions.

As Andrea Funsten, an investor in early-stage startups, notes: "Even as early as the seed stage, an organized data room can make you stand out from the crowd... Left me so impressed and eager to move fast on the deal." This demonstrates how a well-prepared data room, driven by strong data storytelling, can accelerate the fundraising process and create competitive advantages.[33]

The investor data room represents the ultimate test of a startup data professional's comprehensive skill set—combining technical expertise with business acumen and communication skills to drive one of the most critical processes in a startup's life cycle.

---

[33] Andrea Funsten, X (formerly Twitter) post, July 9, 2020, https://x.com/AndreaFunsten/status/1281321440818524162.

The Rise of AI Engineers: Collaboration, Not Replacement

Recent developments in AI are creating a new paradigm of collaboration rather than replacement. AI tools are becoming increasingly sophisticated, but rather than eliminating the need for human engineers, they're creating new developer patterns that treat AI as a foundation for how software gets built.[34,35,36]

According to a16z, developers are adopting new patterns that leverage AI as a foundation:[37,38]

1. **AI-Native Version Control:** The focus shifts from tracking every line-by-line code change to ensuring AI-generated changes pass their tests. The source of truth becomes a combination of the prompt and the test that validates the behavior.

2. **From Dashboards to Synthesis:** Static dashboards evolve into conversational interfaces where users can ask natural language questions like "Summarize the error trends across all services" and receive synthesized insights.

3. **Documentation for Agents:** Documentation is increasingly written for both human and machine consumption, with AI coding assistants using up-to-date documentation as grounding context.

[34] Konstantine Buhler, "AI 50: AI Agents Move Beyond Chat," Sequoia Capital, April 10, 2025, https://www.sequoiacap.com/article/ai-50-2025/.

[35] David Cahn, "AI in 2025: Building Blocks Firmly in Place," Sequoia Capital, December 9, 2024, https://www.sequoiacap.com/article/ai-in-2025/.

[36] David Cahn, "The Next Billion Developers," Sequoia Capital, November 15, 2023, https://www.sequoiacap.com/article/the-next-billion-developers-perspective/.

[37] Yoko Li, "Emerging Developer Patterns for the AI Era," a16z, May 7, 2025, https://a16z.com/nine-emerging-developer-patterns-for-the-ai-era/.

[38] Guido Appenzeller, Matt Bornstein, Yoko Li, and Derrick Harris, "Who's Coding Now? AI and the Future of Software Development," a16z podcast, May 16, 2025, https://a16z.com/podcast/whos-coding-now-ai-and-the-future-of-software-development/.

4. **From Templates to Generation:** Instead of starting with rigid templates, developers can describe the application they want to build and have an AI agent scaffold a personalized project.

5. **Multi-agent Orchestration:** Complex tasks are broken down across multiple specialized AI agents working together.

6. **Human-AI Pair Programming:** AI acts as a coding partner, with humans focusing on architecture and strategy while AI handles implementation details.

7. **Continuous Integration of AI:** AI is embedded throughout the development life cycle, from code generation to testing and deployment.

8. **Prompt Engineering as a Core Skill:** Writing effective prompts becomes as important as writing code itself.

9. **AI-Assisted Debugging:** AI helps identify and fix issues faster than traditional debugging methods.[39]

Despite these advances, AI cannot perform crucial parts of a more complex development process without human involvement. This includes intuitive UX design, assessing code quality and security, and ensuring AI-generated code complies with business requirements and regulatory policies. Most importantly, it's people who understand nuance or can put things into business context. It's people who recognize budget constraints and understand how code integrates into the entirety of a project. It's people who communicate with non-technical stakeholders.

For startup data professionals, this means AI can be leveraged in several practical ways:

- **Offloading Routine Tasks:** Using AI to generate boilerplate code, create standard visualizations, or draft documentation.

- **Accelerating Learning:** Leveraging AI to quickly understand unfamiliar codebases or technologies.

---

[39] Edo Liberty, Harrison Chase, and Sarah Wang, "Bringing Production-Ready GenAI to the Enterprise," a16z podcast, August 7, 2024, `https://a16z.com/podcast/pinecone-langchain-solving-for-ai-in-the-enterprise/`.

- **Enabling Cross-Domain Work:** With AI assistance, data professionals can more confidently contribute to areas outside their core expertise.

- **Automating Data Pipelines:** AI helps with code generation for ETL processes and data transformations.

- **Enhancing Data Analysis:** AI assists in pattern recognition and automated insight generation.

The human remains the architect of the inquiry, but AI is a powerful assistant that makes the process faster and more efficient. This collaboration is particularly valuable in startup environments where data professionals are often generalists capable of handling diverse challenges across the entire data life cycle.

Learning Resources

To build the technical foundation required for a startup data role, here is a curated list of recommended resources. This list focuses on the core tools: SQL for data querying, Python and R for analysis, and Git for version control. The next chapters will also include additional learning resources for more specialized areas of expertise like data engineering, data science, machine learning, and artificial intelligence. This list is a comprehensive guide to high-quality resources. If you are getting started from scratch, I recommend choosing one or two from each category that best fit your learning style to build a solid foundation rather than trying to complete them all.

SQL: The Language of Data

A deep understanding of SQL is non-negotiable. These resources include interactive platforms and practical books.

Interactive Platforms

- **SQLZoo:** An interactive platform with tutorials and exercises that allow you to write queries and see immediate results directly in your browser. It covers fundamental concepts and moves into advanced topics.[40]

- **W3Schools:** A text-based tutorial that is useful for quick reference, with a "Try it Yourself" editor for testing queries instantly.[41]

- **Codecademy:** Offers a beginner-friendly, interactive SQL course that covers essentials like filtering, aggregations, and joins through hands-on projects.[42]

Books

- ***Practical SQL: A Beginner's Guide to Storytelling with Data:*** It is an approachable, fast-paced guide for beginners that teaches how to use the programming language SQL to analyze data, find patterns, and tell stories with it.[43]

- ***SQL Pocket Guide: A Guide to SQL Usage:*** It serves as an on-the-job reference guide, providing syntax examples to help data analysts, scientists, and engineers quickly look up how to perform specific tasks.[44]

---

[40] "SQL Tutorial," SQLZoo, accessed July 9, 2025, `https://sqlzoo.net/`.

[41] W3Schools, "SQL Tutorial," W3Schools, accessed July 9, 2025, `https://www.w3schools.com/sql/`.

[42] "Grow in your career and unlock new opportunities by learning in-demand skills in AI, data, coding, cybersecurity, and more." Codecademy, accessed July 9, 2025, `https://www.codecademy.com/`.

[43] Anthony DeBarros, *Practical SQL: A Beginner's Guide to Storytelling with Data,* 2nd ed. (No Starch Press, 2022).

[44] Alice Zhao, *SQL Pocket Guide: A Guide to SQL Usage,* 4th ed. (O'Reilly Media, 2021).

Python: The Startup Swiss Army Knife

Python's versatility makes it an ideal language for startup environments. The following resources cater to beginners and those looking to deepen their skills.

Interactive Courses and Tutorials

- **Codecademy's Learn Python 3**: A highly regarded interactive course that is ideal for beginners who learn best by doing. The platform also offers a wide variety of courses, skill paths, and career paths for beginners, intermediate, and advanced users.[45]

- **DataCamp:** A platform that helps you learn Python through interactive, expert-curated courses that emphasize a practical, learn-by-doing approach with real-world projects.[46]

- **The Official Python Website:** The best place to start, with official documentation, beginner-friendly tutorials, and in-depth guides.[47]

Books

- ***Python for Data Analysis: Data Wrangling with pandas, NumPy, and Jupyter***: A practical guide by pandas creator Wes McKinney that teaches essential data wrangling, manipulation, and analysis skills by providing hands-on case studies using core Python libraries.[48]

R: The Specialist's Toolkit for Statistical Depth

For roles requiring deep statistical analysis, R is a powerful tool. These resources will help you master its capabilities.

---

[45] "Learn Python 3," Codecademy, accessed July 9, 2025, `https://www.codecademy.com/learn/learn-python-3`.

[46] "Learn data and AI skills," DataCamp, accessed July 9, 2025, `https://datacamp.com`.

[47] "Python.org," Python.org, accessed July 9, 2025, `https://www.python.org`.

[48] Wes McKinney, *Python for Data Analysis: Data Wrangling with pandas, NumPy, and Jupyter*, 3rd ed. (O'Reilly Media, 2022).

Interactive Courses and Tutorials

- **DataCamp:** A platform that helps you learn R programming through interactive, expert-curated courses that focus on a practical, hands-on approach with real-world datasets and projects to build your data analysis and statistical skills effectively.[49]

- **The R Project Official Website:** The best place to download R and access official documentation and manuals.[50]

Books

- *The Book of R: A First Course in Programming and Statistics*: A comprehensive and beginner-friendly guide that uses a hands-on approach with hundreds of examples to teach the fundamentals of R programming, statistical analysis, and data visualization.[51]

- *R for Everyone: Advanced Analytics and Graphics*: This book teaches users how to leverage the R programming language for data analytics, statistical modeling, and creating effective visualizations, covering topics from basic R functionality to advanced machine learning techniques and report generation.[52]

Version Control with Git

Mastery of Git is essential for managing code, collaborating with others, and future-proofing your work.

---

[49] "Learn data and AI skills," DataCamp, accessed July 9, 2025, https://datacamp.com.

[50] The R Foundation, "The R Project for Statistical Computing," R-project.org, accessed July 9, 2025, https://r-project.org.

[51] Tilman M. Davies, *The Book of R: A First Course in Programming and Statistics* (No Starch Press, 2016).

[52] Jared Lander, *R for Everyone: Advanced Analytics and Graphics*, 2nd ed. (Addison-Wesley, 2017).

Free Online Courses

- **Udacity's Version Control with Git:** A free course that covers the fundamentals, from creating your first repository to understanding commits, branches, and merges.[53]

- **GitHub Learning Resources:** GitHub itself provides a list of online courses and resources to help you get started with both Git and the GitHub platform.[54]

Books

- ***Version Control with Git: Powerful Tools and Techniques for Collaborative Software Development:*** It offers a step-by-step guide to mastering the Git version control system, covering everything from fundamental tasks like tracking, branching, and merging to advanced techniques such as managing conflicts, handling submodules, and using Git for both centralized and distributed projects[55]

Summary: The Future-Ready Data Professional

The startup data professional must embrace a **do-it-yourself approach**, solving problems with limited time, skills, and budget. This doesn't mean working in isolation—quite the opposite. It means collaborating effectively with other team members, leveraging AI when appropriate, and constantly learning across domains.

This chapter has explored the essential programming skills that define successful data professionals in startup environments. We've seen how the landscape is shifting from fears of AI replacement to a new paradigm of **human-AI collaboration**, where programming skills become more valuable, not less. **SQL remains the timeless foundation** of data work, serving as the universal language for data retrieval and increasingly incorporating AI capabilities through platforms like BigQuery ML.

---

[53] "Version Control with Git," Udacity, accessed July 9, 2025, `https://www.udacity.com/course/version-control-with-git--ud123`.

[54] "Git and GitHub Learning Resources," GitHub Docs, accessed July 9, 2025, `https://docs.github.com/en/get-started/start-your-journey/git-and-github-learning-resources`.

[55] Prem Ponuthorai and Jon Loeliger, *Version Control with Git: Powerful Tools and Techniques for Collaborative Software Development*, 3rd ed. (O'Reilly Media, 2022).

The choice between **Python and R** depends on your specific context—Python excels in startup environments due to its versatility and cloud integration, while R may be a good choice in specialized statistical domains like pharmaceuticals and finance. We established that startups require **versatile generalists** rather than narrow specialists, where the first data hire must seamlessly transition between roles—from data analyst to infrastructure architect, from compliance officer to storyteller.

We examined how **data storytelling and visualization** transform raw numbers into compelling business narratives. The most effective data professionals understand that creating charts is only part of the equation—the real value lies in weaving data, narrative, and visuals together to drive decisions and influence outcomes. This demands not just technical proficiency but also business acumen, communication skills, and the ability to **prioritize ruthlessly** in resource-constrained environments.

Through the lens of **building investor data rooms**, we saw how all these skills converge in high-stakes scenarios. This practical application demonstrates how technical expertise, business understanding, and storytelling combine to create tangible value—potentially worth millions in funding decisions. The data room creation process showcases the unique combination of

- SQL proficiency for data extraction

- Python/R skills for analysis and visualization

- Data storytelling abilities for narrative creation

- Business understanding for investor-focused metrics

We discovered that **AI tools are reshaping rather than replacing** the data professional's role. The emergence of AI agents represents a new form of collaboration where humans focus on architecture, strategy, and business context while AI handles routine implementation tasks. AI cannot perform crucial parts of a more complex development process as well as humans.

Success in this role doesn't come from narrow specialization but from mastering five key principles:

- **Prioritize Ruthlessly:** Focus on high-impact work that moves key metrics.

- **Embrace Imperfection:** Find solutions that solve most of the problem quickly rather than perfect solutions slowly.

- **Learn Constantly:** Develop enough knowledge across domains to be effective, then dive deeper when necessary.

- **Communicate Clearly:** Translate complex data concepts into actionable insights for non-technical stakeholders.

- **Leverage AI Effectively:** Use AI tools to augment capabilities while maintaining human judgment and business context.

The ability to **communicate clearly**—translating complex data concepts into actionable insights for non-technical stakeholders—separates great data professionals from merely competent ones. In startup environments, this communication skill becomes even more critical as data professionals tend to regularly interface with founders, investors, and cross-functional teams who may have limited technical backgrounds.

Leveraging AI effectively represents the future of data work, where professionals use AI tools to augment capabilities while maintaining human judgment and business context. The most successful data professionals treat AI as a powerful assistant rather than a replacement, using it to offload routine tasks, accelerate learning, and enable cross-domain work.

In the startup world, the most valuable data professional isn't the deepest specialist but the **most adaptable generalist**—someone who can navigate ambiguity, leverage the right tools for each job, and ultimately deliver business impact through data. With the rise of AI tools, this role becomes even more powerful, enabling data professionals to test more ideas, automate routine work, and focus on the creative, strategic tasks that truly drive business value.

The programming skills we've explored in this chapter—**SQL proficiency, Python/R expertise, data storytelling, and AI collaboration**—form the foundation for individual effectiveness. However, as startups scale, the focus must shift from ad hoc analysis to **systematic data infrastructure**. In the next chapter, we'll explore how to evolve from writing individual queries and scripts to building robust data pipelines, implementing automated orchestration, and creating scalable data engineering solutions. We'll examine how the SQL and Python skills you've developed become the building blocks for comprehensive data engineering systems that can support a growing organization's needs.

CHAPTER 8

Data Engineering and Orchestration

> *Their data team hired us to scale their data warehouse back in 2016. We told them the way that they were doing stuff was bad and we needed to move it over to dbt. So they asked for a demo. Over the next week, we refactored all of their existing pipelines and brought them over to dbt. And they were elated.*

—Tristan Handy,[1]

Co-founder of dbt Labs

This is how dbt Labs, then a consultancy named Fishtown Analytics, landed its first real client—the mattress startup Casper. In this story lies the core theme of this chapter: great data tools are often forged in the fires of startup innovation, built to solve immediate and practical problems. Dbt, now an industry standard, began as a tool built by a startup for other startups, and its journey highlights the power of a pragmatic, needs-driven approach to data engineering.

Building on this theme, this chapter transitions from theory to practice. Whereas Chapter 3 established the broader landscape of data engineering and orchestration— exploring **what tools and strategies are available**—this chapter provides a hands-on guide focused on implementation. We will move beyond the theoretical to explore **how a startup can leverage these powerful tools**, providing actionable steps and real-world examples to build a data infrastructure that delivers value quickly and scales with your business. Our journey will cover

---

[1] "How dbt Labs Built a $4.2B Software Business out of a Two-Person Consultancy," First Round Review, accessed September 29, 2025, https://review.firstround.com/dbt-labs-path-to-product-market-fit/.

P. Sidoruk, *From Data to Dollars*, https://doi.org/10.1007/979-8-8688-1898-1_8

- **The Startup Data Engineering Mindset:** We will begin by establishing a pragmatic philosophy for data engineering, rooted in the Pareto Principle, also known as the 80/20 rule. This means embracing the "80% solution"—a mindset focused on achieving 80% of the desired outcome with just 20% of the total effort required for a "perfect" solution. For a startup, this is not about cutting corners; it is about strategically prioritizing tasks that deliver the most business value as quickly as possible.

- **Orchestration with Apache Airflow:** You will learn how to move from manual tasks or chaotic cron jobs to coordinated workflows, using Airflow to manage dependencies, ensure data freshness, and create an observable, reliable data platform.

- **Transformation with dbt:** We will take a deep dive into dbt, exploring how it brings software engineering best practices to SQL. You will learn practical strategies for data modeling, from structuring your project in layers to building agile, domain-aligned models.

- **Ensuring Data Quality and Avoiding Pitfalls:** We will address the critical challenge of building trust in your data with a pragmatic approach to quality and governance and identify common mistakes startups make when implementing these tools.

- **Real-World Architectures:** Through case studies of such startups as Preset, Billie, and Monzo and bigger companies like Airbnb and Spotify, this section demonstrates how to apply these tools in practice to build scalable, reliable, and business-focused data platforms.

This journey from a simple script to a scalable data stack begins not with perfect code but with a specific way of thinking. Let's explore the core principles that define the startup data engineering mindset.

The Startup Data Engineering Mindset

At its core, the startup data engineering mindset is about embracing *Minimal Viable Data Engineering*—building just enough infrastructure to solve current problems while maintaining the flexibility to adapt in the future. This philosophy stands in stark contrast to the traditional enterprise approach, which often involves extensive upfront planning for scale, comprehensive governance, and large, specialized teams. For a startup, where one person often wears many hats, the goal is to start lean, prove value quickly, and iterate.

This leads to the adoption of the 80% Solution philosophy, a pragmatic trade-off that prioritizes getting a functional solution into production quickly over striving for a perfect one. I cannot overstate how critical this is. In more than one startup where I joined as the first data person, there were immediate, ambitious ideas for projects that, in retrospect, were not all necessary. Alternative solutions existed that would have satisfied the immediate business needs in a fraction of the time, freeing us up for projects with far greater potential.

To put this philosophy into practice, let's step into a familiar scenario. Imagine you are the first data professional hired at a promising startup. The requests are coming from all directions. Marketing wants reports that require merging Google Analytics data with other sources, which means getting GA data into your BigQuery warehouse. Meanwhile, finance and some of your key clients need reports based on flawless revenue numbers that involve sophisticated logic. This is a critical area, as the entire business is based on this data. You have a feeling this will most likely require getting started with dbt and Airflow. To top it off, someone from the product team suggests building a custom, complex data pipeline with non-critical user behavior data from several sources, even though no one has any idea if there is any business value in this.

This is not just a hypothetical exercise. It is a high-stakes reality. I have seen firsthand how, after a few months, the business model changes or the money runs out. When that happens, you realize it was not worth spending so much time perfecting an elaborate system when a simplified approach would have been completely sufficient.

With limited time and resources, you cannot tackle everything at once. This is where a prioritization matrix that weighs impact against effort becomes your most valuable tool. This framework helps you sort through the noise, focus your energy where it will generate the most value, and build momentum.

Figure 8-1 divides data engineering tasks into four strategic quadrants:

1. **Quick Wins** (High Impact, Low Effort): The marketing request is a perfect example of a Quick Win. Setting up a Google Analytics to BigQuery data transfer is a relatively low-effort task that unblocks high-value work for a key team. These should be your immediate priorities to build momentum and demonstrate value. Other

examples include creating basic reports with existing tools for
frequently requested numbers or implementing simple data
exports for critical business metrics.

2. **Critical Projects** (High Impact, High Effort): The need for flawless
   revenue data is a quintessential Critical Project. Building a
   sophisticated payment and revenue pipeline with 100% accuracy
   using tools like dbt and Airflow is a high-effort, high-impact
   investment that is foundational to the business's long-term
   success. These are the projects you must invest in heavily.

3. **Small Backlog Tasks** (Low Impact, Low Effort): These are useful
   but non-urgent tasks that can be handled during downtime or
   between larger projects. They add incremental value but are not
   priorities. Examples include improving data documentation,
   applying minor data quality fixes to non-critical datasets
   or implementing optional integrations that provide only
   marginal value.

4. **Time Wasters** (Low Impact, High Effort): That suggestion to build
   a complex pipeline for non-critical user behavior data falls directly
   into the Time Wasters quadrant. It represents a high-effort task
   with low or unproven impact—a classic distraction for an early-
   stage startup that should be avoided. Other examples to avoid
   include over-engineering simple data processes that a basic script
   could handle or building a "perfect," complex architecture for a
   small or non-critical use case.

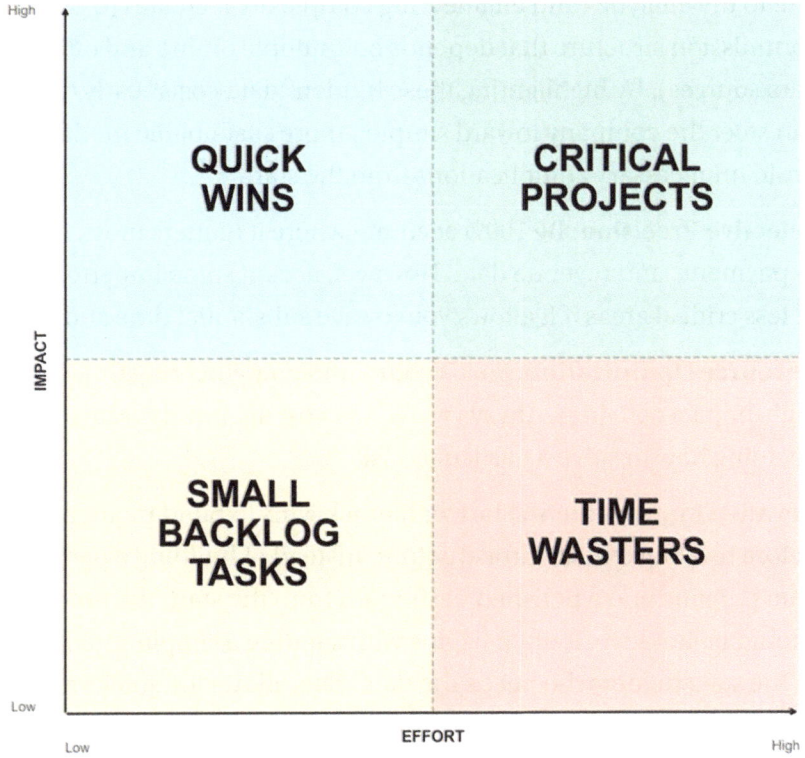

Figure 8-1. *Data Engineering Prioritization Matrix for Startups*

Figure 8-1 embodies the pragmatic approach startups need for data engineering, built on a set of core principles:

- **Assess the Impact:** Ask "*Why should we do this?*" Before building anything, a startup data engineer must translate a general question into specific, answerable sub-questions and understand what actions will be taken based on the results. If the data will not lead to an imminent and significant action, the request may not be worth pursuing at this time.

- **Shape the Business, Not Just the Data:** As a data professional, you can proactively influence business strategy. If, at the idea stage for a new product or commission system, you foresee immense problems with complex reporting and convoluted calculations, speak up. You can convince founders that a certain model may be too costly simply due to the analytics and engineering complexity it creates (e.g., a commission structure that depends on multiple online and offline data sources). By highlighting these hidden "data costs" early, you can steer the company toward simpler, more sustainable models and avoid unnecessary complications from the start.

– **Selective Precision:** Be 100% accurate where it matters most, such as in payments and revenue data. However, accept some imperfection in less critical areas if it allows you to save substantial time and effort.

– **Resource Optimization:** Focus your limited engineering time on high-impact activities. Always weigh the cost against the value when deciding how to solve a particular task.

– **Iterative Improvement:** Start with quick wins to build momentum before tackling critical infrastructure. Instead of building a perfect data pipeline and a polished dashboard from the start, the process should be iterative. It often begins with sending a simple spreadsheet to the stakeholder who needs the data. This allows for quick validation and ensures the data is genuinely useful before more time is invested.

– **Automate Only When Necessary:** Automation and dashboard building should only occur after a dataset has proven its value and is frequently requested. This discipline prevents wasting time on solutions that do not address a real business need.

– **Avoid Perfectionism:** Do not over-engineer solutions for your current scale. Build for what you need now, with an eye toward future flexibility.

A final, crucial principle involves proactively managing the *"Time Wasters"* quadrant. When stakeholders propose ideas that seem to be high effort with unproven value, do not dismiss them outright. Instead, view it as an opportunity to collaborate and redefine the task.

Your goal is to understand the business question behind the request and find a more efficient way to address it. By reframing the problem, you can often transform a potential time waster into a *"Small Backlog Task"* or, in the best-case scenario, a *"Quick Win."*

For instance, consider the earlier suggestion to build a complex pipeline for user behavior data—a classic *"Time Waster"* due to its high effort and uncertain impact. Rather than rejecting it, you could ask the product team what they hope to learn. Perhaps a quick, one-time export of raw, imperfect sample data would suffice for an initial analysis in a spreadsheet. This approach allows the stakeholder to explore the dataset's potential value with minimal engineering effort. It demonstrates a willingness to help while strategically reducing the scope to a manageable size, ensuring that engineering resources remain focused on proven priorities.

This strategic approach to prioritization reveals a fundamental truth: as you begin executing on Quick Wins and Critical Projects, you will inevitably face a coordination challenge. Manual processes and isolated scripts—perfectly adequate for getting started—quickly become bottlenecks as your data ecosystem grows. Eventually, marketing's Google Analytics data, finance's revenue pipeline, and other sources like product's user behavior data all need to run in harmony, not as independent, fragile processes hoping for the best.

This is precisely the moment when orchestration transforms from a nice-to-have into a business necessity. The "cron jobs and prayers" approach no longer scales, and the cost of silent failures becomes too high. To move from this state of chaos to one of coordination, you need a conductor for your data orchestra.

Orchestration with Airflow: From Chaos to Coordination

As we discussed in Chapter 3, we described several options when it comes to workflow orchestration. We identified Apache Airflow as one of the most popular and powerful choices. In this chapter, we transition from the "what" to the "how," providing a practical guide to implementing Airflow in a startup environment.

The need for a robust orchestration tool often becomes apparent as a company's data processes mature. The journey is a familiar one: many startups begin with manual work and spreadsheets, a method that is simple but not scalable. Others adopt easy-to-use sync tools that pipe data from various sources into a warehouse, with many also providing data transformation capabilities. While convenient, these can be costly and may not offer the 100% accuracy required for mission-critical metrics.

A common intermediate step is a fragile system held together by scheduled scripts and hope—a strategy known colloquially as "cron jobs and prayers." When the inevitable silent failures of this approach begin to threaten data integrity, teams start searching for a more reliable and observable solution. This chapter provides a practical guide for those teams, showing how to move from chaos to coordination.

Many of the most impactful tools in the modern data stack were born out of necessity within fast-growing companies, but their journeys often begin with radical simplicity. As we touched on earlier, Airbnb's beginnings exemplify this. Their first data scientist's job was simply to query the production MySQL database and share basic stats with the team. This mirrors the path of numerous successful startups that initially prioritize immediate business needs over technical perfection. For these companies, building a sophisticated data stack is a gradual evolution, not an overnight revolution. They start with the simplest solution that delivers value and iteratively adopt more advanced tools as complexity grows and new challenges arise.[2]

From this simple start, complexity grew, and the need for a more robust solution became clear. Apache Airflow is a prime example of a tool born from that evolution. It was started by Maxime Beauchemin at Airbnb in October 2014 to solve the company's challenges with increasingly complex workflows. It is an open source platform created to programmatically author, schedule, and monitor data workflows. At its core, Airflow allows data engineers to define complex data pipelines as code, specifically using Python to create Directed Acyclic Graphs (DAGs). This configuration-as-code approach provides the flexibility and scalability necessary to manage the intricate dependencies of modern data engineering tasks.[3,4,5]

The story of Airflow's creation underscores its relevance for the startup ecosystem. Beauchemin is also the original creator of Apache Superset (we will elaborate more on it in Chapter 9), a data exploration and visualization platform. This journey from

---

[2] Riley Newman, "At Airbnb, Data Science Belongs Everywhere," Airbnb Engineering & Data Science Blog, July 7, 2015, `https://medium.com/airbnb-engineering/at-airbnb-data-science-belongs-everywhere-917250c6beba`.

[3] "Apache Airflow," Apache Airflow, accessed September 29, 2025, `https://airflow.apache.org/`.

[4] "What is Apache Airflow?," Qubole, accessed September 29, 2025, `https://www.qubole.com/the-ultimate-guide-to-apache-airflow`.

[5] "Architecture Overview," Apache Airflow Documentation, accessed September 29, 2025, `https://airflow.apache.org/docs/apache-airflow/stable/core-concepts/overview.html`.

an internal startup project to an industry-standard tool highlights a common pattern: powerful data solutions often emerge from the high-pressure environment of startups, where they are created to solve immediate and practical problems.[6,7,8,9,10]

The value of an orchestrator like Airflow becomes even clearer when examining the experience of its creator. After his time at Airbnb, Beauchemin founded Preset, a company that provides Apache Superset as a fully managed, enterprise-grade cloud service. As a company whose product is Business Intelligence, establishing an exemplary data foundation from day one was a strategic necessity. Preset's initial data stack was a collection of different SaaS tools designed for specific functions:

- **Data Sources Tools:** Like many startups, Preset quickly started with several out-of-the-box solutions that immediately could become their data sources. Beauchemin names Lever for recruitment data, HubSpot for info about their customers (CRM). Segment as a transport layer for Superset events that then were funneled into the data warehouse. Fivetran was used to sync data sources like HubSpot to the data warehouse.

- **Data Warehouse:** All raw data was centralized in Google BigQuery.

- **Data Transformation:** The team used dbt to execute SQL-based transformations, converting raw data into structured, analytics-ready models.

- **Reverse ETL:** Hightouch was employed to sync processed data from the warehouse back into operational systems, such as sending customer segments to HubSpot for marketing campaigns.

---

[6] "Maxime Beauchemin," Airflow Summit, accessed September 29, 2025, `https://airflowsummit.org/speakers/maxime-beauchemin/`.

[7] Maxime Beauchemin, "A startup's data journey and its growing need for orchestration," Airflow Summit 2022, presentation slides, May, 2022, `https://airflowsummit.org/slides/2022/h4-DataJourney-Maxime.pdf`.

[8] Maxime Beauchemin, "The tale of a startup's data journey and its growing need for orchestration," Apache Airflow YouTube Channel, June 15, 2022, `https://www.youtube.com/watch?v=ER_o1iA6DNO`.

[9] "Airflow," Airbnb Projects, accessed September 29, 2025, `https://airbnb.io/projects/airflow/`.

[10] "Project," Apache Airflow Documentation (version 1.10.7), accessed September 29, 2025, `https://airflow.apache.org/docs/apache-airflow/1.10.7/project.html`.

As more systems were added to the tech stack, the initial harmony of a few powerful, easy-to-use tools began to break down. Beauchemin uses an analogy from the world of music to describe this evolution. He compares a startup's early data stack to a small jazz band, where a few members can coordinate with relative ease. However, as new "instruments"—in this case, data tools—are added, the group swells into a large orchestra. Without a conductor, what should be a symphony quickly devolves into a cacophony.

This was precisely the situation at Preset. Different systems loaded data into the warehouse on their own uncoordinated schedules. This meant that when dbt transformation jobs ran, it was unclear if all the necessary source data was present, leading to non-deterministic outcomes and a critical lack of confidence in the data. It became obvious that a conductor—an orchestrator—was essential to manage the chaos and restore order.

To solve this chaos and establish a reliable data platform, Preset introduced Airflow. As Beauchemin argues, an orchestrator becomes essential as a startup's data complexity grows. It serves several crucial functions:

- **Coordination and Sanity:** In a landscape where new, specialized data tools can be added with ease, an orchestrator is the essential "glue" that holds the stack together. It manages the compounding complexity, ensuring that all components work in harmony.

- **Intelligent Scheduling with Sensors:** Airflow's sensors are a critical feature for building dependable pipelines. They allow a workflow to pause and wait until a specific condition is met—such as a data load from a source system completing—before proceeding. This guarantees that transformations run only on complete, up-to-date data, preventing inconsistent results.[11]

- **Robust Logging and Observability:** Unlike stateless tools that simply execute a task, Airflow provides a persistent and detailed log of every pipeline run. These logs are indispensable for debugging, auditing, and understanding precisely what happened.

---

[11] "Sensors," Apache Airflow Documentation, accessed September 29, 2025, https://airflow.apache.org/docs/apache-airflow/stable/core-concepts/sensors.html.

- **Flexibility for Edge Cases:** While the modern data tools automate many common use cases, every business has unique requirements, homegrown systems, or edge cases that demand custom code. Airflow provides the ideal, centralized platform to run what Beauchemin calls the "little scripts that glue things together," which are often vital for business operations.

- **A Gateway to Data Science:** As a startup matures, it will inevitably move toward more advanced analytics. Airflow offers a natural home for integrating and scheduling sophisticated workloads like machine learning model training or parameterized notebooks, connecting data science outputs to the broader data ecosystem.

Beauchemin concludes that as the modern data stack has matured with a growing ecosystem of specialized tools, Airflow's primary role has evolved. With out-of-the-box solutions now handling many common data ingestion and transformation tasks, Airflow is increasingly used as an "orchestra director" that triggers and manages specialized external systems. However, this does not mean Airflow no longer performs computation itself. A core strength of Airflow remains its flexibility to run custom code, which is essential for handling unique business logic, ML pipelines, or edge cases not covered by standard tools.

Therefore, its modern role is best described as a versatile control plane: it acts as the director for a suite of external tools while retaining the power to be a hands-on performer when custom computation is needed. This lean, powerful model allows even a small startup to build a world-class, reliable data infrastructure. However, this power and flexibility come with complexities and potential trade-offs that, if overlooked, can become significant hurdles.

Common Airflow Pitfalls for Startups

While Airflow can bring order to chaos, its power comes with significant trade-offs. For a resource-constrained startup, the decision to adopt Airflow requires a clear-eyed assessment of its total cost of ownership. Some of the common pitfalls that startups encounter include

- **Underestimating Infrastructure Costs:** Although Airflow is open source and free to download, the cost of running it can be surprisingly high. It demands dedicated infrastructure for its core components. A small, local setup might be cheap initially, but costs can escalate as your data pipelines grow in number and complexity. Scaling Airflow to handle large volumes of concurrent tasks often requires more powerful machines or a distributed setup (e.g., using Kubernetes), leading to increased cloud bills.[12]

- **Underestimating Engineering Time Costs:** Another significant *hidden cost* of Airflow is the engineering time required to deploy, maintain, and debug it. While a basic setup may seem straightforward, configuring and maintaining an efficient, production-ready custom Airflow instance can be a challenging and time-consuming task. The more complex your needs are, the more engineering hours your setup will require—not just for initial deployment but for recurring maintenance, debugging, optimizing, and scaling. This ongoing effort can drain valuable engineering hours that could otherwise be spent on projects that deliver direct business value.[13]

- **A Steep Learning Curve:** Airflow is a complex framework with its own concepts and operational quirks that can be challenging for developers to master. A lack of deep knowledge about Airflow's architecture and configuration is a common root cause of implementation challenges. This steep learning curve can slow down development, particularly for startups or small teams where engineers often wear multiple hats and cannot afford the significant time investment required to become Airflow experts.[14]

---

[12] "The Hidden Costs of Running Apache Airflow," Prefect Blog, June 1, 2024, `https://www.prefect.io/blog/hidden-costs-apache-airflow`.

[13] "The Hidden Costs of Running Apache Airflow," Prefect Blog, June 1, 2024, `https://www.prefect.io/blog/hidden-costs-apache-airflow`.

[14] Jake Roach, "Dagster vs Airflow: Comparing Top Data Orchestration Tools for Modern Data Stacks," DataCamp Blog, September 15, 2024, `https://www.datacamp.com/blog/dagster-vs-airflow`.

Ultimately, you must weigh your individual situation and constraints when choosing an orchestration tool. Analyze your requirements and make your own decision, as you may find that other tools are a better fit. Some teams mitigate the operational burden by using a fully managed service like Astronomer or Google Cloud Composer, which are built on Apache Airflow. Others may find that serverless orchestrators like AWS Step Functions or Azure Durable Functions are better alternatives, as they significantly reduce infrastructure overhead.[15]

This discussion of orchestration reveals an important insight: while a tool like Airflow excels at coordinating when data workflows happen, it does not solve the problem of what happens within those workflows. Airflow is the conductor, ensuring tasks run in the correct order, but it is not the musician playing the notes. For a data team, a critical part of the performance is transforming raw data after it has been loaded into the data warehouse.

You can have a perfectly scheduled pipeline, but if the SQL logic that cleans, joins, and models your data is scattered across files with no version control, testing, or documentation, you're still building on a foundation of sand. To create a reliable data platform, your orchestrator needs a partner that specializes in this transformation step. As we saw in Preset's stack, that specialist is dbt. Let's examine how this tool revolutionizes the way startups build and manage their business logic in SQL.

Data Transformation: The dbt Revolution for Startups

For a startup, dbt is a game-changer because it applies the rigor of software engineering to the familiar world of SQL. This empowers even a solo analyst to build professional-grade data transformation pipelines that are reliable, maintainable, and scalable, without requiring a dedicated data engineering team. It operates on the modern principle of ELT (Extract, Load, Transform), focusing exclusively on the transformation step after raw data is already in your warehouse. This approach allows you to work with

---

[15] Steve Swoyer, "To Build or to Buy? DIY Orchestration with Airflow vs. A Fully Managed Service," Astronomer Blog, May 24, 2022, `https://www.astronomer.io/blog/build-vs-buy-managed-airflow-service/`.

the tool that is simple and popular—SQL. You simply write the business logic as standard SELECT statements, and dbt handles the task of turning that code into reliable, analytics-ready data models.[16]

Beauchemin highlights several key advantages of dbt that make it particularly well suited for startups:

- **Simplicity and Low Barrier to Entry:** Beauchemin emphasizes how easy dbt is to get started with. As a simple, stateless command-line tool, it does not require setting up a dedicated service. In the early days of Preset, their entire data transformation process consisted of a team member manually running dbt run daily from their laptop—a lightweight approach that delivered value immediately without infrastructure overhead.

- **The Power of Incremental Adoption and Full Refreshes:** He points out that dbt is incrementally adoptable. For a startup with small datasets, the ability to simply run a full refresh and recompute the entire data warehouse from scratch is a powerful simplification. This approach avoids the premature complexity of building intricate incremental load logic, which Beauchemin notes can be addressed later as the data grows.

- **A Clear Path to Scalability:** While dbt is easy to start with, Beauchemin also valued its clear path toward a more mature architecture. He knew that as Preset's needs grew, dbt's compatibility with Airflow would allow them to integrate their transformation jobs into a more robust orchestration framework. This provided the confidence that dbt was not a short-term fix but a foundational tool that could scale with the company.[17]

---

[16] "What is dbt?," dbt Developer Hub, accessed September 29, 2025, `https://docs.getdbt.com/docs/introduction`.

[17] Maxime Beauchemin, "The tale of a startup's data journey and its growing need for orchestration," Apache Airflow YouTube Channel, June 15, 2022, `https://www.youtube.com/watch?v=ER_o1iA6DN0`.

Though dbt was created by a startup to solve the problems of other startups, it has since become the industry standard for data transformation. Its robust and scalable framework is now used by thousands of companies of all sizes, from early-stage startups to large enterprises, proving its effectiveness across a wide spectrum of data challenges.[18]

dbt Implementation Strategy

Adopting dbt is a significant step toward building a mature data practice. The implementation strategy should prioritize speed, flexibility, and maintainability. As outlined in Chapter 3, your first decision is whether to use the open source dbt Core or the managed dbt Cloud service. For an early-stage startup, this choice is less about features and more about a fundamental trade-off: engineering time vs. subscription cost.

The first option mirrors the experience at Preset, where Beauchemin's team initially ran dbt Core with minimal overhead—literally just a manual run command on a laptop to process the day's data. It is a scrappy but effective way to get started when the team is small and data needs are relatively simple. As their needs evolved, they transitioned to a simple CI/CD pipeline using Jenkins. This path is ideal for startups that have in-house technical skills and prioritize minimizing costs while maintaining full control and flexibility over their setup.[19]

On the other hand, dbt Cloud offers a powerful, managed solution that is often the right choice for teams wanting to move fast without getting bogged down in infrastructure. It provides a user-friendly, all-in-one platform that handles CI/CD, job scheduling, and observability out of the box. This turnkey approach allows anyone on the data team, regardless of their engineering expertise, to build, test, and deploy data models reliably.

Ultimately, the decision comes down to a strategic choice: invest engineering time for control and lower direct costs with dbt Core, or invest in a subscription to accelerate time-to-value and reduce operational overhead with dbt Cloud.

---

[18] "How dbt Labs Built a $4.2B Software Business out of a Two-Person Consultancy," First Round Review, accessed September 29, 2025, `https://review.firstround.com/dbt-labs-path-to-product-market-fit/`.

[19] Maxime Beauchemin, "The tale of a startup's data journey and its growing need for orchestration," Apache Airflow YouTube Channel, June 15, 2022, `https://www.youtube.com/watch?v=ER_o1iA6DNO`.

Data Modeling for Startups: A Pragmatic Approach

Data modeling within an early-stage startup is fundamentally different from the practices common in large corporations. Startups rarely require endless meetings with a wide array of stakeholders to debate modeling needs across various business units, nor is it necessary to deliberate on every minute detail that may or may not prove valuable. Instead, the primary objective is to deliver a functional solution quickly to prove its value and address a specific, immediate business need. This approach is a response to the unique challenges startups face, such as rapidly changing business requirements, ambiguous metric definitions, and constant pressure to deliver insights quickly. In such a dynamic environment, a rigid, top-down modeling approach is destined to fail. Therefore, startups must adopt a more agile and modular methodology.[20]

Effective data modeling requires an incremental approach. Attempting to address everything at once is a recipe for failure. A common pitfall for startups is the ambition to build a single, monolithic data model. Such a model, which aims to answer every conceivable question, is notoriously slow to build, difficult to maintain, and quickly becomes obsolete as the business evolves. A more effective strategy is to build modular, business-domain-aligned data models. Instead of one large, all-encompassing model, this approach involves creating smaller, independent models for each distinct business area, such as marketing, sales, or product. This methodology offers several key advantages:

- **Speed:** Each model can be developed and deployed independently, delivering value faster.

- **Maintainability:** When a business process changes, you only need to update the relevant model, reducing the risk of breaking downstream dependencies.

- **Clarity:** By aligning models with business domains, the logic is easier for stakeholders to understand and trust.

This modular approach embraces the reality of a startup's constantly evolving data landscape. It enables the data model to adapt and grow alongside the business, transforming it from a potential bottleneck into a strategic asset.

---

[20] Paddy Alton, "Data Modelling for Startups (Part III)," Apolitical Engineering, Medium, May 15, 2023, https://medium.com/apolitical-engineering/data-modelling-for-startups-part-iii-aca31e0f4176.

Building Your First Models: Staging, Intermediate, and Marts

A best-practice dbt project is organized into layers, which promotes modularity, code reuse, and clarity. The objective is to efficiently transform raw data loaded into your warehouse into a structured format ready to meet business needs, such as powering a dashboard. One of the most popular and effective structures involves three key layers:

1. **Staging Models:** This is the first layer of transformation, focusing exclusively on cleaning and standardizing raw data from your sources. Each staging model should correspond one to one with a source table. Typical operations at this stage include renaming columns to be more intuitive (e.g., *consumer_id* instead of *c_id*), casting data types, and performing basic data cleaning. This process creates a solid, reliable foundation for all downstream models. Since many startups have test or fake data in their production environments, you can filter these records out in this step. Staging models are particularly useful for startups where business reality—and thus source data—changes dynamically. You can handle adjustments to these changes here, insulating the rest of your project.

2. **Intermediate Models:** In this layer, you begin to apply more complex business logic and join different concepts together. These models are not typically exposed directly to end users. Instead, they serve as reusable building blocks for your final data marts. For example, you might create an intermediate model that combines session and user data into a comprehensive user_sessions model. By creating these modular components, you avoid repeating the same complex SQL logic in multiple places.

3. **Data Marts:** These are the final, analytics-ready models exposed to your end users and business intelligence tools and are optimized for reporting. Each data mart should be designed to serve a specific business purpose and includes fact and dimension tables. Put simply, fact tables store quantitative business metrics, like sales figures or transaction counts, while dimension tables store descriptive attributes, such as product details, customer

information, or dates, to provide context for the facts. For instance, a *daily_user_activity* model for product analytics may include a fact table (*fact_user_events*) containing numerical values like app open counts, and dimension tables for contextual information such as activity type (*dim_activity_type*), date (*dim_date*), and geography (*dim_geography*).

By gradually adopting data marts, you make it much easier to meet future business requests. The structure of data marts, based on fact and dimension tables, allows you to assemble numerous different dashboards with relative ease. When first getting started, you may be perfectly fine using only the staging and data mart layers. However, as data complexity grows, adding an intermediate layer becomes extremely useful, as it significantly increases the readability and reusability of your code. Of course, your project can have more than the three layers described above. The key takeaway is that by following this layered approach, even a solo data practitioner can build a scalable, maintainable, and trustworthy data transformation pipeline that will serve the startup as it grows.[21]

Common Startup dbt Pitfalls

While dbt is an incredibly powerful tool, the same pressures that make startups agile can also lead to anti-patterns that create technical debt. Being aware of these common pitfalls can help you build a dbt practice that is sustainable and scalable. Beyond the previously discussed mistake of building monolithic models, several other pitfalls can undermine a dbt project in a startup environment:

- **SQL Limitations:** Despite its power, SQL has inherent limitations. Not all data transformations can be easily expressed in pure SQL, especially when dealing with non-tabular data formats like Excel files or other sources that are not natively queryable by the data warehouse. In these cases, you may need to design a more advanced architecture that complements dbt with additional tools and processes to handle data ingestion and transformation beyond SQL. Understanding this helps prevent the misconception that all

---

[21] Christine Berger, "Data Modeling Techniques for More Modularity," dbt Blog, February 21, 2025, https://www.getdbt.com/blog/modular-data-modeling-techniques.

data problems can be solved solely with SQL and dbt. This is an important consideration when planning your data platform architecture.

- **Trying to Do Everything at Once:** A frequent mistake is attempting to build the "perfect" data setup from day one. Instead of focusing on delivering gradual, immediate wins, startups get bogged down creating an overly engineered solution for every potential future use case. Adopting dbt should be an iterative process. Prioritize quick wins that demonstrate value and embrace the 80/20 rule—build what's good enough for now, with the understanding that you will refactor and improve as the business matures.

- **Over-complicating Staging Models:** The purpose of the staging layer is to be a clean, one-to-one reflection of your source data. A critical mistake is performing joins or complex logic at this stage. This makes the layer brittle and difficult to modify, which is especially problematic in a startup where source schemas can change frequently. Keep this layer as simple as possible; joins and business logic belong downstream in your intermediate and mart models.[22]

- **Not Aligning Models with Business Domains:** When data models are structured around technical convenience rather than business concepts, they become a black box to the rest of the company. A model named *daily_user_activity* is far more intuitive to everyone than *reporting_final_v3_temp*. If your stakeholders cannot understand what your models represent, they will not trust them. Structure your dbt project to mirror the structure of your business (e.g., with folders for marketing, finance, and product) and use clear, business-centric naming conventions.

- **Neglecting Essential Tests:** In the rush to deliver, it is easy to skip testing. While you do not need to write thousands of tests to validate scenarios that will probably never happen, you must be prepared for the most likely failure modes.

---

[22] Nicholas Thomson, "Understanding dbt: Basics and Best Practices," Datadog Blog, September 5, 2025, https://www.datadoghq.com/blog/understanding-dbt/.

- **Using dbt to Fix Production Data Issues:** It can be tempting to use dbt transformations as a band-aid for poor data quality originating from source systems. However, dbt should not be the permanent last line of defense for these application-level issues. Your app developers are responsible for testing the features they release and ensuring the integrity of the data they produce. You should advocate for data quality at the source and set the expectation that the data team is not responsible for patching avoidable, upstream problems within dbt.

These pitfalls reveal a deeper truth: dbt's greatest strength—its ability to empower anyone who knows SQL to build data models—also creates its greatest challenge. When more people can contribute, how do you ensure the resulting data is trustworthy? The very flexibility that makes dbt so accessible can lead to an erosion of confidence if not paired with a deliberate approach to quality.

This is why mastering dbt is not just about writing transformation scripts. It is about building a culture of trust from day one. The good news is that dbt provides the very tools needed to implement this culture pragmatically, allowing startups to build confidence without the bureaucratic overhead of complex data governance.

Data Quality: Trust Without Bureaucracy

For a startup, data is only valuable if it is trusted. When stakeholders doubt the accuracy of the numbers they see, dashboards become useless, and data-driven decision-making grinds to a halt. However, startups cannot afford the heavyweight, bureaucratic data governance frameworks of large enterprises, especially when facing rapid schema changes, limited time for testing, and a constant trade-off between speed and accuracy.

The goal is to build confidence through pragmatic, targeted, and automated measures that fit the fast-paced startup environment. To navigate these challenges, it is best to adopt a prioritized approach to data quality, much like building a pyramid. Instead of trying to do everything at once, you must first build a solid foundation of essential tests, gradually adding more sophisticated checks as your team and data mature (see Figure 8-2).

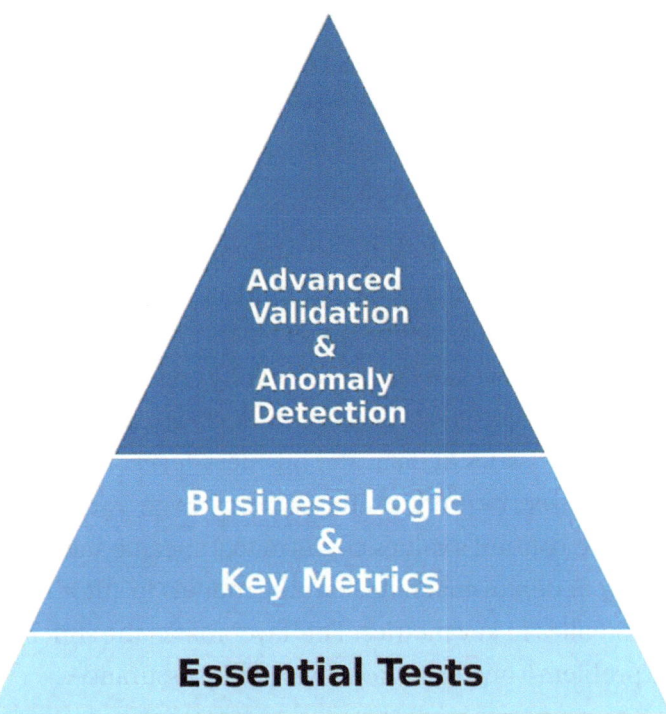

Figure 8-2. *The Data Quality Pyramid for Startups*

Figure 8-2 illustrates the data quality pyramid. Efforts should be focused on building a solid foundation before moving to more advanced quality checks.

1. **The Foundation: Essential Tests:** The pyramid's base represents the foundational, non-negotiable tests for your most critical data models. Instead of attempting to test every column in every model at once, start by applying these essential checks to primary keys and the columns that drive key business metrics. dbt's four built-in data tests are perfect for this stage: uniqueness, not-null constraints, accepted values, and referential integrity.[23]

2. **The Middle Layer: Business Logic and Key Metrics:** With a solid foundation in place, the next layer involves validating core business logic and key metrics. Here, you can use dbt's custom data tests to enforce specific business rules, such as ensuring a

---

[23] "Add data tests to your DAG," dbt Documentation, accessed September 29, 2025, https://docs.getdbt.com/docs/build/data-tests.

user's first activity date does not precede their registration date. This layer is also where you should implement simple alerts for data freshness and volume—for instance, notifying the team if a revenue table has not been updated in 24 hours or if the daily event count drops by more than 50%.[24]

3. **The Top: Advanced Validation and Anomaly Detection:** This is the most advanced layer of the pyramid and should only be tackled once the other layers are in place. This includes implementing sophisticated logic to identify anomalies that signal problems at two levels: technical issues, such as a broken data pipeline causing issues in data, and business-level events, such as fraudulent users or product abuse. At this stage, you might validate that a column's values conform to a specific statistical distribution or deploy machine learning models to automatically detect anomalies in key metrics. This layer marks the shift from reactive problem-solving to proactive quality assurance.

To accelerate these efforts, open source dbt packages like *dbt_expectations* offer a wide range of pre-built tests. By focusing on the base of the pyramid first, even a one-person data team can establish a culture of data quality and build the trust necessary for the organization to become truly data driven.[25]

This structured approach provides a clear road map for implementation, but theory only becomes valuable when tested in the real world. The companies we will examine next—from the Berlin-based fintech Billie to the larger enterprise Spotify—demonstrate how organizations have navigated the journey from startup chaos to mature data operations.

What makes their stories so instructive is seeing how they balanced the pragmatic startup mindset with the need for reliability as they scaled. Their experiences reveal both the power and the practical challenges of implementing tools like Airflow and dbt in high-stakes business environments, turning abstract principles into concrete results.

---

[24] Austin Chia, "A Comprehensive Guide to dbt Tests to Ensure Data Quality," DataCamp, July 1, 2025, https://www.datacamp.com/tutorial/dbt-tests.

[25] Madison Schott, "Testing Data Pipelines with dbt-expectations: A Beginner's Guide," Airbyte Blog, June 7, 2023, https://airbyte.com/blog/testing-with-dbt-expectations.

How Companies Are Building Their Data Stack with Airflow and dbt

The combination of dbt for transformation and Apache Airflow for orchestration has become a cornerstone of the modern data stack, trusted by companies of all sizes. The following examples illustrate how leading companies have leveraged these tools to build scalable and reliable data platforms.

Companies Leveraging Both Airflow and dbt

The pairing of Apache Airflow and dbt represents a natural partnership in the modern data stack—Airflow serves as the "orchestra director," while dbt acts as the "transformation specialist." As we explored earlier in this chapter when discussing Preset's journey, this combination addresses the fundamental need for coordination and reliability as data complexity grows. When both tools have become industry standards, many startups no longer see the need to reinvent the wheel with custom solutions. Instead, they leverage this proven architecture to accelerate their time to market while maintaining scalability and reliability.

This approach offers several compelling advantages for growing companies. Airflow provides sophisticated orchestration capabilities, managing complex dependencies between different systems and ensuring data freshness through its sensor functionality. Meanwhile, dbt brings software engineering best practices to SQL transformations, enabling version control, automated testing, and clear documentation. Together, they create a robust foundation that can scale from startup to enterprise while maintaining operational excellence.

Billie: The journey of the Berlin-based fintech Billie closely follows the pragmatic principles adopted by Preset. Initially, the team manually maintained their data pipelines, a "do-it-yourself" approach that became unscalable and expensive as the company grew. Recognizing the high cost of engineering time, they adopted a hybrid "buy vs. build" model. They chose Fivetran to automate data ingestion from standard sources like Google Analytics, LinkedIn, and Salesforce. This freed up their engineering efforts to be dedicated exclusively to select use cases, building custom pipelines with Apache Airflow only where granular control was essential. These custom pipelines handled data from their critical, proprietary production database, which required complex transformation logic. All data is centralized in a Snowflake data warehouse, where dbt is used for all transformations. Airflow acts as the master orchestrator, triggering both Fivetran syncs and the custom dbt jobs. Billie's story provides a powerful blueprint for startups: strategically buy out-of-the-box services and build only where your business need is unique, using tools like Airflow and dbt to create a scalable and efficient data platform.[26,27]

Monzo: The UK digital bank Monzo offers a powerful case study in how Airflow and dbt can scale to meet exponential growth. As the bank's active user base exploded from 1.6 million in 2019 to nearly 6 million by 2022, its team simultaneously grew from just 4 to over 130 analysts and engineers. This dual-front expansion placed immense strain on its data infrastructure.[28]

Companies Using dbt Without Airflow

While the combination of Airflow and dbt is powerful, it is not a universal solution. The decision to use dbt without Airflow is driven by several strategic factors, ranging from a startup's need for simplicity to an enterprise's reliance on established, cloud-native tools. Understanding these alternative architectures is key to choosing the right path for your organization. These scenarios fall into several common patterns:

- **Using dbt Cloud's Native Scheduler:** For many startups and teams that prioritize speed and want to minimize infrastructure overhead, dbt Cloud's built-in scheduler is a perfect fit. It allows teams to orchestrate their dbt jobs without the complexity of setting up and managing a separate tool like Airflow.

---

[26] "Billie's need for speed and ease of use met by Fivetran," Fivetran Case Studies, accessed September 29, 2025, `https://www.fivetran.com/case-studies/billies-need-for-speed-and-ease-of-use-met-by-fivetran`.

[27] "Billie saves up to 20% warehousing cost with Apache Airflow and Fivetran," Fivetran Case Studies, accessed September 29, 2025, `https://www.fivetran.com/fr/case-studies/billie-airflow-case-study`.

[28] "Airflow at Monzo: Evolving our data platform as the bank scales," Apache Airflow YouTube Channel, October 13, 2023, `https://www.youtube.com/watch?v=1b6Uu4MOExY`.

- **Manual or CI/CD Triggers:** In the earliest stages, a simple approach is often best. As we saw with the early days at Preset, a simple dbt run command executed manually can be sufficient when data needs are straightforward. As needs evolve, many teams graduate to triggering dbt runs via CI/CD pipelines using tools like GitHub Actions or Jenkins, integrating data transformation directly into their software development life cycle.

- **Leveraging Cloud-Native Orchestrators:** As we touched on in Chapter 3, a company's broader cloud strategy often dictates its choice of tools. Teams heavily invested in a specific cloud ecosystem frequently opt for native orchestration services to run their jobs. This approach allows them to manage all their cloud resources within a single, unified environment.

- **Relying on In-House or Legacy Tools:** Some large tech companies developed their own powerful orchestration platforms long before Airflow became the industry standard. Having invested years in building and refining these internal systems, they choose to integrate dbt into their existing, battle-tested solutions rather than migrating to a new tool.

On the other hand, the data stack is not a static decision. As companies' data needs evolve and new solutions become available on the market, many organizations choose to re-architect their platforms. Let's examine how this played out in a real-world scenario:

Spotify: The music streaming giant offers a compelling example of a large organization that has embraced dbt for its transformation layer. Operating at a massive scale, Spotify uses dbt extensively within its Google BigQuery data warehouse to empower its analytics teams. For data orchestration, however, the company chose a different path than Airflow, implementing Flyte to manage its complex workflows. Spotify found Flyte attractive for its centralized service, efficient client SDK, robust caching, and scalability. A key factor was also the relative ease of migrating from their previously used tools, Luigi and Flo. Spotify's architecture highlights a crucial point: dbt's value in bringing software engineering discipline to the transformation layer can be realized independently of any single orchestration tool. This allows companies to pair dbt with the orchestrator that best fits their specific technical requirements and existing infrastructure.[29,30,31]

Companies Using Airflow Without dbt

While the partnership between Airflow and dbt has become a popular standard, many of the world's most innovative tech companies operate powerful data platforms where dbt is not the primary tool for transformation. These organizations often use Airflow as a master orchestrator for a diverse set of tools, specialized processing engines, and complex, non-SQL-based workflows. This architectural choice is typically driven by three main factors:

- **Need for Non-SQL Transformations:** Many advanced use cases, particularly in machine learning and scientific computing, require complex logic written in Python or Scala that goes beyond the capabilities of SQL.

- **Pre-existing Custom Frameworks:** Several tech giants built sophisticated, in-house data platforms and ETL frameworks long before dbt rose to prominence. Having invested years in these battle-tested systems, they choose to integrate them with Airflow rather than re-platform.

- **Extreme Scale and Optimization:** At a massive scale, companies may build custom solutions using distributed computing engines like Spark or Flink to achieve performance optimizations that are highly specific to their infrastructure and workloads.

The following examples illustrate how leading companies have built powerful, Airflow-centric data stacks tailored to their unique needs.

---

[29] "Coalesce 2024: Needle in the (data) stack: How Spotify powers Salesforce," dbt Labs YouTube Channel, October 16, 2024, `https://www.youtube.com/watch?v=vPwsQpjN9mO`.

[30] Guillaume Perchais, "Why We Switched Our Data Orchestration Service," Spotify Engineering Blog, March 14, 2022, `https://engineering.atspotify.com/2022/3/why-we-switched-our-data-orchestration-service`.

[31] Junaid Effendi, "Spotify Data Tech Stack," Substack, August 16, 2025, `https://www.junaideffendi.com/p/spotify-data-tech-stack`.

- **Airbnb:** As the birthplace of Apache Airflow, Airbnb naturally relies on it extensively for data orchestration. Operating at a scale that involves processing billions of events daily and managing petabytes of data, the company's data transformation requirements are highly specific and extend beyond what standard tools like dbt are designed to handle. Consequently, Airbnb's data stack is a complex ecosystem of internally developed tools, including its powerful metrics platform, Minerva.[32,33]

- **Laurel:** For certain companies, the core challenge lies not in transforming tabular data but in orchestrating complex machine learning operations (MLOps). Laurel, an AI-powered timekeeping solution, provides a clear example. The company uses Airflow to manage a suite of machine learning models that require regular retraining to maintain their accuracy and relevance. In this context, Airflow's strength is its ability to serve as a versatile, Python-native orchestrator for a modular system encompassing model retraining, evaluation, and deployment. This use case demonstrates how Airflow can act as the backbone for complex, code-heavy workflows where traditional, SQL-based transformation is only one component of a much larger process.[34,35]

Looking back at the architectures of companies like Preset and Spotify, it is tempting to focus only on the tools they chose. But the most important takeaway is not what they used, but why they made those choices. Their decisions were driven by business context—balancing engineering resources at Billie, integrating with existing systems at Spotify, and achieving rapid time-to-value at Preset.

---

[32] "Democratizing Metrics at Airbnb - Minerva 2.0 and Beyond," CelerData (Powered by StarRocks) YouTube Channel, September 22, 2023, `https://www.youtube.com/watch?v=s4xwqQb5v4k`.

[33] Junaid Effendi, "Airbnb Data Tech Stack," Substack, August 14, 2024, `https://www.junaideffendi.com/p/airbnb-data-tech-stack`.

[34] "Customizing LLMs: Leveraging technology to tailor GenAI using Airflow," Apache Airflow YouTube Channel, November 17, 2024, `https://www.youtube.com/watch?v=T8Leid9EoUI`.

[35] "Best practices for orchestrating MLOps pipelines with Airflow," Astronomer Documentation, accessed September 29, 2025, `https://www.astronomer.io/docs/learn/airflow-mlops`.

Understanding this strategic layer is the key to building a successful data platform. The following lessons are not about technology for its own sake, but about how to align your data infrastructure with the core mission of your startup.

Lessons Learned

This chapter's exploration of data engineering for startups offers several critical takeaways for startups and data leaders. The central theme is a pragmatic, business-focused approach that prioritizes progress over perfection and strategic tool selection over following trends.

- **Embrace Industry-Standard Tools, But Choose Wisely:** Apache Airflow and dbt have achieved widespread adoption for good reason—they solve fundamental problems that virtually every growing startup faces. Airflow emerged from Airbnb's need to orchestrate complex workflows, while dbt was created by a consultancy helping startups like Casper scale their data operations. Their popularity is not accidental. It is the result of solving real startup problems at scale. However, the modern data engineering landscape offers an abundance of options. Beyond Airflow, you will find orchestrators like those mentioned in Chapter 3 and many others. Similarly, while dbt is an excellent SQL-first tool that numerous startups should consider implementing, it does not satisfy every data transformation need. The key insight is not that you must use the most popular tools, but that you should understand why they became popular and whether those reasons align with your specific needs.

- **Navigate the Configuration Landscape Strategically:** The tools landscape is not just wide—it is deep. Even after choosing Airflow and dbt, you face numerous configuration decisions. Should you use dbt Core with manual runs (as Preset initially did) or invest in dbt Cloud for managed infrastructure? Do you deploy Airflow on Kubernetes, use a managed service like Astronomer, or start with a simple setup on a single server? These are not academic questions— they represent fundamental trade-offs between engineering time, operational overhead, and direct costs. The pragmatic approach is to start simple and evolve. Preset began with a team member running dbt manually on their laptop—a lightweight approach that delivered immediate value. As their needs grew, they layered in more sophistication. This evolution from simple to complex, driven by actual business needs rather than architectural idealism, exemplifies the startup data engineering mindset.

- **Prioritize Pragmatic Solutions over Perfect Architecture:** The journey to mature data infrastructure is iterative, not revolutionary. It is perfectly acceptable—often preferable—to begin with manual processes, simple scripts, and imperfect solutions. The 80% solution philosophy allows startups to move quickly, prove value, and adapt as requirements become clearer. This pragmatism extends to tool selection itself. While dbt and Airflow offer powerful capabilities for custom and complex logic, sometimes a simple out-of-the-box solution like Fivetran better serves your immediate needs. As Billie demonstrated, they strategically chose Fivetran for out-of-the-box data ingestion while building custom Airflow pipelines only where their business required unique logic and complete control. The key is matching tool sophistication to problem complexity.

- **Adapt Your Setup to Your Unique Context:** Every startup operates within unique constraints—limited engineering resources, specific industry requirements, existing technical investments, and evolving business models. The most effective data stack is not the one that follows best practices most faithfully, but the one tailored to your current reality. Consider how different companies approached the same tools: Monzo scaled dbt and Airflow to handle explosive user growth, while Spotify chose dbt but paired it with Flyte instead of Airflow to integrate with their existing, large-scale engineering infrastructure. Neither approach is inherently superior. Both reflect thoughtful adaptation to specific circumstances.

- **Build for Today, Design for Tomorrow:** Think of your data architecture like a set of Lego bricks. The modularity of modern tools provides a critical advantage for startups by allowing you to avoid building a single, monolithic system that is difficult to understand, maintain, and modify. Instead, by focusing on independent, interchangeable modules, you gain the flexibility to adapt quickly. When a business process changes, you only need to swap out or update the relevant "brick"—whether it is a dbt model for a specific business domain or a particular Airflow DAG—without the risk of breaking the entire structure. This modular approach empowers you to start with the simplest viable implementation and add sophistication incrementally. You may decide to begin with manual dbt runs, graduate to CI/CD triggers, and then evolve to full orchestration as complexity demands it. This requires resisting the temptation to over-engineer for hypothetical future needs, while building a system that can evolve gracefully as your understanding deepens and your business grows. The goal is not to predict the future perfectly but to build a flexible foundation that can be easily reconfigured to meet new challenges.

With this strategic foundation in place, the journey now shifts from theory to practice. Mastering the tools and techniques discussed in this chapter is the critical step in bringing a resilient data infrastructure to life. To guide you on this path, the following section provides a curated list of essential resources to help you deepen your expertise and build hands-on skill.

Additional Resources

This chapter was designed to give you a strong, practical foundation for tackling the most common data engineering and orchestration challenges faced by startups. You now have the essential toolkit to begin building value.

The journey into data engineering is a continuous one. While our deep dive into Airflow and dbt equips you with the core of a modern batch-processing stack, the industry is vast and constantly evolving. As you advance, you will encounter challenges requiring a broader toolkit—many of which were introduced in Chapter 3. This includes real-time data streaming with tools like Apache Kafka and alternative orchestration platforms such as Dagster.

To help you deepen your understanding of this chapter's topics and begin exploring that wider world, I highly recommend the following resources. They provide invaluable, in-depth knowledge directly from leading practitioners in the field.

Online Platforms

- For hands-on, interactive learning, consider exploring dedicated online platforms. Websites like Datacamp offer structured courses and tutorials that are excellent for building foundational skills in both Airflow and dbt.

Books

- *Fundamentals of Data Engineering: Plan and Build Robust Data Systems* by Joe Reis and Matt Housley[36]

- *Data Engineering Design Patterns: Recipes for Solving the Most Common Data Engineering Problems* by Bartosz Konieczny[37]

- *Analytics Engineering with SQL and dbt* by Rui Pedro Machado and Helder Russa[38]

- *Unlocking dbt: Design and Deploy Transformations in Your Cloud Data Warehouse* by Dustin Dorsey and Cameron Cyr[39]

- *Data Pipelines with Apache Airflow* by Julian de Ruiter and Bas Harenslak[40]

- *Apache Airflow Best Practices* by Dylan Intorf, Dylan Storey, and Kendrick van Doorn[41]

---

[36] Joe Reis and Matt Housley, *Fundamentals of Data Engineering: Plan and Build Robust Data Systems* (O'Reilly Media, 2022).

[37] Bartosz Konieczny, *Data Engineering Design Patterns: Recipes for Solving the Most Common Data Engineering Problems* (O'Reilly Media, 2025).

[38] Rui Pedro Machado and Helder Russa, *Analytics Engineering with SQL and dbt* (O'Reilly Media, 2023).

[39] Cameron Cyr and Dustin Dorsey, *Unlocking dbt: Design and Deploy Transformations in Your Cloud Data Warehouse* (Apress, 2023).

[40] Julian de Ruiter and Bas Harenslak, *Data Pipelines with Apache Airflow* (Manning Publications, 2021).

[41] Dylan Intorf, Dylan Storey, and Kendrick van Doorn, *Apache Airflow Best Practices* (Packt Publishing, 2024).

Official Documentation

- **Apache Airflow:** The official documentation is the definitive source for understanding Airflow's architecture, core concepts like DAGs and operators, and extensive provider integrations. It is an essential resource for planning and building a production-grade Airflow environment. For practical application, I highly recommend Airflow's official YouTube channel. It features a wealth of case studies and best practices from the Airflow Summits. These talks are particularly useful for seeing how more mature companies solve complex data challenges, offering valuable insights as your startup's needs grow and your team expands.[42,43]

- **dbt:** The dbt documentation is a comprehensive and well-structured guide covering everything from getting-started tutorials to advanced techniques. Whether you need to master macros, implement a robust testing strategy, or learn data modeling best practices, this is your go-to resource for mastering the dbt workflow.[44]

Summary

This chapter navigated the crucial landscape of data engineering and orchestration, offering a pragmatic road map for startups to build a robust and scalable data foundation. We began by establishing the core principle of a "startup data engineering mindset," which prioritizes progress over perfection. We emphasized an iterative approach, where beginning with manual processes and focusing resources on high-impact areas is not just acceptable but strategic. This philosophy, coupled with careful "build vs. buy" decisions, allows a small team to deliver value quickly without becoming bogged down in over-engineering.

---

[42] "Apache Airflow," Apache Airflow, accessed September 29, 2025, `https://airflow.apache.org/`.

[43] Apache Airflow YouTube Channel, accessed September 29, 2025, `https://www.youtube.com/channel/UCSXwxpWZQ7XZ1WL3wqevChA`.

[44] "What is dbt?," dbt Developer Hub, accessed September 29, 2025, `https://docs.getdbt.com/docs/introduction`.

At the heart of our discussion were two industry-standard tools: Apache Airflow and dbt. We explored how Airflow acts as the essential orchestrator—the "orchestra director"—that brings order to the modern data stack by coordinating disparate tools and ensuring reliable, observable pipelines. We then detailed how dbt revolutionizes the transformation layer, empowering teams to apply software engineering best practices like version control, automated testing, and documentation directly to their SQL-based data models. Through practical strategies, real-world company examples, and the case study of Billie, we demonstrated how to implement these tools effectively. We outlined a layered approach to data modeling and introduced a pragmatic Data Quality Pyramid to build trust without creating bureaucracy.

By following the principles in this chapter, you can construct a reliable system that transforms raw data into clean, structured, and analytics-ready datasets. However, data's true value is only unlocked when it is placed in the hands of decision-makers. In Chapter 9, we will bridge this final gap, exploring the tools and strategies that bring your data to life through dashboards, visualizations, and analytics.

CHAPTER 9

Business Intelligence Platforms

The onslaught of AI and niche data tools has made BI more complicated, not better.

—Colin Zima,[1]

CEO of Omni, a BI startup

When you hear *Business Intelligence*, you probably think of dashboards. But today's BI landscape is a sprawling, complex ecosystem of tools that goes far beyond charts, making it incredibly challenging to navigate. Colin Zima's statement cuts to the heart of the problem: a flood of AI-powered tools has promised to make analytics easier, but for a startup, it has often created a paradox of overwhelming choice.

You have a mountain of data, a mandate to be "data driven," and a dizzying array of tools all claiming to be the answer. So where do you even begin? For startups already stretched thin on resources, this explosion of options presents both unprecedented opportunities and daunting challenges.

In the startup world, data-driven decision-making isn't just a goal. It's a survival mechanism. Yet the very tools designed to bring clarity have created their own problem: *BI debt*, a digital graveyard of unused dashboards and proliferating tools that drain time and money.

---

[1] Colin Zima, "A $69 Million Series B for our Third Birthday," Omni Analytics blog, March 13, 2025, `https://omni.co/blog/omni-series-b-funding`.

© Piotr Sidoruk 2026
P. Sidoruk, *From Data to Dollars*, https://doi.org/10.1007/979-8-8688-1898-1_9

Think of this chapter as your practical road map. It's designed to arm you with the clarity to choose the right BI tools for your startup. After reading it, you will be able to

- **Make Sense of the Chaos:** Confidently distinguish between the hype and reality of the modern BI landscape, from open source champions and low-cost cloud solutions to powerful AI-driven platforms.

- **Build Your Decision Playbook:** Apply a simple, powerful framework to select a platform based on your startup's unique stage, budget, and goals.

- **Learn from Real-World Scenarios:** Move beyond theory by seeing which tools are right for common startup situations, from an early-stage company to one that's scaling.

- **Avoid the Common Traps:** Recognize the hidden costs and critical mistakes many founders make, ensuring you choose a tool that will become a long-term asset, not a liability.

As we explored in Chapters 4-6, startups need to track metrics that matter. Business intelligence platforms are the foundation for transforming those critical metrics into intuitive dashboards with compelling visualizations. These platforms enable you to monitor key performance indicators, share insights across your team, and communicate progress to stakeholders—all while maintaining the agility and cost-consciousness that startup success demands.

Introduction to BI Tools

The business intelligence landscape has evolved dramatically from the era of heavy and slow enterprise solutions that dominated the early 2000s. I still remember the beginnings of my professional career and the heavy interfaces of tools such as IBM Cognos and Oracle Discoverer. During these times, the main challenge was transitioning from Excel to dedicated BI platforms. For many business users, who were accustomed to the immediate flexibility of spreadsheets, this laborious and restrictive process felt like it created more problems than it solved. These legacy systems required extensive IT involvement, suffered from poor user interfaces, and frequently delivered static reports that were outdated by the time they reached decision-makers.[2,3]

BI tools are in a constant state of flux, with vendors rolling out new features, interfaces, and AI-driven capabilities at a dizzying pace. By the time you read this chapter, the specifics of any given platform may have changed—sometimes substantially. However, the core purpose of this chapter is not to catalog features that will soon be outdated but to help you develop a clear, adaptable framework for evaluating BI solutions on your own terms. Startups today often find themselves at a crossroads: some are drawn toward cutting-edge platforms rich in AI-powered automation and promise, eager to push the boundaries of insight and efficiency. Others consciously reject the complexity and cost of "all-in-one" solutions, instead opting for streamlined, affordable tools that focus on the essentials—tracking mission-critical metrics without unnecessary bells and whistles. Whichever direction you lean, your ability to critically assess what's on offer, ask the right questions, and prioritize your startup's real needs over hype will be the key to making business intelligence work for you.[4]

The Evolution of BI Platforms

The transformation of business intelligence tools reflects broader technological shifts in how we store, process, and interact with data. The traditional era of BI, exemplified by tools like IBM Cognos, SAP BusinessObjects, and Oracle BI, was characterized by on-premises deployments, extensive data modeling requirements, and IT-controlled report generation. These platforms operated on the principle of preaggregated OLAP cubes and required specialized expertise to modify or create new reports.[5]

---

[2] "Business Intelligence is at a Breaking Point, New Report Finds," Business Wire, April 7, 2025, https://www.businesswire.com/news/home/20250407843252/en/Business-Intelligence-is-at-a-Breaking-Point-New-Report-Finds.

[3] Richie Cotton and Colin Zima, "The Next Generation of Business Intelligence with Colin Zima, CEO at Omni," DataFramed podcast, DataCamp, June 15, 2025, https://www.datacamp.com/podcast/the-next-generation-of-business-intelligence.

[4] Alireza Sadeghi, "The Evolution of Business Intelligence: From Monolithic to Composable Architecture," Practical Data Engineering, December 18, 2024, https://www.pracdata.io/p/the-evolution-of-business-intelligence-stack.

[5] Patrycja Zajac, "History of Business Intelligence," 10 Senses, accessed August 2, 2025, https://10senses.com/blog/history-of-business-intelligence/.

The shift began in earnest around 2010 with the emergence of cloud data warehouses and self-service BI tools. Companies like Tableau revolutionized the market by introducing intuitive drag-and-drop interfaces that enabled business users to create visualizations without extensive technical training. This democratization of analytics coincided with the growth of cloud computing, making powerful analytical capabilities accessible to organizations of all sizes.[6]

The current era, which began around 2020, is defined by the integration of artificial intelligence and machine learning into BI platforms. Tools like ThoughtSpot's Spotter and conversational analytics capabilities are transforming how users interact with data, enabling natural language queries and AI-driven insights discovery. However, as Colin Zima observed, this proliferation of AI-enhanced tools has also introduced new complexity that startups must navigate carefully.[7]

This new complexity arises not just from the sheer number of tools available but from the blurring lines between different types of analytical experiences. As we discussed in Chapter 3, the modern BI landscape is no longer a monolith of traditional dashboards. Instead, it has fractured into at least three distinct categories that startups must understand:

- **Dashboards** for core metrics monitoring

- **Augmented Analytics** platforms that promise to uncover hidden insights automatically

- **Embedded Analytics** that integrate data directly into the product experience

Navigating this landscape requires a startup to look beyond a single tool and instead develop a cohesive strategy that determines which of these analytical outputs—or which combination—will best serve their internal teams and their customers at each stage of growth.

---

[6] Mike Schiff, "That Was the Year That Was: Major Data Warehousing Events of 2010 (and Predictions for 2011)," TDWI, December 15, 2010, https://tdwi.org/articles/2010/12/15/dw-events-2010.aspx.

[7] Rodrigues, Nilima (2025). From Dashboards to Data Science: How AI is Transforming BI Platforms. INTERNATIONAL JOURNAL OF COMPUTER ENGINEERING & TECHNOLOGY. 16. 394–413. 10.34218/IJCET_16_03_026.

Why Startups Need Business Intelligence Platforms

The transition from spreadsheet-driven analysis to sophisticated BI platforms represents more than just a technological upgrade—it's a fundamental shift in how startups operationalize their data strategy. In Chapters 4-6, we established the critical importance of identifying and tracking the right metrics for startup success. BI platforms transform these metrics from static numbers in spreadsheets into dynamic, interactive insights that drive action across your organization.

Modern startups generate data at unprecedented scales. Every user interaction, transaction, marketing campaign, and operational process creates digital footprints that, when properly analyzed, reveal patterns crucial for growth. Traditional Excel-based analysis, while familiar and accessible, quickly becomes inadequate as data volumes grow and the need for real-time insights intensifies.

BI platforms address several specific challenges that startups face in their data journey. First, they enable data democratization—allowing non-technical team members to access and interpret business insights without requiring extensive SQL knowledge or data science expertise. This capability is particularly valuable for resource-constrained startups where wearing multiple hats is the norm rather than the exception.

Second, modern BI tools provide real-time or near-real-time insights that enable rapid decision-making. In the startup environment, the ability to detect trends, identify problems, or capitalize on opportunities as they happen can mean the difference between success and failure.

Third, BI platforms offer scalable visualization capabilities that grow with your business. What begins as simple charts tracking user acquisition can evolve into sophisticated dashboards monitoring multiple business units, customer segments, and operational metrics. The best BI tools accommodate this growth trajectory without requiring complete platform migrations.

The Pragmatic Choice Framework

Selecting the right BI platform requires balancing multiple competing priorities. Unlike enterprise organizations with dedicated IT teams and substantial budgets, early-stage startups must find tools that deliver immediate value while remaining flexible enough for rapid growth.[8]

---

[8] Adam Finer, "How to Choose a BI Tool For Your Business," Adam Finer - Learn BI Online YouTube Channel, October 2, 2024, https://www.youtube.com/watch?v=NS8NVrIeqo4.

Rather than getting lost in feature comparisons or vendor marketing, approach your selection with a structured framework that accounts for your unique constraints. This framework begins not with features but with fundamental questions about your audience, resources, budget, and core requirements.

1. Start with Your Audience and Core Use Case

Before evaluating any tool, first define who you are building for and what you need analytics to accomplish.

- **Who is the end user?** Are you building dashboards for internal teams to track basic metrics? Do you need to provide secure portals for stakeholders and investors? Or is your goal to embed analytics directly into your product for customers?

- **What is the primary job to be done?** Do you simply need to run SQL queries against your data warehouse and create simple charts? Or do you require advanced features like the ability to use Python or R within the tool? Clearly differentiate between must-haves and nice-to-haves.[9]

2. Assess Your Technical Resources and Existing Stack

An honest assessment of your team's capabilities and current infrastructure will significantly narrow your options.

- **Can you support a self-hosted solution?** Open source tools like Metabase offer powerful features without licensing fees, but they require technical know-how to deploy, manage, and scale. If you lack the engineering resources, a managed solution may be a better fit, even if it comes at a higher price.

- **How easily does it integrate with your data stack?** Tools that connect seamlessly with your existing data warehouse can dramatically reduce setup time. For example, if you are running on AWS, Amazon QuickSight is a natural fit. If your data lives in Google

---

[9] "Create Power BI visuals using R," Microsoft Learn, February 28, 2025, https://learn.microsoft.com/en-us/power-bi/create-reports/desktop-r-visuals.

BigQuery, Looker Studio could be the quickest path to basic reporting. However, avoid choosing a tool solely for its convenience; ensure it will scale with your future needs.

- **Is it easy for everyone to use?** The best tool is one that gets adopted. Consider the ease of use for both technical and non-technical users to ensure it empowers your entire team.

3. Evaluate the True Cost

For most startups, cost is the most pressing concern. Look beyond the sticker price to understand the total cost of ownership.

- **What is the cost structure?** Does the pricing model remain sustainable as you scale your team and data volume? Try to avoid tools with pricing that becomes prohibitive as you grow.

- **What is the total cost of ownership?** Open source isn't truly free. You must account for the cost of hosting and, more importantly, the engineering time required for setup and maintenance. Sometimes, a low-cost managed solution has a lower total cost.

4. Define Your Feature and Scalability Requirements

Once you have clarity on the above, you can confidently assess features.

- **Does it meet your visualization needs?** Ensure the tool supports the specific chart types, formatting, and interactive features your use cases demand.

- **Do you need embedded analytics?** If you plan to integrate BI into a customer-facing application, this is a must-have. Evaluate the quality of the embedded features—are they intuitive and powerful enough for your end users?[10]

---

[10] Ivan Seow, "Embedded analytics: A step-by-step guide to unlocking your data's potential," ThoughtSpot, November 27, 2024, `https://www.thoughtspot.com/data-trends/embedded-analytics`.

- **Do you really need AI-driven features?** Many BI tools now heavily advertise their augmented analytics capabilities. For an early-stage startup without massive datasets, these "shiny features" may not provide immediate value. Ask what you can concretely gain from them before paying a premium.

Remember, you can always change tools as your requirements evolve. It is often better to start with a simple, low-cost tool that meets your immediate needs than to invest heavily in a complex solution you are not ready for.

Common Traps and Practical Trade-Offs

Once you have a shortlist of tools, the final decision often comes down to navigating real-world pressures and practical compromises. Here are the most common challenges to anticipate.

The Trap of Social Proof

Avoid the mistake of selecting a tool based purely on social proof. Just because a competitor uses Tableau or an investor insists that Mode is the best solution doesn't mean either is right for your startup's unique needs and constraints.

For instance, a startup may be urged to adopt a tool like Mode as its first BI tool based on an investor's recommendation. A closer look, however, might reveal that Mode is not a classic BI tool, but a data workspace platform—a distinction we will explore later in this chapter. Because it is a different category of tool, an objective evaluation may find it fails to meet the startup's core business intelligence requirements.

Always conduct your own thorough evaluation—including a proof-of-concept with your own data—before making a long-term commitment.

The Trade-Offs of Open Source

While traditional or open source BI tools might better fit your requirements and budget, it's important to recognize their trade-offs. These tools often come with a more limited feature set and the potential for bugs.

My own experience with Metabase is a case in point. The platform is known to have a substantial backlog of unresolved issues; at the time of writing, there are thousands of open issues on its GitHub repository. However, most users will only ever encounter a small fraction of these, and if your use case is basic, you may not run into any bugs at all.[11]

Metabase's visualization capabilities also have limitations compared to premium BI solutions, but for many teams, the substantial cost savings make these compromises worthwhile. For startups with straightforward needs, choosing Metabase can be a deeply satisfying decision despite its imperfections.

The Importance of Managing Expectations

This is why clear communication is so critical. Whether you select an open source solution or a premium platform, you must manage your team's expectations.

- If you choose an open source tool like Metabase, let your team know that occasional quirks are part of the package in exchange for the savings.

- If you opt for a premium platform, ensure everyone understands the ongoing investment required to maximize its value.

Being upfront about these trade-offs prevents disappointment and keeps the entire team aligned on your BI strategy and the role your chosen tools will play.

A Day in the Life: Choosing the Right BI Tool

To make the framework tangible, let's walk through a common scenario. Imagine you are in charge of data at *Coffee Mugs*, a small, early-stage ecommerce startup. Your business operates in the creator economy: you partner with social media influencers who promote your custom-made mugs and drive traffic to your website. Your technical infrastructure is lean—you use Google Analytics for web data, Stripe for payments, and all your information is centralized in a BigQuery data warehouse.

---

[11] "Metabase Issues," GitHub, accessed August 3, 2025, `https://github.com/metabase/metabase/issues`.

You have recently joined the startup and you need to choose your first BI tool for reporting. An initial discussion with your team brings the classic challenges to the surface:

- **The Voice of Social Proof:** A co-founder suggests Power BI is "probably the best tool" because he heard about it from other founders. After a brief discussion, you realize the startup he's referring to is much larger and already operates heavily within the Microsoft ecosystem.

- **The Influence of Corporate Experience:** A colleague who recently joined from a large corporation insists that Looker is "just great." However, her experience comes from the big-company world, where teams can afford expensive and complex enterprise tools.

- **The Long-Term Vision:** Another team member suggests focusing on an embedded analytics solution to provide stats to your influencer partners. The team agrees this is a valuable idea for the future, but not a priority to invest in this year.

- **The Trend Chaser:** The CTO, fresh from a conference, pushes for an AI-driven approach, suggesting you invest in a cutting-edge augmented analytics solution. He even shares a popular article from Andreessen Horowitz on the "unified data architecture" that names several of these tools.[12]

The conversation becomes a whirlwind of brand names and buzzwords. To cut through the noise, the CEO offers a practical constraint: "Let's find an open-source tool if possible to minimize costs. We can afford to pay for a tool after our next funding round."

This directive brings clarity. You realize that to move forward, you need to conduct a structured comparison based on your startup's actual requirements. You decide to apply the pragmatic choice framework to organize your evaluation, starting with the answers you already have.

---

[12] Matt Bornstein, Jennifer Li, and Martin Casado, "Emerging Architectures for Modern Data Infrastructure," Andreessen Horowitz, accessed August 3, 2025, https://a16z.com/emerging-architectures-for-modern-data-infrastructure/.

1. **Start with Your Audience and Core Use Case**

   You already know the main goal is to assemble basic dashboards for internal teams to track key metrics. Nothing more is needed for now. You don't need advanced features or integration with R or Python. While those are nice to have for you as a data professional, they aren't necessary to meet stakeholders' current needs. The job is simply to connect to your data warehouse and enable automated reporting of basic stats.

2. **Assess Your Technical Resources and Existing Stack**

   It's clear you work with Google Cloud, with BigQuery as the single source of truth. The question "Can you support a self-hosted solution?" is relevant—yes, if necessary, but you would prefer not to. As a focused data professional with numerous responsibilities, you want to avoid unnecessary operational work. The most critical question is "How easily does it integrate with your data stack?" Looker and Looker Studio come to mind as natural candidates, but you remain open to other options. The final consideration "Is it easy for everyone to use?" is a key part of the exploration to follow.

3. **Evaluate the True Cost**

   You know your most likely winner will be either a low-cost tool or a self-hosted open source solution. However, you will explore the pricing of all options to understand the trade-offs. This will also equip you to provide your colleagues with objective arguments for why you ultimately choose tool X over the ones they proposed.

4. **Define Your Feature and Scalability Requirements**

   You've already established that embedded analytics isn't necessary yet, but it's a definite nice-to-have for the future. Therefore, it's worth exploring tools that have this feature for future planning. Similarly, when asking "Do you really need AI-driven features?", the answer is no, it's not a requirement. But to stay current, you need to grasp what these cutting-edge tools offer and what they cost.

With these framework questions answered, you can now analyze the market in a structured way. You identify four primary categories of tools from the team's discussion to explore:

1. The Open Source Champions

2. The Augmented Analytics Specialists

3. The Low-Cost, Easy-Entry Players

4. The All-in-One Enterprise Platforms

With our pragmatic choice framework in place, let's now dive deeper into the four main categories of tools you identified, exploring the specific pros and cons of each for a startup like *Coffee Mugs*.

1. The Open Source Champions

For startups with engineering talent and a desire to control their own data stack while minimizing costs, open source BI tools are a powerful starting point. These platforms offer flexibility and a strong community but require more hands-on effort to deploy, manage, and scale.

– Here are some tools you may have heard of: **Metabase:** Renowned for its simplicity and ease of use, Metabase is a popular choice for enabling self-service analytics among non-technical users. While the company showcases numerous startup success stories on its website, a deeper look reveals that the free, open source edition has a more limited feature set compared to its paid, managed versions.[13,14]

– **Apache Superset:** For those needing more advanced and highly customizable visualizations, Superset is a strong contender. However, it is often described as a tool "built by engineers, for engineers,"

---

[13] "How companies use Metabase," Metabase, accessed August 2, 2025, https://www.metabase.com/case-studies.

[14] "Metabase Pricing," Metabase, accessed August 2, 2025, https://www.metabase.com/pricing/.

which can make it challenging to set up and less approachable for non-technical team members.[15,16]

–   **Redash:** Similar to its peers, Redash allows you to connect to your data sources and visualize key metrics. Like other open source tools, setting up a Redash instance requires technical effort, as your team will be responsible for maintaining the infrastructure and performing regular upgrades.[17,18]

–   **Lightdash:** This tool integrates directly with dbt (Data Build Tool), allowing you to define and manage metrics alongside your data transformation code. This ensures consistency from the data warehouse all the way to the dashboard. Lightdash can be self-hosted or used via a paid plan that includes additional features.[19,20]

–   **Grafana:** While your engineering team might mention Grafana as a BI tool due to its varied visualizations, it's crucial to understand that Grafana is primarily designed for real-time monitoring and observability. It's favored by DevOps and engineering teams for tracking application and infrastructure performance rather than for traditional BI analytics.[21,22]

[15] "Apache Superset," Apache Superset Website, accessed August 2, 2025, `https://superset.apache.org/`.

[16] Chris Nguyen, "Tool Evaluation Series: Superset," Medium, January 6, 2024, `https://datacorner.medium.com/tool-evaluation-series-superset-e7774f64898e`.

[17] "Stasher Redash Case Study," Redash, accessed August 3, 2025, `https://redash.io/case-studies/stasher/`.

[18] "Setting up a Redash Instance," Redash, accessed August 3, 2025, `https://redash.io/help/open-source/setup/`.

[19] Abdelmounim Boufous, "How Stake built a successful self-serve data culture using Lightdash," Lightdash, accessed August 3, 2025, `https://www.lightdash.com/customer-story/stake`.

[20] "Pricing & Plans," Lightdash, accessed August 3, 2025, `https://www.lightdash.com/pricing`.

[21] "Grafana OSS," Grafana Website, accessed August 3, 2025, `https://grafana.com/oss/grafana/`.

[22] "Use Grafana as a BI Tool," Grafana Community Forum, accessed August 3, 2025, `https://community.grafana.com/t/use-grafana-as-a-bi-tool/76284`.

- **Cube:** As your research deepens, you'll discover tools like Cube that also aren't traditional BI platforms. Instead, Cube positions itself as a "headless BI" solution—an open source semantic layer that provides consistent data models across your infrastructure. It's designed to work in conjunction with visualization tools like Metabase or Superset and is used by major companies like SpaceX and Walmart. Like other open source projects, it can be self-hosted for free or accessed through paid tiers with more features. Its D3 AI-driven analytics platform enables users to query data through natural language.[23,24,25]

2. The Augmented Analytics Specialists

As you continue your research, you drift farther from traditional BI and discover a category of tools for startups wanting to leverage AI for automated insights. These platforms are designed to sift through massive datasets to find patterns, anomalies, and key business drivers that a human analyst might otherwise miss.

Unlike dashboarding tools, solutions like Anodot, Outlier, and Sisu connect to your data warehouse and proactively monitor your metrics. Your investigation into these tools, however, reveals a critical lesson about the volatile world of AI startups:

- First, you explore **Outlier**, a tool you saw referenced in the popular a16z "unified data architecture" diagram. What you find is shocking: Outlier has been shut down and no longer exists. It's a stark reminder of how short the lifespan of an AI tool can be, even with industry recognition.[26,27]

---

[23] Adnan Rahic, "Self-Service Analytics with Metabase and Cube," Cube Blog, October 27, 2022, https://cube.dev/blog/self-service-analytics-with-metabase-and-cube.

[24] "Cube.js Repository," GitHub, accessed August 3, 2025, https://github.com/zhjuncai/cube.js/.

[25] "Cube," Cube Website, accessed August 3, 2025, https://cube.dev/.

[26] Matt Bornstein, Jennifer Li, and Martin Casado, "Emerging Architectures for Modern Data Infrastructure," Andreessen Horowitz, accessed August 3, 2025, https://a16z.com/emerging-architectures-for-modern-data-infrastructure/.

[27] "Outlier AI," LinkedIn, accessed August 3, 2025, https://www.linkedin.com/company/outlier-ai/?originalSubdomain=br.

Next, you look into Sisu, another solution from the same diagram that focuses on automated diagnostics. Here, you find another surprise: Sisu was acquired by Snowflake, which announced plans to discontinue the stand-alone product and integrate its technology into the broader Snowflake ecosystem.[28,29]

Lesson Learned: The Risk of Stand-Alone AI Tools

For a startup without a large data science team, these specialized tools can act as a powerful force multiplier when used alongside traditional BI solutions. However, the examples of Outlier and Sisu reveal a significant risk: the augmented analytics market is highly volatile.

AI startups in this space face a constant **displacement risk**—the fear that their core technology could be rendered obsolete overnight by a more powerful open source model or a new feature from a major tech giant. (We will explore this concept further in Chapter 12.)

This reality should lead you to ask a critical question: is it safer to choose a more complex, all-in-one platform that already has these AI features built in rather than relying on a stand-alone tool that could disappear?

A Note on AI for Dashboard Creation

It's important to separate the risk of specialized AI tools from the value of AI in the analytics process. While stand-alone platforms can be precarious, using AI to automate the creation of dashboards within established BI tools can be incredibly useful.[30]

---

[28] "Sisu Data LinkedIn Post," LinkedIn, accessed August 3, 2025, `https://www.linkedin.com/posts/sisu-data_weve-spent-the-last-five-years-working-on-activity-7119690491054485505-v2vL/`.

[29] Ben Horowitz, "Sisu," Andreessen Horowitz, November 12, 2023, `https://a16z.com/sisu/`.

[30] "AI+BI Analytics," Strategy Software, accessed August 3, 2025, `https://www.strategysoftware.com/pl/strategyone/ai-analytics`.

For readers interested in a deep dive, I recommend books like *Powered-AI Business Intelligence: From Concepts to Implementation with Power BI and Copilot: The Complete Guide to Transforming Your Enterprise with AI-Optimized Analytics*. They provide comprehensive overviews, case studies, and step-by-step implementation advice for leveraging AI-optimized analytics in a more stable ecosystem.[31]

3. The Low-Cost, Easy-Entry Players

For many startups, the simplest and most cost-effective path forward is the BI tool offered by their primary cloud provider. These tools are designed for seamless integration and often include generous free or low-cost tiers that are perfect for early-stage needs. Let's explore the most popular options within the Google, Amazon, and Microsoft ecosystems.

- **Looker Studio** (Google): In our hypothetical scenario, Looker Studio is the most natural place to start. Formerly known as Google Data Studio, this completely free tool is incredibly easy to use and integrates seamlessly with the Google Cloud Platform (GCP) services your fictional startup already uses, like BigQuery and Google Analytics. While it lacks the advanced functionality of more powerful platforms, it is often the perfect choice for creating and sharing simple dashboards with minimal effort. With your existing setup, you are just a few clicks away from your first dashboard, with no additional deployment or maintenance required.[32,33]

- **Amazon QuickSight** (AWS): For startups running on AWS, QuickSight is the native BI service. It offers a pay-per-user pricing model that is highly attractive for startups, as costs scale with team growth. It's a full-featured tool that supports dashboards, augmented analytics with Amazon Q, and embedded analytics, though the advanced AI features come with a higher price tag. QuickSight has

---

[31] Steven Favour Bright, Cherydan Lucas, et al., *Powered-AI Business Intelligence: From Concepts to Implementation with Power BI and Copilot: The Complete Guide to Transforming Your Enterprise with AI-Optimized Analytics* (Independently published, 2025).

[32] "Looker Studio," Google, accessed August 4, 2025, `https://lookerstudio.google.com/`.

[33] "Looker Studio," Google Cloud, accessed August 4, 2025, `https://cloud.google.com/looker-studio`.

made significant progress over the years. It used to be quite simple at the very beginning. Today, it is a robust platform that continues to add new features and attract a growing user base.[34,35]

- **Power BI** (Microsoft): As a leader in the BI market, Power BI offers a powerful suite of tools that are deeply integrated into the Microsoft ecosystem (Azure, Office 365). Its desktop version is free, and its paid tiers are competitively priced, making it an accessible yet highly scalable option.[36]

After exploring other tools like Zoho Analytics and IBM Cognos, you conclude that none are a better fit for your immediate needs than the options you have already explored. The path of least resistance—starting with a low-cost, easy-entry player—is the clear winner.[37,38]

4. The All-in-One Enterprise Platforms

This final category includes the most powerful and feature-rich BI platforms on the market. With a correspondingly high price tag, they are best suited for later-stage startups or large corporations with dedicated data teams and complex requirements. These tools offer comprehensive solutions covering everything from data modeling and governance to advanced dashboards, augmented insights, and embedded analytics.

- **Looker:** You learn that companies choose Looker over Looker Studio when they need advanced data modeling, deep analytical capabilities, and enterprise-grade governance. It excels at handling large, intricate datasets where robust security and scalability are

---

[34] "Amazon QuickSight," Amazon Web Services, accessed August 4, 2025, `https://aws.amazon.com/pm/quicksight/`.

[35] "Amazon QuickSight Pricing," Amazon Web Services, accessed August 4, 2025, `https://aws.amazon.com/quicksight/pricing/`.

[36] "Power BI Pricing," Microsoft, accessed August 4, 2025, `https://www.microsoft.com/en-us/power-platform/products/power-bi/pricing`.

[37] "Zoho Analytics Pricing," Zoho, accessed August 4, 2025, `https://www.zoho.com/analytics/pricing.html`.

[38] "IBM Cognos Analytics Pricing," IBM, accessed August 4, 2025, `https://www.ibm.com/products/cognos-analytics#choose-the-right-plan`.

paramount. You can integrate Looker with Gemini to allow users to ask your data the same way they would ask an analyst.[39]

— **Other Enterprise Platforms:** A quick review of other leaders like Tableau, **ThoughtSpot**, **Qlik Sense**, **Domo**, and **Sisense** confirms your suspicion: you simply don't need all the "shiny features" that come at such a high cost.

Lesson Learned: Start Simple, Stay Focused

Your research brings you to a clear conclusion. Your startup's current requirements are basic, and your budget is limited. With this clarity, you take a few minutes to build your first dashboard in Looker Studio and present it to your colleagues, outlining the pros and cons of each approach you explored.

This structured comparison also serves another crucial purpose: managing stakeholder expectations. By walking your co-founders and investors through the evaluation, you demonstrate that you've respected everyone's input and made a data-informed decision, not just a personal one. It provides an objective basis for your choice and aligns the entire team.

By applying the simple pragmatic choice framework, you successfully navigated the paradox of choice. The final decision became a logical outcome of the process:

1. **Use Case:** The need for simple internal dashboards immediately clarified that powerful, complex enterprise platforms were overkill.

2. **Stack and Resources:** Your reliance on BigQuery and limited engineering time made a native, cloud-based tool like Looker Studio far more practical than a self-hosted open source option.

3. **Cost:** The low-budget constraint was the primary filter, ruling out any tool with bigger recurring subscription fees.

---

[39] Vijay Venugopal, and Kate Grinevskaja, "AI and BI converge: A deep dive into Gemini in Looker," Google Cloud Blog, April 16, 2025, `https://cloud.google.com/blog/products/data-analytics/gemini-in-looker-deep-dive`.

4. **Features:** Future needs like embedded analytics and augmented AI were considered but not prioritized, allowing you to solve today's problem without over-investing in tomorrow's technology.

The framework didn't just lead you to a tool. It gave you a clear, defensible story to tell your team, turning a potentially contentious decision into a moment of strategic alignment.

The result of this analysis is summarized in Figure 9-1. It may seem counterintuitive that Looker Studio, the winning tool, has just one checkmark. This outcome proves a critical lesson for startups: the "best" platform is not the one with the most features but the one that solves your immediate problem effectively.

It is important to note that this comparison is intentionally adjusted to your specific requirements. For example, while Looker Studio has an embedded analytics feature, it was not checked off here because it lacks the specific row-level security and visualization capabilities required for this use case. The final chart reflects a pragmatic assessment against the startup's unique needs, not an objective list of every feature a tool offers in general.

	Classic Dashboards	Embedded Analytics	Augmented Analytics	Open Source	
Metabase	✓	✓	✗	✓	
Superset	✓	✓	✗	✓	
Redash	✓	✓	✗	✓	
Grafana	✓	✓	✗	✓	
Lightdash	✓	✓	✗	✓	
Cube	✗	✗	✓	✓	
Anodot	✗	✗	✓	✗	
Outlier	✗	✗	✓	✗	
Sisu	✗	✗	✓	✗	
Looker Studio	✓	✗	✗	✗	⭐ free of charge easiest setup
QuickSight	✓	✓	✓	✗	
Power BI	✓	✓	✓	✗	
Looker	✓	✓	✓	✗	
All-in-One: Others	✓	✓	✓	✗	

Figure 9-1. *BI Tool Capability Comparison for the Coffee Mugs Startup*

Data Workspace Tools

While *Coffee Mugs'* needs were met with a simple dashboarding tool, the Business Intelligence tools landscape is much broader than what we've explored. So far, we have elaborated on the general data analysis and output platforms that are popular choices for dashboards, embedded analytics, and augmented analytics. This next category of tools is not for every startup. Data workspace tools are best suited for companies that have at least one data-savvy team member (but typically more) who need to go beyond pre-built dashboards and perform deeper, more collaborative exploratory analysis.

If a BI dashboard is the polished annual report shared with the board, a Data Workspace is the collaborative workshop where the analysts build it—complete with their raw data, calculations, and internal notes.

These cloud-native workbenches typically fuse the SQL editor, notebook kernel, version control, and lightweight app builder into a single browser tab, letting an analyst query the warehouse, refine results in Python or R, and publish an interactive dashboard without changing context. Because the code, commentary, and visualizations share one real-time document, multiple users can edit, comment, and iterate together, collapsing the time from first question to shareable insight.[40]

Mode, Hex, and Deepnote illustrate the spectrum:

- **Mode** begins with a SQL editor, pipes results into Python/R notebooks, and schedules or embeds finished dashboards—designed for data teams that still live heavily in queries but need polished reports. However, it's much different from typical "BI tools for dashboarding" as it offers a lot of advanced tools for collaborative data team work.[41]

- **Hex** openly states on their website that they offer much deeper analytical capabilities than just dashboards. They position themselves as the "AI-powered analytics workspace built for teams driving faster answers and better decisions."[42]

---

[40] Matt Bornstein, Jennifer Li, and Martin Casado, "Emerging Architectures for Modern Data Infrastructure," Andreessen Horowitz, accessed August 3, 2025, https://a16z.com/emerging-architectures-for-modern-data-infrastructure/.

[41] "Mode," Mode, accessed August 4, 2025, https://mode.com/.

[42] "Hex," Hex, accessed August 4, 2025, https://hex.tech/.

– Similar to Hex, **Deepnote** positions itself as *"the AI-powered data workspace."* It's a collaborative data science and analytics platform that combines notebooks and AI-driven features—enabling teams to connect to data; analyze with Python, SQL, or R; visualize results; and share insights all in one place.[43]

For startups, such workspaces compress the analytics cycle: they sit directly on the cloud warehouse, avoiding data egress and duplicate governance, and turn notebooks into shareable assets that embed in internal wikis or even customer-facing portals. The result is faster decisions, lower tooling overhead, and a flexible foundation that scales from exploratory analysis to lightweight data apps. However, these tools are typically designed for startups with more advanced tasks and bigger teams than small, early-stage startups like our fictional use case examined in this chapter.

Specialized BI Tools for Startups

The modern startup landscape increasingly demands sophisticated financial intelligence that goes beyond traditional reporting. Founders operating in venture-backed environments may need analytics platforms that mirror the perspective of their investors and provide the metrics-driven insights crucial for fundraising success and operational optimization.

The Rise of VC-Centric Analytics Tools

Another interesting type of tool I have encountered when working with startups and investors is specific BI tools used by VCs. In such cases, VCs might ask you to integrate your data sources (e.g., your Stripe account) or send them data extracts from your database (like specific user activity data). Some VCs might even encourage you to use their tools for your own internal analytics.

Headline's Deepdive represents such a kind of specialized business intelligence platforms designed specifically for startup founders. Unlike traditional BI tools that focus on operational dashboards and general business metrics, Deepdive provides venture

---

[43] "Deepnote," Deepnote, accessed August 4, 2025, https://deepnote.com/.

capitalist-grade financial analysis that helps founders understand their business through the lens of potential investors. The platform automatically processes transaction data to deliver cohort analysis, unit economics calculations, and capital efficiency metrics.[44]

What distinguishes Deepdive from conventional analytics is its focus on fundraising-critical metrics such as payback ratios, customer retention dynamics, and capital efficiency benchmarks. The platform transforms complex datasets into investor-ready visualizations and provides benchmarking against Headline's proprietary database of venture-backed companies. This approach enables founders to identify potential red flags before entering due diligence processes and better communicate their business performance in the language investors understand.[45]

The platform addresses a common disconnect between internal business dashboards and investor expectations. While companies typically track operational metrics like MRR growth and customer acquisition costs, investors evaluate businesses using different frameworks that emphasize sustainable unit economics and capital deployment efficiency. Deepdive bridges this gap by providing both perspectives within a single platform, allowing founders to prepare for investor conversations more effectively.[46,47]

On the other hand, Andreessen Horowitz's Guide to Growth Metrics is a simple tool available on their website that provides interactive benchmarking for B2B companies based on a16z's proprietary dataset of successfully scaled companies. The tool segments benchmarks by revenue scale, software type, and go-to-market motion, allowing founders to compare their performance against relevant peer groups. The platform emphasizes metrics that correlate with long-term success rather than vanity metrics that can mislead decision-making.[48]

[44] "Deepdive," Headline, accessed August 4, 2025, https://deepdive.headline.com/.

[45] "Why We Quit Excel for Analysis," Headline YouTube Channel, April 4, 2024, https://www.youtube.com/watch?v=OvOzDbhgjKc.

[46] "Deepdive Learn," Headline, accessed August 4, 2025, https://deepdive.headline.com/learn.

[47] Haje Jan Kamps, "How Headline is using AI to make better investment decisions," TechCrunch, November 20, 2023, https://techcrunch.com/2023/11/20/headline-deep-dive/.

[48] "Guide to Growth Metrics," Andreessen Horowitz, accessed August 4, 2025, https://a16z.com/growth/guide-growth-metrics/.

Traditional SaaS Analytics Platforms: The Foundation Layer

Another category of specialized analytics platforms has emerged over the years that help subscription startups analyze their business. Platforms like ChartMogul and Baremetrics each serve different segments of the market with varying degrees of analytical sophistication.

ChartMogul offers comprehensive cohort analysis, customer segmentation, and extensive customization capabilities. The platform excels in providing detailed subscription metrics analysis and supports complex revenue recognition scenarios. ChartMogul publishes reports that may be interesting to a lot of subscription businesses as they allow them to benchmark their performance with other similar businesses.[49]

Baremetrics uses data from over 800 startups partnering with them to provide live benchmarks to help startups benchmark their performance. The platform gives insights into cancellation reasons, and it helps analyze what pricing models work best. Baremetrics is particularly well suited for early-stage companies that need immediate insights without extensive setup complexity.[50]

We have discussed several kinds of BI tools, each serving a distinct purpose. While platforms like ChartMogul or Deepdive often provide pre-packaged insights, the general-purpose BI tools we've covered give you a blank canvas. This freedom is powerful, but it's also where the real challenge begins. Turning raw data into meaningful metrics is one thing; presenting those metrics in a way that drives action is another entirely.

This brings us to the art and science of building classic dashboards—reports that are so clear and relevant they become essential to your team's workflow.

Creating Actionable Reports and Dashboards

I've seen the dashboard graveyard too many times: a collection of reports launched with enthusiasm only to be abandoned weeks later. This phenomenon of "dashboard abandonment" is a persistent challenge, especially in startups where rapid pivots and changing priorities can quickly make yesterday's critical metrics irrelevant.

---

[49] "ChartMogul Insights," ChartMogul, accessed August 4, 2025, `https://chartmogul.com/insights/`.

[50] "Baremetrics Open Benchmarks," Baremetrics, accessed August 4, 2025, `https://baremetrics.com/open-benchmarks`.

When to Build a Dashboard

Building a new dashboard can be a tricky task, often revealing deeper issues with your data before you even begin. You might discover that critical events aren't being tracked at all or that the data from a recent release is a mess because testing was poorly executed (startup reality!). This is why creating a dashboard is a commitment that goes far beyond design. Every report you create requires initial development time, ongoing maintenance, and user support. Before you begin, ask these fundamental questions to ensure the investment will generate lasting value.

1. **Start with Clear Business Questions:** The most successful dashboards emerge from specific, actionable business questions rather than generic desires to "monitor performance." Instead of building a dashboard because you think you should track certain metrics, identify specific decisions the dashboard will inform. For example, rather than creating a general "customer metrics" dashboard, focus on answering specific questions like "Which customer acquisition channels are most cost-effective?" or "What product features correlate with higher retention rates?"

2. **Validate Ongoing Need:** Before investing in dashboard development, confirm that the underlying business question will remain relevant for the foreseeable future. Startups often shift strategies, pivot business models, or change focus areas— dashboards tied to deprecated strategies quickly become digital debris. Engage with intended users to understand not just their current information needs but how those needs might evolve over the next 6-12 months.

3. **Assess User Capacity:** Consider whether intended users have the bandwidth and motivation to regularly engage with new dashboards. Teams already overwhelmed with existing reporting may not adopt additional analytics tools, regardless of their quality or relevance. Sometimes, enhancing existing dashboards or communication processes proves more valuable than creating new ones.

The Pre-build Checklist

Once you've confirmed a dashboard is necessary, define the requirements with this pre-build checklist. Getting clear answers upfront prevents confusion and ensures the final product is useful.

- **Objectives:** What is the single, primary purpose of this dashboard?

- **Metric Definitions:** Is the exact calculation for every metric clearly defined and agreed upon?

- **Data Sources:** Where will the data come from? Is it reliable?

- **Refresh Rate:** How often does the data truly need to be updated (daily, weekly, monthly)? Be sure to differentiate between "must-haves" and "nice-to-haves."

- **Access and Security:** Who needs to see this dashboard, and what are the security requirements?

Finally, design for your audience. Make the layout intuitive and present only the necessary data. If there's any room for misinterpretation, add clear text notes, run a short training session, or record a quick walkthrough video. The ultimate goal is to enable true self-service analytics, where the dashboard empowers users to find answers on their own.

Essential Dashboard Categories for Startups

Drawing from the metrics framework established in Chapters 4-6, startups typically require dashboards across several critical functional areas. Each category serves distinct audiences and decision-making processes, requiring tailored approaches to design and implementation.

Consider our hypothetical startup: *Coffee Mugs*. Effective dashboard organization requires balancing comprehensiveness with usability, ensuring that users can quickly find relevant information without being overwhelmed by unnecessary details.

1. **Executive Dashboards** provide high-level overviews of business health and progress toward strategic objectives. In our case, we could get started with the north star metric and supporting metrics discussed in Chapter 4. Executive dashboards may update less frequently—often weekly or monthly—but require

exceptional clarity and context to support strategic decision-making. They should answer questions like "Are we on track to meet our quarterly goals?" and "What are the most significant opportunities or risks facing the business?"

2. **Operational Dashboards** monitor day-to-day business processes and enable rapid response to emerging issues or opportunities. These dashboards update more frequently—sometimes in real time—and serve teams responsible for ongoing operations. Marketing teams might monitor campaign performance, customer acquisition costs, and lead generation metrics. Product teams might track user engagement, feature adoption, and performance metrics. Sales teams might monitor pipeline health, conversion rates, and customer interactions. All these teams could have their own OKR dashboards to track day-to-day progress on objectives and key results.

3. **Financial Dashboards** track revenue, expenses, cash flow, and other financial metrics critical for startup survival and growth. These dashboards often serve dual purposes: internal management and external stakeholder communication. Investors, board members, and leadership teams all require clear visibility into financial performance, making these dashboards some of the most scrutinized in startup environments.

4. **Customer Analytics Dashboards** provide insights into user behavior, satisfaction, and life cycle progression. For startups focused on product-market fit and growth, understanding customer patterns can inform everything from product development to marketing strategy. These dashboards might track user acquisition, activation, retention, and revenue metrics across different customer segments or cohorts. In the case of *Coffee Mugs*, we could build a Consumer Funnel Dashboard to understand which creators or which marketing campaigns are most effective in acquiring new customers.[51]

---

[51] Ray Yu, Derek Rodenhausen, Yotam Ariav, Trevor Sponseller, and Clémentine Remy, "It's Time for Marketers to Move Beyond the Linear Funnel," Boston Consulting Group, January 17, 2025, https://www.bcg.com/publications/2025/move-beyond-the-linear-funnel.

BI vs. Product Analytics

The distinction between business intelligence and product analytics may confuse startup teams, particularly as both categories involve dashboards, data visualization, and user behavior tracking. While there is indeed overlap in functionality, these two domains serve fundamentally different purposes and typically require different toolsets and approaches.

Fundamental Differences in Scope and Purpose

Business intelligence encompasses the broader organizational perspective, analyzing data across multiple business functions to inform strategic and operational decisions. BI platforms typically aggregate data from various sources—sales systems, marketing platforms, financial applications, customer support tools—to provide comprehensive views of business performance. The primary goal is to understand overall business health, identify trends across departments, and support executive decision-making through historical analysis and reporting.[52]

Product analytics focuses specifically on how users interact with digital products, analyzing behavior patterns, feature usage, and user journeys to optimize product design and user experience. Product analytics tools typically work with event-stream data generated by user interactions within applications, websites, or digital platforms. The primary goal is to understand user behavior, improve product features, and optimize user engagement and retention. Product analytics tools like Mixpanel may both be used as a tool to visualize data and as one of the data sources that could be combined with other data sources (e.g., financial data) in the data warehouse and then visualized in a BI tool.

Data Sources and Architecture

The architectural differences between BI and product analytics reflect their distinct purposes and use cases. BI systems typically operate on structured data stored in data warehouses, drawing from multiple business systems and requiring ETL or ELT

---

[52] Ali Baghshomali, "Business Intelligence vs Product Analytics," Metabase Community Post, February 10, 2025, https://www.metabase.com/community-posts/what-is-the-difference-between-product-analytics-and-business-intelligence.

processes to aggregate and clean data from diverse sources. As discussed in Chapter 3, BI implementations often leverage the data infrastructure components we explored, including data warehouses, transformation tools, and structured reporting processes.

Product analytics platforms, by contrast, work primarily with real-time event streams generated by user interactions. These systems track clicks, page views, feature usage, session duration, and other behavioral metrics as they occur. Product analytics tools like Amplitude, Mixpanel, or PostHog are optimized for analyzing time-series data and typically support cohort analysis, funnel analysis, and user segmentation based on behavioral patterns.

User Audiences and Use Cases

The target audiences for BI and product analytics tools differ significantly, reflecting their distinct purposes and data requirements. BI tools primarily serve executives, managers, analysts, and business stakeholders who need comprehensive views of organizational performance. These users typically ask questions like "What was our customer acquisition cost last quarter?" or "How do marketing channel performances compare across different regions?"

Product analytics tools serve product managers, UX designers, growth marketers, and product development teams who focus on optimizing user experiences and product features. These users ask questions like "Which features correlate with higher user retention?" or "Where in our onboarding flow do users typically drop off?"

When Startups Need Both

Many startups eventually require both BI and product analytics capabilities, but the timing and prioritization depend on business model, growth stage, and strategic focus. Early-stage startups with simple business models might satisfy most analytical needs with a single BI platform. However, as products become more sophisticated and user bases grow larger, dedicated product analytics tools often become necessary to understand user behavior at sufficient granularity.[53]

---

[53] Tim Flack, "I've helped build 5 startups: Here's how product analytics cuts costs and headaches," Mixpanel Blog, February 24, 2025, https://mixpanel.com/blog/ive-helped-build-5-startups-heres-how-i-use-product-analytics-to-cut-costs-and-headaches/.

- **SaaS startups** typically need both categories relatively early in their development. BI tools help track business metrics like monthly recurring revenue, customer acquisition costs, and churn rates, while product analytics tools provide insights into feature usage, user engagement, and product-market fit indicators.

- **Ecommerce startups** might prioritize BI tools initially to track sales performance, inventory management, and marketing effectiveness but eventually require product analytics to optimize website user experience, cart abandonment rates, and conversion funnel performance.

- **Mobile app startups** often begin with product analytics tools to understand user behavior and optimize app experience, later adding BI capabilities to track business performance, monetization metrics, and operational efficiency.

The key is recognizing that BI and product analytics complement rather than compete with each other. The most successful startups develop integrated approaches that leverage insights from both domains to inform comprehensive growth strategies. As we'll explore further in Chapter 10, product analytics deserves dedicated attention due to its specialized requirements and critical importance for product-led growth strategies.

Learning Resources

For readers eager to go deeper with the BI tools discussed in this chapter, it's crucial to leverage a broad spectrum of learning resources—and to stay current. Official documentation and hands-on tutorials from each platform remain the most reliable starting point. Tableau offers engaging content through Tableau Public, Microsoft Power BI gives comprehensive beginner-to-advanced guides at Microsoft Learn, and Metabase provides detailed resources for both self-hosted and cloud versions.

To make the best technology choices, regularly consult up-to-date industry analyses. These reports not only highlight current leaders and challengers in the BI space but also break down the strengths and weaknesses of each platform as they evolve:

- **Gartner: Magic Quadrant Report** and Analytics and Business Intelligence Platforms Reviews and Ratings[54]

- **Forrester Wave Report:** Business Intelligence Platforms[55]

Exploring documentation for several platforms—including but not limited to Metabase, Tableau, Power BI, and Looker—will help you compare usability and feature sets firsthand. Many modern vendors offer free trials, enabling you to experiment before making any commitment.

For those who prefer structured, in-depth learning, numerous books are available for every major BI tool. The titles listed below are just a few prominent examples; a wide range of other books exists for different tools and levels of expertise:

- **Looker Studio Users:** *Data Storytelling with Google Looker Studio: A hands-on guide to using Looker Studio for building compelling and effective dashboards* is a hands-on guide for creating engaging dashboards.[56]

- **Looker Users:** *Business Intelligence with Looker Cookbook: Create BI solutions and data applications to explore and share insights in real time* offers practical, real-world applications and tips for building effective BI in the Looker environment.[57]

- **Microsoft Ecosystem Users:** Consider titles such as *The Definitive Guide to DAX: Business Intelligence for Microsoft Power BI, SQL Server Analysis Services, and Excel.* This book dives deep into the DAX language and methodologies that power many advanced analytics solutions.[58]

[54] "Analytics and Business Intelligence Platforms Reviews," Gartner, accessed August 4, 2025, https://www.gartner.com/reviews/market/analytics-business-intelligence-platforms.

[55] "The Forrester Wave™: Business Intelligence Platforms, Q2 2025," Forrester Research, 2025, https://www.forrester.com/report/the-forrester-wave-tm-business-intelligence-platforms-q2-2025/RES182218.

[56] Sireesha Pulipati, *Data Storytelling with Google Looker Studio: A hands-on guide to using Looker Studio for building compelling and effective dashboards* (Packt Publishing, 2022).

[57] Khrystyna Grynko, *Business Intelligence with Looker Cookbook: Create BI solutions and data applications to explore and share insights in real time* (Packt Publishing, 2024).

[58] Marco Russo and Alberto Ferrari, *The Definitive Guide to DAX: Business Intelligence for Microsoft Power BI, SQL Server Analysis Services, and Excel*, Second Edition (Microsoft Press, 2019).

- **For Universal Principles of Dashboard Design:** *Information Dashboard Design: Displaying Data for At-a-Glance Monitoring* is a classic reference that's valuable regardless of which BI platform you choose.[59]

You'll find many other relevant books tailored to a variety of platforms and use cases. By making continual learning a habit and staying attuned to both new releases and authoritative market analyses, you'll ensure your BI journey is well informed, strategic, and adaptable to the ever-changing landscape.

Summary

My goal in this chapter was to cut through the noise of the crowded business intelligence market and give you a durable framework for making a confident decision for your startup—the pragmatic choice framework. We began by tackling the central paradox of modern BI: while the explosion of powerful, AI-driven tools offers incredible potential, it has also created overwhelming complexity and the risk of costly "BI debt."

Throughout our discussion, I guided you through a strategic process for navigating this landscape.

- **First, we demystified the market** by exploring the four key categories of BI tools relevant to our hypothetical early-stage startup use case: open source champions, low-cost cloud players, augmented analytics specialists, and all-in-one enterprise platforms.

- **Next, you were equipped with a decision-making playbook** focused on asking the right questions—prioritizing your startup's unique stage, resources, and integration needs over hype or secondhand recommendations.

- **Finally, we confronted the common traps**, from team biases to hidden costs, ensuring the platform you choose becomes a genuine asset for growth, not a source of frustration.

---

[59] Stephen Few, *Information Dashboard Design: Displaying Data for At-a-Glance Monitoring*, Second Edition (Analytics Press, 2013).

Ultimately, selecting a BI platform is not just a technical task but a core strategic decision. The right tool empowers your entire team to monitor performance, share insights, and act with the data-driven agility that modern success demands. A well-chosen BI platform isn't just a cost center. It's an engine for finding product-market fit, reducing churn, and increasing revenue. By using the principles from this chapter, you can choose a solution that will scale with you, turning your data into a powerful engine for growth.

With your BI platform established to provide a high-level view of your company's overall health—tracking everything from sales funnels to operational efficiency—our focus must now shift to the very heart of your startup: the product itself. How are customers actually using what you've built? Which features lead to delight and retention, and where are the points of friction? Answering these questions requires a different, more granular set of tools and a specialized approach.

In Chapter 10, we will transition from the company-wide lens of BI to the user-focused world of product analytics. I'll show you how to instrument your application to capture meaningful user actions, turning raw clicks and interactions into the actionable insights that will guide your product road map and fuel sustainable growth.

CHAPTER 10

Product Analytics and Event Tracking

How do you make something for a million people? I don't know where to start. But if you pick one person, study them, and take their journey, you can actually build something really personal. You can design something and keep iterating until they love it. Don't stop improving it until that person loves it, and you're not allowed to move to the second person until the first person loves it. Then you get the second person and keep iterating until they love it. And so on.

—Brian Chesky,[1]

Co-founder of Airbnb

Product analytics and event tracking are fundamentally about enabling you to understand these user journeys. How much time do users spend with your product? What steps do they take until they reach their "aha moment"? Where do they stop engaging? Where do they churn and never return? Can they even find that feature your team worked so hard to build?

Product analytics and event tracking are the tools that allow you to scale this "handcrafted" understanding. They are the microscope that lets you observe thousands of user journeys at once. As we established in Chapter 9, while BI platforms provide the wide-angle lens to view overall business health, product analytics offers the microscope

---

[1] Brian Chesky, "Airbnb cofounder Brian Chesky on how to design a product for a million people," Startup Archive YouTube Channel, February 1, 2024, `https://www.youtube.com/watch?v=2LtfuAEOBLE`.

© Piotr Sidoruk 2026
P. Sidoruk, *From Data to Dollars*, https://doi.org/10.1007/979-8-8688-1898-1_10

needed to examine the user journey in intricate detail. It moves our focus from company-wide metrics to the clicks, taps, and scrolls that define the user experience. This chapter is dedicated to mastering that microscope.

This chapter will go beyond the basics of event tracking. It will provide a startup-focused playbook for choosing the right tools, designing a resilient tracking plan, blending quantitative data with qualitative insights, and avoiding the common pitfalls that lead to data mistrust.

What you'll learn in this chapter:

- **Qualitative research methods** that reveal the "why" behind user behavior
  - Session replays and heatmaps for visualizing user struggles
  - Survey design principles that avoid misleading insights
  - User interview techniques using *The Mom Test* framework
  - When and how to use focus groups effectively
- **Product analytics tool selection** using a systematic approach
  - Overview of popular tools categorized by different criteria
  - Four-step pragmatic framework for choosing the right platform
  - Case study: *Empathic* therapy app's tool selection process
  - Startup programs and cost optimization strategies
- **Event tracking implementation** that delivers clean, actionable data
  - Building tracking plans that capture what you actually need
  - Naming conventions and data quality best practices
  - Common pitfalls and how to avoid them
  - Case study: 5% to 35% conversion improvement through proper tracking

- **Frameworks and templates** you can implement immediately

  - Event tracking plan templates with sample data

  - Tool evaluation decision matrices

  - Research workflow from quantitative insights to qualitative validation

By the end, you'll have both the analytical mindset and practical tools to turn user insights into product decisions that drive growth.

Beyond the Numbers: Integrating Qualitative Insights

We were trying to figure out how to provide an amazing experience for hosts in New York. We were in Mountain View (SF). And Paul Graham said something to me that I'll never forget. He said: 'Do things that don't scale.' And it's counterintuitive. What that meant for us was we were able to go door-to-door, meeting every user, trying to provide an amazing experience.

—Brian Chesky,[2,3]

Co-founder of Airbnb

Building on Paul Graham's advice, the next step is learning to pair quantitative analytics with qualitative research so the startup team understands not just what users do but why they do it. We will go through the following qualitative methods for product teams:

- **Session Replays and Heatmaps:** The bridge between quantitative and qualitative insights. When your funnel shows a drop-off, watch session replays to see users struggle in real time.

---

[2] "Do things that don't scale - Paul Graham's advice to Brian Chesky (CEO of Airbnb)," LinkedIn post, accessed August 12, 2025, `https://www.linkedin.com/posts/useintro_do-things-that-dont-scale-paul-grahams-activity-7203441747358621697-eINE/`.

[3] Paul Graham, "Do Things that Don't Scale," Paul Graham Essays, July 2013, `https://paulgraham.com/ds.html`.

- **Surveys:** May be very useful for startups, and getting started can be as simple as one-question, behavior-triggered in-app surveys. For example, if a user skips a key feature, immediately ask why to capture feedback while the context is fresh.

- **User Interviews:** The gold standard for deep empathy. This is how you "study one person" to understand their core motivations and frustrations.

- **Focus Groups:** Useful for gauging reactions to new concepts or designs in a group setting.

Understanding Data from Session Replays and Heatmaps

Product analytics funnels are excellent at telling you what is happening—for example, that 70% of users drop off between signing up and completing their user profile. But to understand why they are dropping off, you need to see the product through their eyes. This is where session replays and heatmaps act as the perfect bridge between quantitative and qualitative data. They allow you to watch users interact with your product, providing visual context to the numbers.

Session Replays: Watching the User's Journey

A session replay is a video-like recording of a real user's interactions with your app or website. You can see every click, scroll, mouse movement, and text input, allowing you to retrace their exact journey. When your analytics show a problem, session replays are your go-to tool for diagnosis. Instead of guessing why users are failing to convert, you can watch recordings of users who dropped off at that specific step to see them encounter the bug or confusing interface element in real time.[4,5]

---

[4] "Session Replay," PostHog, accessed August 13, 2025, https://posthog.com/session-replay.

[5] "What user behavior analysis tools for CRO, such as heatmaps and session recordings, offer the best insights?," Merge.rocks, accessed August 17, 2025, https://merge.rocks/resources/websites-playbook/what-user-behavior-analysis-tools-for-cro-such-as-heatmaps-and-session-recordings-offer-the-best-insights.

This capability is invaluable for

- **Bug Reproduction:** Developers can see the exact sequence of events that led to an error, drastically reducing the time spent trying to reproduce it.[6]

- **Understanding User Frustration:** Observe behaviors like "rage clicks"—repeatedly clicking on an element that isn't working—or erratic mouse movements that indicate confusion.[7]

- **Validating Product Decisions:** See firsthand whether users are discovering and engaging with a new feature as you intended.[8,9]

The main value of session replay tools lies in their ability to uncover deep, nuanced insights into customer behavior—an especially critical capability in the early days of a startup. When you're still refining your product design, every click, scroll, and hesitation tells a story about what resonates, what frustrates, and where opportunities for improvement lie. As your product and user base evolve rapidly, these replay recordings give you a front-row seat to real user experiences, allowing you to iterate more thoughtfully and confidently. In the "scrappy" phase—when user counts are still manageable—analyzing even a handful of replays can yield breakthroughs in usability, conversion, and feature adoption that would be impossible to detect through aggregate metrics alone. By harnessing these qualitative insights early, you set the foundation for a product that truly meets your customers' needs from day one.

---

[6] "Session Replay," Statsig, accessed August 13, 2025, `https://www.statsig.com/session-replay`.

[7] "How to read and analyze session replays?," Capturly, accessed August 13, 2025, `https://capturly.com/guides/how-to-read-and-analyze-session-replays/`.

[8] "OpenReplay," OpenReplay, accessed August 14, 2025, `https://openreplay.com/`.

[9] Jane Leung, "How to analyze session recordings," UXCam Blog, June 19, 2024, `https://uxcam.com/blog/how-to-analyze-session-recordings/`.

Heatmaps: Visualizing Aggregate Behavior

While session replays offer a deep dive into individual experiences, heatmaps provide a broad, aggregated view of user behavior across many sessions. They use a color scale (typically from cool to warm) to show which parts of a page receive the most attention. There are several common types of heatmaps, each answering a different question.[10]

Click Maps

Click maps visualize where users click with a mouse on desktop devices or tap the screen on mobile devices. Hotspots reveal the most popular elements, while clicks on non-interactive areas can signal a confusing design.[11]

Let's take a look at the click map below. Figure 10-1 shows where visitors click across a single long page that contains stacked content blocks with text and buttons. Click popularity is usually shown on a color scale ranging from red to blue, with the "hot" red areas indicating the most popular points where users click most frequently. The most intense activity appears on large buttons and on several smaller buttons, indicating where visitors most often choose to interact. Interestingly, lighter clusters appear in the lower part of the page, beneath the buttons, hinting at specific user behaviors.

This map can help you identify which buttons and content blocks truly drive engagement, as well as which visible elements are largely overlooked despite occupying valuable screen space.

---

[10] "The complete guide to heatmaps," Contentsquare, accessed August 14, 2025, `https://contentsquare.com/guides/heatmaps/`.

[11] "Heat Mapping: Visualize User Behavior Like Never Before," Statsig Perspectives, July 6, 2024, `https://www.statsig.com/perspectives/heat-mapping-visualize-user-behavior-like-never-before`.

Figure 10-1. *Click Map*

Move Maps

Move maps track where users move their mouse cursors. Since mouse movement is often correlated with eye tracking, these maps can indicate what users are looking at, even if they don't click.[12,13]

---

[12] "How to use mouse tracking move maps to improve UX and conversions," Contentsquare, accessed August 14, 2025, `https://contentsquare.com/guides/heatmaps/move-maps/`.

[13] Milisavljevic A, Abate F, Le Bras T, Gosselin B, Mancas M and Doré-Mazars K (2021) Similarities and Differences Between Eye and Mouse Dynamics During Web Pages Exploration. Front. Psychol. 12:554595. doi: 10.3389/fpsyg.2021.554595

Figure 10-2 illustrates how mouse movements concentrate and flow across the same page. As with the click map above, the data is shown on a red-to-blue color scale, with "hot" red areas marking the most popular points where cursors linger or pass through repeatedly.

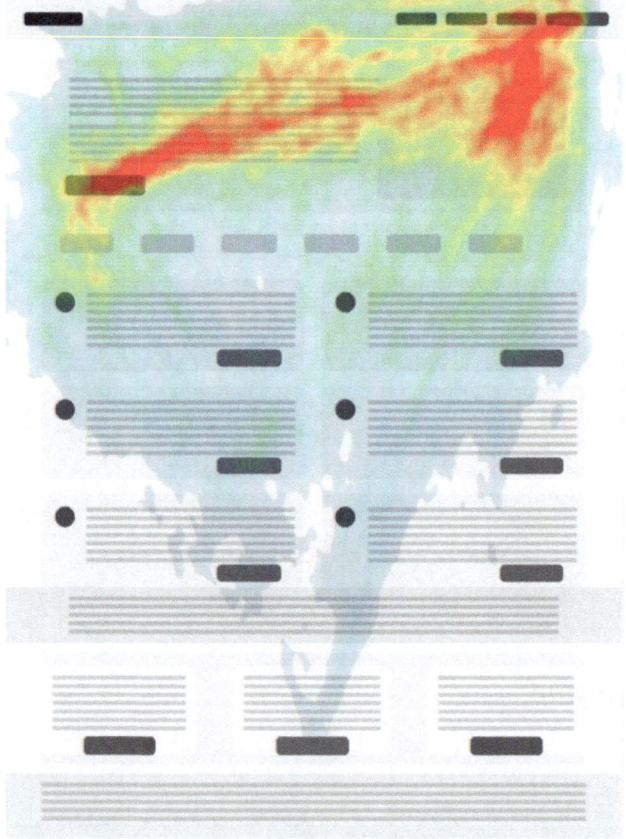

Figure 10-2. *Move Map*

Notice how attention clusters around the buttons in the top section of the website and extends into nearby text areas within the same blocks, forming smooth fields rather than sharp lines—indicating how visitors explore before deciding to click.

Movement occurs not only near the top but also around the middle and lower buttons, showing that users who click in these areas often pause to examine nearby content first. This map can help you identify hesitation or curiosity: broad, diffuse fields can signal comparison or uncertainty, while tight, focused fields suggest clear intent.

Scroll Maps

Scroll maps reveal how far down a page the average user scrolls. This is critical for ensuring that important calls-to-action or key information are placed in the red zones where most users will likely see them.[14]

Figure 10-3 shows how far down the page visitors typically scroll. As with the click and move maps above, the scroll map uses a red-to-blue color scale, with "hot" red areas marking the sections that receive the most attention. The strongest focus appears near the top, with additional attention areas emerging around certain mid-page and lower blocks where users slow down to read or evaluate buttons. This map can help you place key messages and calls to action where attention naturally concentrates, and reconsider or streamline sections that remain "cooler" despite occupying substantial space.

---

[14] "The complete guide to heatmaps," Contentsquare, accessed August 14, 2025, https://contentsquare.com/guides/heatmaps/.

Figure 10-3. *Scroll Map*

By combining these views, you can quickly identify which parts of your interface are effective and which are being ignored.[15]

- **Clicks Confirm Action:** The click map shows which visible buttons truly earn interactions.

- **Movement Explains Why:** The move map reveals where visitors pause and explore around those buttons and text blocks.

- **Scroll Frames Context:** The scroll map shows whether visitors even reach a block—and how long they linger there.

---

[15] "Heatmaps," Hotjar, accessed August 15, 2025, `https://www.hotjar.com/product/heatmaps/`.

If a block attracts attention in the scroll map and shows concentrated cursor movement but generates few clicks, refine its message or improve the clarity of its button. If a block fails to capture attention in the scroll map yet is important, move key content or a button higher on the page, or add a stronger visual break that naturally draws the eye.

From "What" to "Why": A Powerful Workflow

The most effective way to use these tools is to let your quantitative data guide your qualitative investigation. Start with your analytics to identify a drop-off point or a poorly performing metric. Then, use heatmaps to analyze the relevant page for broad patterns. Finally, dive into session replays of users who fit that specific segment to uncover the "why" behind their behavior.

However, it's critical to be mindful of user privacy. Modern tools provide robust features to automatically mask passwords and other personally identifiable information (PII) to comply with regulations like GDPR and CCPA, and these configurations should be a priority. When handled correctly, this workflow provides unparalleled insight, turning user behavior data into actionable product improvements.[16]

Startup's Survey Design

Surveys offer startups a cost-effective, scalable means to capture user attitudes, intentions, and satisfaction levels—insights that quantitative behavioral analytics alone can't reveal—and to measure improvements over time, such as before and after a new feature launch. While user interviews provide depth, surveys provide breadth, helping you validate hypotheses, measure customer satisfaction, or segment your market with data from a larger audience. However, a poorly designed survey doesn't just yield poor data; it can be actively misleading. The "garbage in, garbage out" principle is critical here.[17,18]

---

[16] "Session Replay," Amplitude, accessed August 15, 2025, https://amplitude.com/session-replay.

[17] George Kuhn, "Conducting Market Research for Startups," Drive Research Blog, March 15, 2025, https://www.driveresearch.com/market-research-company-blog/market-research-for-startups/.

[18] Erin Gilliam Haije, "Best practices for conducting user experience surveys," Mopinion, May 17, 2024, https://mopinion.com/user-experience-surveys/.

The key to effective survey design is to move beyond "nice-to-know" questions and focus exclusively on gathering actionable insights that will inform a specific decision.[19]

Core Principles for Impactful Surveys

Before writing a single question, a successful survey starts with a clear strategy. Adhering to a few core principles will ensure the data you collect is reliable and valuable:

- **Define a Clear Objective:** Every survey should have a well-defined goal. Are you trying to identify the primary reason for churn, gather feedback on a newly released feature, determine how to adjust pricing tiers to balance affordability with revenue, or measure brand perception? Every question should directly serve your main objective. If a question doesn't help inform a decision, remove it.[20,21]

- **Keep It Short and Focused:** Respondent fatigue is real. Long or unfocused surveys often result in high drop-off rates, poor-quality answers, and reluctance to start in the first place. Aim for surveys that take five minutes or less to complete, and be transparent about the estimated time commitment from the start. A short, well-focused survey will almost always yield higher-quality data than a lengthy, exhaustive one.[22]

- **Use Clear, Simple Language:** Write questions in plain, straightforward wording and avoid internal jargon, acronyms, or overly technical terms. Tailor your language to the target audience. A useful test is to ask yourself whether someone completely new to your product or organization would understand the question without difficulty.[23]

---

[19] Elizabeth Ferrall-Nunge, "Improve your startup's surveys and get even better data," GV Library, April 4, 2012, https://library.gv.com/improve-your-startup-s-surveys-and-get-even-better-data-7b0272f74c23.

[20] Ryan Stuart, "Survey Design Best Practices: 7 Steps to Create Impactful Customer Surveys," Kapiche Blog, October 18, 2024, https://www.kapiche.com/blog/survey-design.

[21] "How to Write a Startup Customer Survey," Process Street, accessed August 15, 2025, https://www.process.st/how-to/write-a-startup-customer-survey/.

[22] "Survey Design Best Practices: Tips and Examples," Gozen Blog, accessed August 15, 2025, https://gozen.io/blog/survey-design-best-practices/.

[23] "How to Write a Startup Customer Survey," Process Street, accessed August 15, 2025, https://www.process.st/how-to/write-a-startup-customer-survey/.

- **Design for Mobile First:** For many startups—though not all—a significant share of survey responses will come from smartphones or tablets. Make sure your survey layout is clean, responsive, and easy to navigate on small screens. Use large, legible fonts, clearly labeled buttons, and avoid questions that require lengthy typing.[24,25]

Crafting Effective Questions

The quality of your insights is directly tied to the quality of your questions. Biased or confusing questions can invalidate your results.

- **Avoid Leading and Biased Questions:** Frame questions neutrally to avoid influencing the respondent.

  - Frame questions in a neutral way to prevent influencing respondents. For example, instead of asking *"Don't you agree that our new feature is a great improvement?"*, ask "On a scale of 0 to 10, how would you rate our new feature?" Chapter 4 discussed the NPS (Net Promoter Score) question, *"How likely are you to recommend our product to a friend or colleague?"*, which also uses a 0-to-10 scale.

  - In practice, I've often seen product managers create surveys with leading or biased questions about features they personally designed, aiming to confirm their own success. To avoid this pitfall, survey questions should be reviewed by others before the survey is launched.

---

[24] Steve Wigmore, "15 Survey Best Practices for Effective Designs," Kantar, July 16, 2025, https://www.kantar.com/Inspiration/Research-Services/11-survey-design-best-practices-to-increase-effectiveness-pf.

[25] "Survey Design: Survey Design Strategies for Startup Growth," FasterCapital, April 11, 2025, https://www.fastercapital.com/content/Survey-Design--Survey-Design-Strategies-for-Startup-Growth.html.

- **Use a Mix of Question Types:** A good survey balances closed-ended and open-ended questions.[26]

  - **Closed-ended questions** such as multiple-choice items or rating scales provide quantitative data that is straightforward to analyze and compare over time. Rating scales are particularly effective for measuring satisfaction. Common formats include

    - **Likert Scale:** Typically a 5-point scale where 1 = *Strongly disagree*, 3 = *Neither agree nor disagree*, and 5 = *Strongly agree*.[27]

    - **Semantic Differential Scale:** For example, a 7-point scale from *old-school* to *modern* or from *amateur* to *professional*.

    - These rating scales often use an odd number of points (e.g., 5 or 7) to provide a neutral middle option such as *Neither agree nor disagree*.

  - **Open-ended questions** yield qualitative "why" insights by allowing respondents to explain their reasoning in their own words. While highly valuable, they require greater effort—particularly on mobile devices where typing may be less convenient—so their use is typically limited to the most important topics.

- **Ask One Thing at a Time:** Avoid combining multiple topics into a single question, often called a double-barreled question. This can make it unclear which part the respondent is addressing. For instance, rather than asking *"How would you rate our design and ease of use?"*, break it into two separate questions—one about the design and another about ease of use.

---

[26] "10 Tips for Effective Survey Design," LimeSurvey Blog, March 12, 2024, `https://www.limesurvey.org/blog/tutorials/tips-for-effective-survey-design`.

[27] "Likert Scale," ScienceDirect Topics, accessed August 15, 2025, `https://www.sciencedirect.com/topics/psychology/likert-scale`.

- **Start Broad, Then Go Deep:** Structure your survey with a logical flow. Begin with simple, general questions to ease the respondent in before moving to more specific or sensitive topics. Grouping related questions together also helps maintain context and reduces cognitive load.

Launching Your Survey

Once your survey is designed, the final steps are crucial for ensuring its success.

- **Pilot Test Your Survey:** Before sending your survey to your entire audience, test it with a small group of colleagues or trusted users. This "soft launch" helps you identify confusing questions, technical glitches, or logical errors in your survey flow.[28]

- **Consider Anonymity and Incentives:** Allowing respondents to remain anonymous can lead to more honest and candid feedback. In some cases, offering a small, relevant incentive—such as a discount or a free month of service—can help boost participation rates. Just be sure the reward aligns with your audience's interests and is used thoughtfully to avoid biasing responses.[29]

- **Close the Loop:** If possible, follow up with your respondents. Sharing a summary of the findings and explaining how their feedback will be used shows that you value their input and encourages participation in future research.[30]

---

[28] George Kuhn, "Conducting Market Research for Startups," Drive Research Blog, March 15, 2025, https://www.driveresearch.com/market-research-company-blog/market-research-for-startups/.

[29] Jamie Page, "Top Survey Design Tips To Get Better Responses (And Avoid Common Pitfalls)," ScoreApp, January 21, 2025, https://www.scoreapp.com/survey-design-tips-better-responses/.

[30] "Survey Design: Survey Design Strategies for Startup Growth," FasterCapital, April 11, 2025, https://www.fastercapital.com/content/Survey-Design--Survey-Design-Strategies-for-Startup-Growth.html.

Common Traps in Startup Survey Design

While following best practices is crucial, it's equally important to recognize and avoid common traps that can invalidate your research. Many startups, in their eagerness to gather data, fall into pitfalls that render their survey results useless or, even worse, misleading. Here are some of the most common bad practices to watch out for.

The Engagement Echo Chamber

A common pitfall is trying to understand unengaged or churned users by surveying only your most active and loyal customers. This leads to selection bias: the insights you gather come from a group that is, by definition, still engaged—often giving you a distorted view of why others dropped off.

Imagine this scenario: you run a fitness app and identify a segment of users who have churned. You send them a survey, but the majority of replies come from your most active former users—the ones who still check their email and feel connected enough to answer. In their open-ended responses, many say they aren't currently working out because they're injured or sick. If you target this group with promotional discounts in your new anti-churn campaign, it's unlikely to work—they're not leaving for price-related reasons. Meanwhile, the people who might be swayed by a discount never even filled out your survey. The takeaway? Be mindful of who is actually responding, and don't assume their reasons for disengaging reflect the wider audience you're trying to win back.

My own experience has demonstrated this pitfall clearly: users who churned due to lack of engagement were, unsurprisingly, also unwilling to participate in surveys about their experience. Their silence is, in itself, a form of data. Instead of relying solely on survey responses from your happy users, you must cross-reference your findings with quantitative data. Analyze the behavior of non-respondents within your product analytics. Their usage patterns—or lack thereof—often tell a more accurate story than the polite responses of your engaged user base.

Asking Questions You Already Know the Answer To

Surveys should be used to uncover information you cannot get elsewhere. A particularly damaging practice is asking questions that could be answered by a simple query of your own database. This wastes your users' time and signals that you don't understand your own product data.

Consider this real-world example of a poorly designed question I encountered: "Do you know you have free access to 2137 influencers on our marketplace?" At first glance, it might seem like an engaging prompt. In reality, it was problematic for several reasons. One manager believed many app users thought they had access to only a single influencer—the one who originally brought them to the app through their social media profile. Yet the question failed to address that belief and introduced other issues:

- **It Was Factually Inaccurate:** The access required a paid subscription, so framing it as "free" was misleading and could erode user trust.

- **The Data Was Outdated:** At the time the survey was sent, the number of influencers was already higher, making the company look unprofessional and out of touch.

- **It Was Unnecessary:** The manager's "gut feeling" that users didn't understand the marketplace's scale was contradicted by existing product analytics, which showed most customers engaged with multiple influencers. The data disproved the hypothesis before the survey even started.

The takeaway? If your existing data already contains the answer, don't ask the question. Use surveys to close genuine knowledge gaps, not to confirm hunches that your own analytics can already prove or disprove.

The Peril of Misguided Incentives

Earlier, we noted that incentives can help lift survey response rates. But they carry a significant trade-off: **response bias**. When you pay users for their time or throw in a free subscription, you're not just gathering feedback—you're entering into a transaction. That shift subtly changes the dynamic.

Once a reward is on the table, many people feel an unspoken obligation to be *nice*. They may soften their criticism, inflate scores, or hold back on hard truths—especially if they feel the gift deserves a "thank-you" in return. The result? Data that looks great on paper but paints a dangerously optimistic picture.

This risk is especially acute when you're trying to uncover uncomfortable realities about your product—the kind of feedback you need most. If you must use incentives, do so with your eyes open: adjust your analysis to account for possible bias, and scrutinize feedback that seems too positive to be true.

Be clear that any incentive is a thank-you for the respondent's time—not a payment for positive feedback. Keep the reward small enough to avoid skewing who participates or how they answer. An oversized reward can lead to disastrous unintended consequences.

Consider this cautionary tale from my own experience. A manager, excited to gather feedback, offered several months of free service as a survey incentive. The response was massive, but it came at a steep cost. We soon traced a significant dip in revenue directly to our most engaged, long-term subscribers who had, quite rationally, cancelled their paid plans to take advantage of the free offer. We were essentially paying our best customers to give us feedback, which also risked biasing their answers.

This experience highlights a critical rule: never let your incentive cannibalize your revenue or create a conflict of interest for your users. Instead, offer a choice of smaller tokens of appreciation—a gift card they can give to a friend, a brand-related gadget, a modest discount, or perhaps a single free month of service. Varying the incentive can also help attract a more diverse mix of participants.

Ultimately, authentic, unsolicited feedback is far more valuable than opinions purchased with perks. Insights from genuine users—offered freely and without conditions—will almost always lead to better analytics and better product decisions.

User Interviews

Clayton Christensen, a Harvard Business School professor, once discovered why a fast-food chain's milkshake sales were flat. Morning commuters were "hiring" milkshakes to make long drives less boring and keep them full until 10 a.m.—not to satisfy a chocolate craving. The real competition wasn't other milkshakes but bananas, donuts, bagels, Snickers, and coffee. The thick, cup-holder-friendly milkshake simply did the job better. That insight came only from watching real behavior and asking the right questions in user interviews. Armed with this knowledge, the company could create a far better solution: a morning milkshake that was even thicker to last the entire commute, perhaps with small fruit chunks to make the experience more engaging and target the customer's need to alleviate boredom.[31,32]

---

[31] Carmen Nobel, "Clay Christensen's Milkshake Marketing," Harvard Business School Working Knowledge, February 14, 2011, https://www.library.hbs.edu/working-knowledge/clay-christensens-milkshake-marketing.

[32] Clayton Christensen, "Understanding the Job," Edward Capaldi YouTube Channel, March 31, 2016, https://www.youtube.com/watch?v=sfGtw2C95Ms.

While analytics can reveal what users are doing, interviews let you dig into why they do it—often uncovering insights far deeper than surveys can deliver. This is your most direct path to building genuine empathy and seeing the world through the user's eyes: their challenges, motivations, and lived experiences.

In early-stage startups, this work is the essence of "doing things that don't scale." You can't automate empathy—and at this stage, you shouldn't try. Treat these conversations not as an opportunity to pitch your product, but as a chance to learn about the person's life, context, and needs. The richer your understanding, the better your product decisions will be.[33]

A common mistake in interviews is asking leading questions that confirm existing biases. To avoid this, many founders use *The Mom Test*—simple rules for framing questions that yield honest, actionable insights instead of polite praise. Popularized by Rob Fitzpatrick's book *The Mom Test: How to talk to customers & learn if your business is a good idea when everyone is lying to you.* The approach suggests that even a supportive parent won't mislead you if you ask the right questions. The core: focus on concrete past behavior, listen closely, and avoid hypotheticals or opinions about the future.[34,35]

Here are some best practices for conducting effective user interviews:

- **Focus on Their Life, Not Your Idea:** Ask about their workflow, their frustrations, and how they currently solve a given problem. Good questions sound like: "Tell me about the last time you…" or "What is the most difficult part about…?"[36]

- **Avoid Hypotheticals and Feature Requests:** Users are not good at predicting their own future behavior. Instead of "Would you use a feature that did X?", ask "How are you currently dealing with the problem that X solves?" This grounds the conversation in real-world behavior, not speculation. Avoid yes/no questions, as they are

---

[33] "How User Interviews Can Transform Your Startup," FasterCapital, April 3, 2025, `https://fastercapital.com/content/How-User-Interviews-Can-Transform-Your-Startup.html`.

[34] Katie Mulligan, "How To: User Interviews," Pillar VC, accessed August 14, 2025, `https://www.pillar.vc/playlist/article/how-to-user-interviews/`.

[35] Rob Fitzpatrick, *The Mom Test: How to talk to customers & learn if your business is a good idea when everyone is lying to you* (Robfitz Ltd, 2013).

[36] Eric Migicovsky, "How to Talk to Users," Y Combinator YouTube Channel, July 25, 2019, `https://www.youtube.com/watch?v=MT4Ig2uqjTc`.

conversation dead ends. You invest in interviews to explore the rich details of a user's experience, not to collect data points you could have gathered in a survey.[37]

— **Listen More Than You Talk:** Your primary job is to be a detective, uncovering facts and motivations. Let the user's answers guide the conversation and be comfortable with silence, as it often encourages the user to elaborate further.[38]

— **Separate Problem Discovery from Solution Validation:** Use initial interviews to deeply understand the problem space. Emmett Shear, former CEO of Twitch, emphasizes that you must validate solutions by testing for action, not opinion, by either building a minimal "hack" to observe real-world behavior or asking customers for a financial commitment upfront.[39]

While one-on-one interviews give you a deep, personal understanding of an individual user's motivations, they represent just a single perspective at a time. To explore how ideas resonate across a broader cross-section of your audience— and to capture the spark that can come from group dynamics—you can turn to focus groups.

Focus Groups

Another form of deeper qualitative analysis, particularly useful during the discovery phase in startups—when you aim to identify who your target users are and how they collectively perceive your product or concept.

---

[37] "How To Talk To Users | Startup School," Y Combinator YouTube Channel, December 1, 2022, https://www.youtube.com/watch?v=z1iF1c8w5Lg.

[38] Brandon Kindred, "Startup founders guide to user interviews," Medium Design Bootcamp, August 24, 2021, https://medium.com/design-bootcamp/startup-founders-guide-to-user-interviews-a5a372b43db7.

[39] Emmett Shear, "Lecture 16 - How to Run a User Interview (Emmett Shear)," YC Root Access YouTube Channel, November 14, 2014, https://www.youtube.com/watch?v=qAws7eXItMk.

Focus groups bring together a small, curated group of people to discuss a specific topic, product, or concept under the guidance of a moderator. Unlike one-on-one interviews that dive deep into an individual's context, focus groups are designed to generate a breadth of ideas and observe group dynamics.[40]

Focus group interview techniques include

- **Personification:** A method for uncovering subconscious perceptions of your product by asking participants to bring it to life. Prompt them with *"If this brand were a person, what would they be like? Describe their personality, their job, and who their friends are."*

- **Analogies:** This involves asking participants to make a comparison to understand abstract feelings. For example, *"If this app were a type of weather, what kind would it be and why?"*

- **Fantasy Scenarios:** These prompts encourage creative thinking by placing the user in a position of power. A classic example is *"Imagine you just became the owner of this company. What is the very first change you would implement for this product?"*[41]

For startups, focus groups are particularly useful for gauging initial reactions to concepts that have a strong subjective or social component. They are not a replacement for one-on-one interviews when you need to understand a user's specific workflow or deep-seated problems.[42]

---

[40] Laura White, "How to conduct a focus group: Nine simple steps for startups," Transmit Startups, accessed August 15, 2025, https://www.transmitstartups.co.uk/marketing-sales/focus-groups.

[41] "Focus Group Participant," ScienceDirect Topics, accessed August 15, 2025, https://www.sciencedirect.com/topics/computer-science/focus-group-participant.

[42] Jakub Nawrocki, "How to Use Focus Group Interviews to Build Better Digital Products?," Futuremind, November 30, 2023, https://www.futuremind.com/insights/how-to-use-focus-groups-to-build-better-digital-products/.

Consider using a focus group for

- **Branding and Messaging:** How does a group react to a new name, logo, or tagline? The discussion can reveal shared associations, potential misinterpretations, and the right language used to describe the benefits of your product.[43]

- **Broad Concept Exploration:** In the very early stages, a focus group can be a quick way to brainstorm and assess the general appeal of a new idea before investing in deeper research.

- **Understanding Social Dynamics:** If you are building a collaborative or community-based product, observing how a group discusses the topic can provide insights that individual interviews might miss.

However, be aware of the significant pitfalls. **Groupthink** is a major risk, where one or two dominant personalities can sway the opinion of the entire group. This can lead to false consensus and misleading feedback. Effective moderation is critical to ensure all voices are heard and to steer the conversation away from unproductive tangents. For this reason, many startups choose to prioritize individual user interviews, which provide cleaner, more reliable data on user needs and behaviors.[44,45,46]

Now that we've explored the qualitative methods that reveal the "why" behind user behavior, the next question is scale: how do you capture and analyze diverse signals across thousands of users without losing nuance? This is where a well-chosen product analytics platform becomes a force multiplier—connecting individual insights to population-level patterns. Interviews illuminate individual motivations, and surveys surface broader sentiment, but analytics ties it all together by revealing behavioral trends

---

[43] "How to Create a Successful Focus Group for Your Startup Product," MyCTOFriend - Tech startup tips for entrepreneurs YouTube Channel, August 2, 2023, `https://www.youtube.com/watch?v=pxxOCLdwIP8`.

[44] James Simonetta, "Why We Think Focus Groups Suck for Start Ups," Hustle Agency Blog, November 26, 2024, `https://hustle.agency/digital-agency-blog/why-we-think-focus-groups-suck-for-start-ups/`.

[45] Edward Boon, "How to Do Focus Groups On a Budget — For Startups and Small Businesses," Better Marketing, March 29, 2024, `https://medium.com/better-marketing/how-to-do-focus-groups-on-a-budget-for-startups-and-small-businesses-792e8a0719d0`.

[46] Sanjida Satter, "Focus Group vs Interview: Pros, Cons, & When to Use Each," Trymata Blog, accessed August 14, 2025, `https://trymata.com/blog/focus-group-vs-interview/`.

across the entire user base—turning anecdote into evidence and insight into repeatable decisions. The hard part isn't whether to use analytics—it's selecting the right tool from an overcrowded landscape and aligning it with the questions the team needs to answer.

Choosing Product Analytics Tools

Selecting the right product analytics tool can be a daunting task. The market is vast and, much like the business intelligence space, in a constant state of flux. Vendors are continuously adding new capabilities, sunsetting old features, and acquiring competitors, which means the landscape is always shifting. It is a near certainty that by the time you read this, new players will have emerged and existing ones will have evolved.

Similar to business intelligence tools discussed in the previous chapter, leading product analytics platforms are increasingly integrating AI-powered insights. These platforms compete by expanding their feature sets to support both quantitative and qualitative analysis. They now cover broader functional areas, including data privacy compliance and seamless integration with external tools such as customer data platforms (e.g., Twilio Segment) and data export capabilities to data warehouses, data lakes, and other storage systems.

Categories of PA Tools

Categorizing product analytics tools is a tricky task. The lines between platforms are often blurred, as many tools offer features that span multiple categories. The framework below is not intended as a rigid classification but rather as a guide to help you make sense of the market, orient yourself, and narrow the field of potential tools for your startup.

Event-Based Digital Analytics Solutions

This category includes powerful platforms that are purpose built for deep, self-serve analysis of funnels, cohorts, and user retention. They are ideal for product and growth teams that need to answer complex behavioral questions. Independent research firms such as Forrester regularly evaluate this market, and The Forrester Wave: Digital

Analytics Solutions report provides a detailed evaluation of the top vendors in this space classified as *Leaders, Strong Performers,* and *Contenders.* Let's review the solutions highlighted in that report.[47,48]

Leaders

Below, we summarize their findings on the two companies named as *Leaders,* offering a look into their key strengths and ideal use cases:

- **Amplitude:** Amplitude was named both a *Leader* and a *Customer Favorite* in the evaluation. It received the highest score of all vendors in the "Current Offering" category and achieved the highest possible scores in numerous different criteria such as AI-related features, deployment, analysis for UX, UI, and digital product performance. The platform offers a wide range of tools for product and web analytics such as retention, engagement, and conversion analyses. One of its key technical differentiators is its bidirectional integration with data warehouses. That lets you access your data warehouse directly within the Amplitude platform. Forrester appreciates Amplitude's ease of use and states that it is best suited for companies with close alignment between product and marketing teams.[49,50]

  According to the Forrester report, users also appreciated the speed of its query engine and its cost-to-value ratio. Amplitude is well known for its commitment to the startup ecosystem. You can apply for its startup scholarship program. According to their website, over 96% of scholarship applicants get approved, if they meet certain criteria.

---

[47] Thomas Hansen, "The First-Ever Forrester Wave™ for Digital Analytics Solutions Is Here—and We're a Leader," Amplitude Blog, August 4, 2025, `https://amplitude.com/blog/forrester-wave-das-leader`.

[48] Chiara De Gasperin, Oliwia Berdak, Agnes Nkansah, and Frank Harris, "The Forrester Wave™: Digital Analytics Solutions, Q3 2025," Forrester Research, accessed August 15, 2025, `https://reprint.forrester.com/reports/the-forrester-wavetm-digital-analytics-solutions-q3-2025-855ddc02/index.html`.

[49] "Amplitude," Amplitude, accessed August 15, 2025, `https://amplitude.com/`.

[50] "Data Warehouse Native Overview," Amplitude Documentation, accessed August 15, 2025, `https://amplitude.com/docs/data/warehouse-native/overview`.

The program provides a generous free year of its paid Growth plan, which includes a high limit of events per month, along with access to numerous features such as session replay, experimentation, feature flags, and behavioral analytics. This allows early-stage companies to leverage the power of the platform without the enterprise price tag. For those not in the scholarship program, Amplitude still offers a robust free starter plan that is sufficient for many new products. Amplitude blog describes numerous customer success stories from companies of different sizes, including startups such as Correcto and well-known companies such as Vimeo.[51,52,53]

- **Adobe:** Also named a Leader, with Forrester focusing on two products: Adobe Customer Journey Analytics and Adobe Analytics. However, early-stage startups do not typically choose Adobe for an affordable product analytics solution. Unlike Amplitude or Mixpanel, Adobe is not known for offering attractive startup pricing tiers.[54]

  Customer Journey Analytics enhances Adobe Analytics by enabling analysis across multiple channels. The tool is perceived as excellent for analyzing marketing performance. However, it offers limited insight into digital product performance. Compared to some of the competitors, it also lacks features for technical performance and UX/UI analysis, such as session replays.[55]

---

[51] Noa Ziv, "How Vimeo Scaled Self-Serve Analytics to Make Data Accessible to Everyone," Amplitude Blog, August 3, 2023, `https://amplitude.com/blog/vimeo-self-serve-analytics`.

[52] "Amplitude for Startups," Amplitude, accessed August 15, 2025, `https://amplitude.com/startups`.

[53] Abraham López Lee, "From Scholarship to Growth: How Correcto Boosted Activation by 208%," Amplitude Blog, May 14, 2025, `https://amplitude.com/blog/correcto-boosted-activation?siteLocation=Read+Blog+Article`.

[54] "Comparison with Adobe Analytics," Adobe Experience League, accessed August 15, 2025, `https://experienceleague.adobe.com/en/docs/analytics-platform/using/compare-aa-cja/cja-aa-comparison/overview`.

[55] "Contentsquare vs. Adobe Analytics: which is better for your team?," Contentsquare Blog, accessed August 15, 2025, `https://contentsquare.com/blog/contentsquare-vs-adobe/`.

According to Forrester, Customer Journey Analytics may be a good choice for marketers in big companies. It is ideal for those who want to integrate digital and offline data and who already use or plan to adopt other Adobe products. It's a complex solution for bigger data teams, not particularly a good choice for startups that are just getting started with their infrastructure.[56,57]

Strong Performers

In this category, Forrester recognizes vendors with a solid market presence and a compelling product offering, even if they don't meet the full criteria for leadership status. They represent strong choices for many organizations:

– **Mixpanel:** Mixpanel is a popular and user-friendly choice for product-focused companies, serving clients ranging from early-stage startups like copy.ai and Zealy to large enterprises such as Uber and Yelp. Its startup program makes it particularly appealing to new companies, and its design effectively supports cross-functional teams.[58,59]

– **Google:** With Google Analytics 4 (GA4), one of the most popular web analytics tools, it is especially a popular choice in sectors like ecommerce and retail. Numerous startups appreciate the ability to use GA4 free of charge and its deep integration with the Google ecosystem, particularly BigQuery and Google Ads. Many startups decide to use GA4 for web analytics and choose a separate product analytics solution like Mixpanel for app analytics. This approach helps them use pricing tiers data limits more effectively and optimize costs.[60]

---

[56] "Adobe Analytics," Adobe, accessed August 15, 2025, `https://business.adobe.com/products/adobe-analytics.html`.

[57] "Why choose Mixpanel over Adobe Analytics," Mixpanel, accessed August 15, 2025, `https://mixpanel.com/compare/adobe/`.

[58] "Mixpanel," Mixpanel, accessed August 15, 2025, `https://mixpanel.com/`.

[59] "Mixpanel Customers," Mixpanel, accessed August 15, 2025, `https://mixpanel.com/customers/`.

[60] "Google Analytics," Google Marketing Platform, accessed August 15, 2025, `https://marketingplatform.google.com/about/analytics/`.

–   **Pendo:** Distinguishes itself by uniquely combining product analytics with in-app user guidance presenting itself as a Software Experience Management platform. Its core strength lies in enabling product teams to not only measure metrics such as Net Promoter Score, analyze user behavior, but also to directly influence it through its "Guides" feature, making it highly effective for improving feature adoption and onboarding from within the application itself.[61,62]

–   **Contentsquare:** Exemplifies the dynamic consolidation occurring in the analytics market. Through strategic acquisitions of prominent platforms like Heap for product analytics and Hotjar for user feedback, it has rebranded itself as an "experience intelligence" provider. The company aims to merge its powerful visual journey analysis with the robust features of these acquired tools. This positions Contentsquare as an excellent choice for marketing and UX teams seeking a holistic, visual map of user behavior, though the seamless integration of these technologies into a single platform is still a work in progress.[63,64]

–   **Quantum Metric:** Specializes in real-time performance monitoring and friction detection, serving major enterprises such as Lenovo, Western Union, and Lufthansa. Forrester identifies it as a strong choice for analytics, user experience, and engineering teams that need to rapidly diagnose and resolve problems across web and app platforms. Given its enterprise-grade features, it is not a first-choice solution for early-stage startups but is instead better suited for medium to large companies.[65]

---

[61] "Pendo," Pendo, accessed August 15, 2025, `https://www.pendo.io/`.

[62] "A Comprehensive Guide To Net Promoter Score (NPS)," Pendo Glossary, accessed August 15, 2025, `https://www.pendo.io/glossary/net-promoter-score-nps/`.

[63] "Contentsquare," Contentsquare, accessed August 16, 2025, `https://contentsquare.com/`.

[64] "Contentsquare Completes Acquisition of Heap," Contentsquare Blog, December 7, 2023, `https://contentsquare.com/blog/contentsquare-completes-acquisition-heap/`.

[65] "Quantum Metric," Quantum Metric, accessed August 15, 2025, `https://www.quantummetric.com/`.

Contenders

In its analysis, Forrester also identifies *Contenders*—vendors that offer viable solutions but may have a more specialized focus or are still developing the breadth and depth of their platforms compared to the Leaders and Strong Performers:

- **Glassbox:** Tailored for UX/UI and technical teams in regulated industries.[66]

- **Acoustic (Tealeaf):** Founded in 1999. Primarily useful for tech and User Experience teams working in large companies such as PayPal.[67,68]

- **Fullstory:** Best for product and engineering teams looking for User Experience insights without significant development work.[69]

Other Solutions

Except for solutions included in the Forrester report, there are also many other tools specialized in particular features such as

- **Replaying User Sessions: Mouseflow and Sprig** can be used to visualize their exact experience, including clicks, rage clicks, and mouse movements. Invaluable for identifying UI friction and bugs.[70,71]

- **"Autotrack" Events:** Tools that capture every interaction automatically, allowing teams to analyze data retroactively without needing developers to instrument every event upfront. Great for speed and flexibility but can lead to "messy" data if not managed. Some experts believe that the best approach for early-stage startups is to be able to define events in code while having auto-track as a safety net.

[66] "Glassbox," Glassbox, accessed August 15, 2025, `https://www.glassbox.com/`.

[67] "Tealeaf," Acoustic, accessed August 15, 2025, `https://www.acoustic.com/tealeaf`.

[68] "PayPal Case Study," Acoustic, accessed August 16, 2025, `https://www.acoustic.com/resources/case-studies/paypal`.

[69] "FullStory," FullStory, accessed August 16, 2025, `https://www.fullstory.com/`.

[70] "Mouseflow," Mouseflow, accessed August 16, 2025, `https://mouseflow.com/`.

[71] "Sprig," Sprig, accessed August 16, 2025, `https://sprig.com/`.

Userpilot is an example of such a platform designed to help product teams enhance user engagement and drive growth without requiring extensive coding.[72,73]

Open Source Alternatives

Except for paid solutions from numerous vendors, there are also open source solutions that many cost-efficient startups choose over well-known market leaders. However, choosing self-hosted open source solutions means other costs: the cost and time required for engineering (deployment and maintenance) and hosting. Let's briefly go through some of them.

— **PostHog:** An all-in-one platform offering a suite of tools including product analytics, session replay, feature flags, and A/B testing. It is designed for technical users who want to self-host and have full control over their analytics stack. Similar to other popular Open Source solutions, you can choose either the self-hosted free version with a limited set of features or pay for the managed cloud version. You can apply for the PostHog for startups program.[74,75]

— **Matomo:** A popular open source web and app analytics platform that positions itself as an ethical, privacy-focused alternative to Google Analytics.[76]

— **OpenReplay:** An open source, self-hostable session replay suite that provides developers with tools to see what users are doing on their app to troubleshoot issues faster. It combines session replay with product analytics features. It has a free self-hosted version and paid cloud options.[77]

---

[72] "Web Analytics," Satchel, accessed August 16, 2025, `https://satchel.com/web-analytics/`.

[73] "Heap Autocapture," Userpilot Blog, accessed August 16, 2025, `https://userpilot.com/blog/heap-autocapture/`.

[74] "PostHog," PostHog, accessed August 16, 2025, `https://posthog.com/`.

[75] "PostHog for Startups," PostHog, accessed August 16, 2025, `https://posthog.com/startups`.

[76] "Matomo," Matomo, accessed August 16, 2025, `https://matomo.org/`.

[77] "OpenReplay," OpenReplay, accessed August 16, 2025, `https://openreplay.com/`.

If this comprehensive landscape of product analytics tools feels overwhelming, you're not alone. With dozens of vendors offering hundreds of features across multiple categories, many startup teams fall into "analysis paralysis"—spending weeks comparing features instead of gathering user insights. The solution isn't to choose the most powerful tool but to choose the right tool for your current stage and constraints. This is where a structured evaluation framework becomes invaluable.

A Pragmatic Choice Framework for Product Analytics Tools

Choosing a product analytics tool is a critical decision that directly impacts your ability to understand user behavior and guide your product strategy. Just as with BI platforms, the sheer number of options can be overwhelming. Instead of getting lost in a sea of features, use this structured framework to make a pragmatic choice that aligns with your startup's specific goals, resources, and budget.

The framework is built around four key questions. By answering them honestly, you will develop a clear set of requirements that will naturally lead you to the right tool for your current stage.

1. Start with Your Audience and Core Use Case

Before looking at any specific tool, you must first define who it's for and what problems it needs to solve. This initial step prevents you from choosing a powerful tool that your team can't use or a simple tool that doesn't answer your most critical questions.

- **Who are the primary users?** Will the tool be used mainly by product managers and data analysts who are comfortable with complex data exploration? Or do you need a tool that is accessible to UX designers, marketers, and even founders with varying technical skills? The answer will determine whether you should prioritize power and flexibility or ease of use and an intuitive interface.

- **What is the primary job to be done?** Are you just getting started with a simple website where you only need to track basic steps of a consumer funnel (like Landing Page Visit, Checkout Page Visit, or a click on the Purchase button)? For such cases, getting started might be as easy as configuring Google Analytics on your website. Or do you already have an existing website tracked with other tools and now want to track your complex mobile app across iOS and Android? This might involve multiple screens with sophisticated user actions that require manually defined events with specific properties, potentially pulled from external data sources (e.g., your payments database).

2. Assess Your Technical Resources and Existing Stack

Your team's technical capabilities and current infrastructure will significantly narrow down your options. Be realistic about what you can support. Focus on what is possible and necessary now, not in a year from now.

- **Do you have the engineering resources for a self-hosted solution?**
  Open source tools like PostHog offer immense power and control, often with a generous free, self-hosted plan. However, this "free" option comes with the hidden costs of deployment, maintenance, upgrades, and troubleshooting. If your engineering team is already stretched thin, a managed cloud solution might have a lower total cost of ownership.

  - How much developer time can you invest in implementation?

    - **"Autotrack" solutions (like Heap)** automatically capture every click, tap, and pageview without requiring developers to manually implement event tracking. This offers immediate, retroactive insights but can lead to large, messy, and overwhelming amounts of data that are expensive to store, process, and interpret. The autocapture approach may be a good choice at the very beginning when a business is small and its tracking needs are basic. However, as your business and product grow or your needs become more sophisticated, manual event tracking may turn out to be a more effective approach.

    - **Event-Based Tracking tools** (like Mixpanel or Amplitude) allow developers to manually instrument the specific events (and their properties) you want to track. This requires more work upfront, but when executed properly, it results in cleaner, more intentional data and gives you greater control over volume and quality.

- **How well does the tool integrate with your existing stack?** A tool that seamlessly connects to your data warehouse, CRM, or marketing automation platforms will provide a more holistic view of the customer journey and prevent data silos.

3. Evaluate the True Cost and Startup Programs

For the vast majority of early-stage startups, budget is a primary constraint. It's crucial to look beyond the monthly subscription fee to understand a tool's true, long-term cost.

- **What is the pricing model?** Typically, the price of product analytics tools is based on Monthly Tracked Users (MTUs) or the volume of events. Model how these costs will scale as your user base grows. A tool that seems cheap today could become prohibitively expensive in a year. Be mindful of your top-of-funnel traffic. If you have many website visitors but only a small percentage convert into app users, implementing the same analytics solution for both web and app tracking can become unexpectedly costly.

- **Are you taking advantage of startup programs?** For any early-stage company, this is a critical question. Many leading analytics platforms, including Mixpanel and Amplitude, offer generous plans that provide their premium tools for free or at a steep discount, typically for one to two years. These programs can grant you access to premium tools long before you could otherwise afford them, giving you a valuable window to grow your business and revenue.

  However, approaching these programs requires a clear strategy. You must have a plan for when the program ends. The free period will eventually expire, and if you haven't budgeted for the full price, you could face a difficult choice: pay an unexpectedly high bill or endure a painful migration to a cheaper tool. **Don't hesitate to negotiate with providers.** If you present a strong case, you can often secure better discounts or additional features.

This is a calculated risk. While it's true your startup might not survive the year—or it could be acquired—that volatility is precisely why you must be strategic. Startups frequently experiment with different tools, and migrations are a normal part of the ecosystem. The key is to weigh the immense immediate benefit of a powerful, free tool against the potential future cost of a migration.

— **What is the Total Cost of Ownership (TCO)?** For a self-hosted tool, factor in server costs and engineering salaries. For a managed tool, consider the time your team will spend on implementation and training. The cheapest option isn't always the one with the lowest sticker price.

4. Define Your Feature and Scalability Requirements

Finally, once you have clarity on your audience, resources, and budget, you can confidently assess specific features.

— **What are your core visualization and exploration needs?** Ensure the tool can build the analyses that matter most to you, such as funnels, retention curves, and user path reports. Does it offer intuitive dashboards and customizable reports that are easy for the target audience to understand?

— **Do you need advanced capabilities?**

  — **A/B Testing and Experimentation:** Do you need a platform that can also manage experiments to test the impact of new features?

  — **Qualitative Insights:** Are you looking for a tool that combines quantitative data with qualitative insights like session replays and heatmaps?

  — **AI-Driven Insights:** Some tools leverage AI to automatically surface anomalies or opportunities in your data, acting as a force multiplier for a small team. However, is this feature essential if it comes at a significantly higher price?

- **Is the tool built to scale with you?** Carefully consider not only the financial constraints discussed in Step 3 but also any inherent technical limitations. Select a platform capable of handling your increasing data volumes without experiencing performance degradation or sudden cost escalations.

A Day in the Life: Choosing the Right Product Analytics Tool at Empathic

To bring this framework to life, let's step into the shoes of the recently hired data lead at *Empathic*, a new mobile application with a mission to make mental healthcare more accessible by allowing users to schedule therapy sessions with licensed psychologists.

Empathic aims to stand out by integrating both mental and physical well-being. The app already offers a robust suite of features, including tools for tracking and improving diets, reminders for regular eating and hydration, medication alerts, and prompts for daily walks. Additionally, it features multiple dedicated screens for exercise routines and meditation practices.

Given this comprehensive functionality, users can engage in numerous distinct and specific actions across various screens. The app generates revenue through both subscription access (offering several different pricing tiers) and one-time purchases.

Empathic launches with a flurry of positive press, leading to thousands of downloads and a strong initial sign-up rate. However, a critical problem lurks beneath the encouraging top-line numbers: while users are signing up, very few are actually booking their first therapy session. The company's core mission isn't being fulfilled.

The team relies on Google Analytics, which confirms that their marketing efforts are successfully driving downloads. However, it offers no visibility into the crucial user journey inside the app. They know users are entering the front door but have no idea where they get lost or why they leave.

During a tense team meeting to discuss the low booking rates, the conversation becomes a classic startup debate fueled by opinion and anecdote:

- **The Voice of Convenience:** A back-end engineer, concerned about the development road map, suggests, "Let's just stick with Google Analytics 4. It's free and can track in-app events. We just need to spend some time configuring it."

- **The Influence of Social Proof:** The Head of Marketing, who just attended a tech talk, argues, "I've been hearing amazing things about Heap. Their autocapture feature means we could see every user action immediately without needing to define events. We could get answers fast."

- **The Voice of Past Experience:** The founder, who previously worked at a successful telehealth company, counters, "My last company was built on Amplitude. We need a tool that's designed for deep funnel analysis. We're dealing with sensitive user journeys, and we need precise, reliable data, not just a sea of clicks."

The room is divided. To provide clarity and a structured path forward, you step in to apply the pragmatic choice framework, guiding the team through the key questions tailored to *Empathic's* unique situation.

1. **Start with Your Audience and Core Use Case**

   You begin by asking the most important question: "What is the primary job to be done?"

   The team agrees that the primary goal is to fix the broken user activation funnel. The Head of Product defines this as the journey from signing up to successfully booking a first session. The critical path is: User Signed Up → Intake Questionnaire Completed → Viewed Therapist Profile → First Session Booked.

   The primary audience for this tool is the product manager and the UX designer. They need to answer specific, sensitive questions:

   - Where exactly are users dropping off in the booking funnel? Is it during the intake questionnaire, while browsing therapists, or at the final payment step?

   - What percentage of users who view a therapist's profile actually proceed to book?

   - Are users leveraging our filtering system to find the right therapist, or are they overwhelmed by choice?

This initial step immediately makes it clear that a generic tool like Google Analytics is inadequate. The team needs a purpose-built product analytics tool designed for deep behavioral analysis, with strong, intuitive funnel and user pathing visualizations.

2. **Assess Your Technical Resources and Existing Stack**

Your engineering team is small and intensely focused on core features, such as improving video call functionality and ensuring data security. Consequently, you lack the dedicated resources required to deploy, maintain, and secure a self-hosted, open source solution. As the sole data professional on the team, you are already overwhelmed with numerous other priorities, making an open source, self-hosted approach an impractical choice.

The conversation then shifts to the "autotrack vs. manual event tracking" debate.

– **Autocapture Events:** While tempting for its speed, this approach poses a significant risk for *Empathic*. Automatically capturing every interaction could inadvertently collect Protected Health Information (PHI) from the intake forms, creating a major security and compliance liability. Moreover, the app is complex, and the autocapture approach might quickly lead to generating a big amount of messy data.

– **Manual Event Tracking (with tools such as Mixpanel or Amplitude)**: This approach requires more upfront engineering work to manually define and implement each event. It forces the team to be deliberate and intentional about what data is collected, ensuring they only track what is necessary for analysis while protecting user privacy.

The team unanimously agrees that the manual tracking approach is the only responsible choice.

3. **Evaluate the True Cost**

As a mission-driven, early-stage startup, *Empathic's* budget is limited. An expensive enterprise license is not an option.

This is where you highlight the importance of startup programs. You research and present the fact that leading platforms like Amplitude and Mixpanel offer generous free plans for early-stage startups. This allows *Empathic* to use a best-in-class, secure platform without the prohibitive cost.

4. **Define Your Feature and Scalability Requirements**

   Based on *Empathic's* core use case and technical constraints, the following features are deemed must-haves:

   – **Funnel Analysis:** To visualize drop-off at each step of the activation flow.

   – **User Segmentation:** To compare the behavior of users across different pricing tiers or to analyze those who complete the intake questionnaire vs. those who skip it.

   – **HIPAA-Compliant Setup:** A non-negotiable requirement. The platform must adhere to stringent healthcare data privacy standards. Research confirms that Amplitude can be configured for HIPAA compliance, and numerous case studies from healthcare companies with similar needs further validate its suitability.[78]

A valuable nice-to-have feature is qualitative insight tools. While session replays of actual therapy sessions are unequivocally out of the question, the ability to watch anonymized recordings of the booking flow could provide invaluable insights into user confusion or frustration points. The startup program for Amplitude includes Session Replays, notably with the capability to block sensitive data. This functionality appears highly promising.

Furthermore, considering Forrester's assessment of Amplitude's collaborative approach, *Empathic* can likely rely on their guidance for proper HIPAA-compliant configuration. While this level of support is a compelling future benefit, it is not an immediate priority. Overall, Amplitude's broad feature set and user-friendly data exploration capabilities appear well suited to *Empathic's* target audience.

---

[78] "Amplitude for healthcare," Amplitude, accessed August 16, 2025, `https://amplitude.com/industry/healthcare`.

Given *Empathic's* current user base and projected growth for the foreseeable future, the capacity provided by Amplitude's startup program appears more than sufficient to accommodate their scaling needs.

Lesson Learned: Start with What Matters Most

Your structured evaluation leads to a clear decision. Your team applies for and is accepted into Amplitude's startup program, confident in its robust features and commitment to security.

Figure 10-4 illustrates the results of your detailed analysis of several popular analytics tools, assessed through four key criteria relevant to startups. These criteria are represented in the following columns:

- **Easy Setup and Maintenance** reflects how much technical effort it requires to implement and manage the tool.

- **Affordable Expected Future Pricing** captures your projection of the tool's cost-effectiveness over time (e.g., when the startup program ends), taking into account anticipated usage growth and scaling costs. This helps anticipate whether the tool remains financially viable in the long run.

- **Attractive Program for Startups** evaluates special free tiers tailored for startups.

- **Overall Requested Features Match** summarizes how well each tool meets your specific feature requirements based on your startup's unique analytics priorities. This is a subjective but crucial measure reflecting the alignment between the tools' capabilities and your practical needs.

It's important to emphasize that these columns don't present an objective universal ranking but rather your assessment of how well each tool meets the defined criteria in the context of your startup's situation and strategic goals. Unlike broad market reports such as those by Forrester, your focus is practical and tailored, considering unique limitations and priorities.

The graph clearly shows that while Amplitude doesn't have a perfect score across all categories, it stands out as a very promising market leader offering a wide range of features and excellent ease of use, particularly for non-technical users. Its startup program allows free usage for the first year, which is a significant advantage. However, the decision matrix also exposes a major potential challenge: if your budget after this period is not big enough, migrating away from Amplitude could be painful.

This visualization serves as an effective communication tool to discuss the pros and cons of selecting Amplitude over alternatives like Google Analytics or Heap. It provides an objective foundation to address colleagues' suggestions by highlighting strengths and tradeoffs, always through the lens of your startup's specific needs.

	Easy Setup & Maintenance	Affordable Expected Future Pricing	Attractive Program for Startups	Overall Requested Features Match	
Amplitude	✓	✗	✓	✓	★
Mixpanel	✓	✗	✓	✗	
Google Analytics	✓	✓	✗	✗	
Heap	✓	✗	✗	✗	
PostHog	✗	✓	✓	✗	

Figure 10-4. *Product Analytics Tool Choice Decision Matrix*

Choosing the right product analytics tool is a critical decision that moves beyond a simple feature comparison. As we have demonstrated, applying a structured evaluation framework is essential for grounding this choice in your startup's unique context. By starting with your core business objectives—whether that's boosting user activation or refining a key feature—the framework ensures your evaluation is tied to tangible outcomes, not just technical specifications.

Ultimately, this structured approach transforms a potentially overwhelming decision into a strategic exercise. By systematically weighing solutions against your budget limitations and specific needs for data analysis, integration, team workflow, and scalability, you ensure the tool you embed in your stack becomes the powerful microscope we discussed—one that magnifies user journeys rather than just another dashboard that gathers dust.

However, selecting the right analytics platform is only half the battle. *Empathic's* team now had access to powerful funnel analysis and user segmentation capabilities, but these tools are only as valuable as the data flowing into them. This brings us to the core of any successful manual event tracking strategy: a meticulously designed event tracking plan. Such a plan is not just about data quality; it's also crucial for fostering effective teamwork and enabling self-serve analytics.

When you have one shared document detailing how each user action is tracked, the entire team gains a consistent understanding of what the data truly means. Without this clarity, analyzing data properly becomes impossible—you simply can't derive accurate insights if you're unsure what particular events actually represent. Even the most sophisticated analytics platform is rendered useless by poor-quality data, irrelevant event tracking, inconsistent naming conventions, or an overwhelming flood of non-essential data. Let's explore how to build a robust tracking plan that truly transforms your chosen tool into a powerful product growth engine.

The Blueprint: Designing and Managing Your Event Tracking Plan

Effective product analytics hinges on clean, reliable data. This is where the discipline of event tracking becomes paramount, as it operates on the fundamental principle of "garbage in, garbage out." Without a structured approach, your data can quickly become a tangled mess of inconsistent, irrelevant, and untrustworthy information. The antidote to this chaos is a meticulously crafted tracking plan. This document serves as the single source of truth for your analytics, aligning the entire company on what user actions are being measured and, more importantly, why.

1. **Define Clear Business Objectives:** Ask critical questions like "Are users discovering our key feature?" or "What behaviors correlate with long-term retention?" to ensure every tracked event ties back to a strategic goal.

2. **Map the Critical Path:** Identify the essential actions users must take to realize your product's core value (e.g., User Signed Up → Project Created → Teammate Invited). Prioritize these events to generate the most immediately actionable insights.

3. **Enforce a Consistent Naming Convention:** Use a simple schema (e.g., Object_Action such as Song_Played or Document_Shared) to avoid ambiguity and make data intuitive for all stakeholders. Crucially, the most important aspect is to ensure this convention is applied consistently across all platforms and tracking points. Avoid variations like "Song Played" on iOS and "song_played" on Android, which can significantly complicate your data analysis later. Additionally, be mindful of technical limitations, as some databases do not accept specific characters, which will also impact your naming convention choice. For instance, certain systems reject hyphens unless the name is wrapped in special characters like backticks. While various approaches exist, some tools offer a dual system: an internal, consistent event name (e.g., "song_played") and a more business-user-friendly Event Display Name (e.g., "Song Played"). Always consult your chosen tool's documentation for best practices and examples of event tracking plans relevant to different industries.[79]

4. **Enrich Events with Contextual Properties:** Attach relevant metadata (e.g., for Song_Played, include song_title, artist, genre) to turn a single event into a rich dataset capable of revealing nuanced user preferences and behaviors.

Table 10-1 presents a simplified example of an event tracking plan. This template is designed to demonstrate the core components in action, offering a clear, practical starting point. It is important to note that real-world tracking plans are often far more complex, tailored to the specific needs of a business.

For those looking to explore more detailed examples, product analytics vendors like Mixpanel often provide comprehensive templates on their websites. These resources typically include Google Sheet documents that showcase intricate tracking plans for various business models and industries, from ecommerce to SaaS.[80]

---

[79] "Events: Capture behaviors and actions," Mixpanel Documentation, accessed August 16, 2025, https://docs.mixpanel.com/docs/data-structure/events-and-properties.

[80] "Create A Tracking Plan," Mixpanel Documentation, accessed August 16, 2025, https://docs.mixpanel.com/docs/tracking-best-practices/tracking-plan.

The simplified plan below includes the following columns to help you understand the structure:

- **Event Name:** The unique, consistently named identifier for the user action

- **Trigger:** The specific user interaction that fires the event

- **Page:** Where in the product the event occurs

- **Property:** The contextual details collected with the event

- **Sample Values:** Example data for each property

Table 10-1. *Simplified Event Tracking Plan Template*

Event Name	Trigger	Page	Property	Sample Values
Sign-up Completed	User signs up	Sign-up Page	User ID	2137
			Registration Method	Google, Facebook
			Registration Date	YYYY-MM-DDTHH:MM:SS
			Payment Methods Selected	Debit, Crypto
Login Completed	User logs in	Login Page	User ID	2137
			Login Method	Google, Facebook

While the simplified template provides a solid foundation, a truly robust tracking plan often includes additional columns to enhance clarity, streamline implementation, and ensure data quality across the organization. These extra details bridge the gap between product strategy and engineering execution, creating a document that is useful for everyone from marketers to developers. Here are some additional columns you might include in your tracking plan:

- **Event Display Name:** A clean, human-readable name for the event (e.g., "User Signed Up" for the event user_signed_up). This is useful for building reports and dashboards that are accessible to non-technical team members.

- **KPI:** This column explicitly links an event to the Key Performance Indicator (KPI) it influences. For example, tracking Subscription Upgraded directly impacts the "Monthly Recurring Revenue (MRR)" KPI. This connection ensures that every tracked event serves a clear business purpose.

- **Event Definition:** A precise description of what the event measures and the exact conditions under which it is triggered. This eliminates ambiguity and ensures consistency in how data is collected.

- **Property Type:** Categorizes the event properties. For instance, properties could be grouped into types like user properties (e.g., account creation date) or event-specific properties (e.g., playlist name for a Song Played event).

- **Property Definition:** Similar to the event definition, this provides a clear explanation of what each property represents (e.g., for a plan type property, the definition might be "The user's subscription tier at the time of the event").

- **Data Type:** Specifies the technical format for each property's value, such as String, Integer, Boolean, or Timestamp. This is critical for preventing data quality issues during implementation.

- **Platform:** Indicates on which platform(s) the event is implemented. An event like Push Notification Opened would be specific to iOS and Android, whereas Form Submitted might be tracked on Web.

- **Implementation Status:** A tracking column, often broken down by platform (Web, iOS, Android, Server), to monitor the progress of engineering work. It typically uses values like Not Started, In Progress, or Completed.

- **Method Call:** An example code snippet showing how the event should be implemented by developers. This serves as a practical reference and helps ensure consistency.

- **Developer Notes:** A space for any technical considerations, implementation nuances, or other important notes for the engineering team.

Ultimately, the columns you include should be adapted to the specific challenges and workflow of your startup. The most effective tracking plans are living documents, not rigid templates. For instance, in startups where the app has numerous screens that evolve quickly, I have found it invaluable to add a Screenshot or Visual Reference column. A quick visual of the UI element or screen where an event is triggered can instantly clarify what an event means, overcoming the common pitfall of short, ambiguous descriptions that become confusing over time.

Similarly, in environments where multiple people are implementing events, adding an Owner column is a simple and effective way to assign responsibility. This immediately tells you who to approach with follow-up questions, whether you are clarifying a confusing definition or investigating an event that is sending unexpected data. The goal is to create a practical, shared resource that maximizes clarity and alignment, ensuring your tracking plan serves as a robust foundation for your analytics strategy.

While these tracking plan principles seem straightforward on paper, the reality of implementation often reveals unexpected challenges. Many startups begin with the best intentions—clear naming conventions, focused event lists, and detailed documentation—only to find their data becoming unreliable within months. Understanding these common pitfalls before you encounter them can save weeks of debugging and prevent the erosion of data trust across your organization.

Common Pitfalls and How to Avoid Them

Implementing event tracking may seem straightforward, but several common pitfalls can compromise the quality and utility of your data. Here's how to anticipate and avoid them:

- **Vague or Inconsistent Naming:** Without a clear convention, you might end up with ProjectCreated, created_project, and new_project_creation all tracking the same action. This makes analysis chaotic and unreliable.

    - **How to Avoid:** Establish and enforce a clear naming convention from day one. The Object_Action format (e.g., Project_Created) is a robust standard. Document this in a shared tracking plan that is the single source of truth for developers, product managers, and analysts.

- **Tracking Too Much or Too Little:** Tracking every single click creates a noisy, unmanageable dataset leading to analysis paralysis and high costs. Conversely, tracking too few events creates critical blind spots in the user journey.

  - **How to Avoid:** Start with a minimal set of events tied directly to your key performance indicators. Map out the critical path for a user to find value in your product and track the steps along that path. You can always add more events later.

- **Neglecting Event Properties:** An event like User Signed Up is useful, but it's far more powerful when you know how they signed up. Properties add the necessary context to answer deeper business questions.

  - **How to Avoid:** For each event, consider what context is needed. For User Signed Up, properties like the signup method (e.g., email, Google) or user cohort date are invaluable for analysis.

- **Poor Data Quality and Validation:** "Garbage in, garbage out" is the cardinal rule of data. Broken or inaccurate tracking renders your analytics platform useless and erodes trust in data across the company.

  - **How to Avoid:** Implement a rigorous validation process. It is the developer's responsibility to ensure correct event implementation. They can use instrumentation debugging tools in a development environment to verify events before production deployment. Additionally, you can set up validation mechanisms at the data warehouse level to identify anomalies and ensure data integrity. Periodically audit your data to confirm it's collected as expected.

- **Ignoring the "Why":** Quantitative data shows you what users are doing (e.g., "70% of users drop off at this step"), but it rarely explains why they are doing it.

  - **How to Avoid:** As discussed, pair your quantitative data with qualitative methods. If you see a drop-off point in your analytics, watch session recordings of users who failed at that step, analyze heatmaps, design a targeted survey, or conduct user interviews to understand the friction.

Case Study: Implementing an Event Tracking Plan

To see how these principles work in the real world, let's walk through a practical, end-to-end example with our fictional startup, *Empathic*. As discussed previously, the team agrees that their primary goal is to fix a broken user activation funnel. The Head of Product defines this as the complete journey from signing up to successfully booking a first session. This clarity translates into a focused critical path: User Signed Up → Intake Questionnaire Completed → Viewed Therapist Profile → First Session Booked.

Table 10-2 presents how this could look in their hypothetical event tracking plan.

Table 10-2. *Simplified Event Tracking Plan for Emphatic*

Event Name	Trigger	Page	Properties and Sample Values
User Signed Up	When the user successfully submits the sign-up form and verifies their email	Sign-up page	Sign-up method: "google" utm source: "newsletter"
Intake Questionnaire Completed	When the user clicks the "See my matches" button on the final step of the intake form	Intake questionnaire	Time to complete seconds: 223 Questionnaire version: "v2"
Viewed Therapist Profile	When the user clicks on a therapist's card from the recommendations list to view their detailed profile	Therapist recommendations	Therapist ID: "20050402" List position: 3 Source page: "recommendations"
First Session Booked	When the user clicks the "Confirm Booking" button and sees the confirmation screen	Booking confirmation	Therapist ID: "20050402" Session cost: 80.00 Booking flow variant: "deferred_payment"

With this sharp focus, the engineering team was tasked with instrumenting only those four critical events. Within two weeks, the data revealed a shocking and immediately actionable insight. A funnel visualization showed that while 70% of new users completed the intake questionnaire and 50% viewed a therapist's profile, only 5% of the original cohort proceeded to book a session (see Figure 10-5). The data undeniably pointed to a massive leak at the final step.

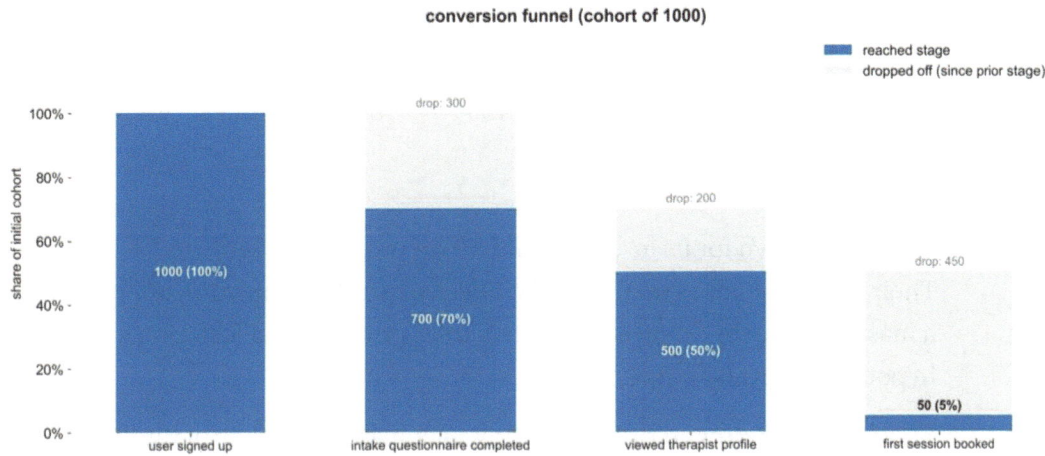

Figure 10-5. *Case Study: Conversion Funnel*

Armed with this insight and cross-referencing it with other analyses, you formed a clear hypothesis: the upfront payment screen was creating excessive friction. Having gained access to Amplitude's A/B testing capabilities, you swiftly designed an experiment to confirm this. (In Chapter 11, we will delve deeper into the principles of experimentation and A/B testing.)

Your team promptly developed a new user flow, allowing users to book their first session without immediate payment, instead being charged 24 hours before the appointment. After ensuring proper configuration and validation of Amplitude's A/B testing functionality, you launched the test with the new design.

The results were transformative. The new flow dramatically increased the First Session Booked conversion rate from a mere 5% to 35%—a sevenfold improvement. For the first time, *Empathic's* core activation metric surged, and this success soon cascaded into a corresponding improvement in their crucial Day 30 retention numbers.

Empathic's success profoundly underscores the true value of a disciplined analytics framework. It didn't just facilitate tool selection. It catalyzed critical conversations about priorities, focused your team on addressing their single most important problem, and transformed a vague sense of underperformance into data-driven progress.

Learning Resources

To further deepen your understanding of product analytics and related disciplines, the following resources offer a wealth of knowledge, from high-level strategic frameworks to practical, hands-on techniques.

Foundational Concepts

- **Reforge:** Known for its in-depth content on product and growth. Their articles, such as their guide to effective product analytics, offer a masterclass in building a sound data infrastructure and fostering a hypothesis-driven culture.[81]

- **Lenny's Newsletter:** Written by Lenny Rachitsky, this is one of the most popular and respected newsletters for product and growth professionals. It covers invaluable product growth advice, analytics strategies, and comprehensive guides on using data to achieve product-market fit.[82]

Tool-Specific Guides

- For practical, hands-on guidance, the best resources are the official learning centers of the tools themselves. The documentation, articles, and tutorials offered by companies like Amplitude, Mixpanel, and PostHog are essential for mastering their platforms. Furthermore, their company blogs are excellent sources for case studies, best practices, and expert commentary on broader product analytics topics.[83]

---

[81] "Slow Down to Speed Up Your Product Analytics," Reforge Blog, accessed August 16, 2025, `https://www.reforge.com/blog/three-steps-to-effective-product-analytics`.

[82] "Lenny's Newsletter," Lenny Rachitsky, accessed August 16, 2025, `https://www.lennysnewsletter.com/`.

[83] "Amplitude Blog," Amplitude, accessed August 16, 2025, `https://amplitude.com/blog`.

Book Recommendations

- *Qualitative Data Analysis: A Methods Sourcebook* provides a comprehensive exploration of how to systematically analyze and interpret qualitative data.[84]

- *Designing Quality Survey Questions* is a practical guide for crafting surveys that avoid bias and elicit clear, actionable responses.[85]

- *Interviewing Users: How to Uncover Compelling Insights* teaches the art and science of conducting effective user interviews to understand the true needs and motivations driving customer behavior.[86]

- *Focus Groups: A Practical Guide for Applied Research* serves as an excellent handbook for planning and executing focus groups, a powerful method for gauging collective reactions to new concepts and designs.[87]

Summary

In this chapter, we shifted our focus from the wide-angle lens of business intelligence to the powerful microscope of product analytics. You learned how to understand the intricate journeys your users take, anchored by the core principle of "doing things that don't scale" to build a product people truly love. We established that product analytics and event tracking are the tools that allow you to apply this handcrafted, user-centric approach at scale, giving you a detailed view of the clicks, taps, and scrolls that define the user experience.

We then explored the toolkit for gathering deep user insights. You learned how to pair quantitative data with qualitative methods to understand the "why" behind the "what." We covered powerful product analytics tools, including

---

[84] Matthew B. Miles, A. Michael Huberman, and Johnny Saldaña, *Qualitative Data Analysis: A Methods Sourcebook* (SAGE Publications, 2019).

[85] Sheila Robinson, and Kimberly Firth Leonard, *Designing Quality Survey Questions* (SAGE Publications, 2018).

[86] Steve Portigal, *Interviewing Users: How to Uncover Compelling Insights* (Rosenfeld Media, 2023).

[87] Richard A. Krueger, and Mary Anne Casey, *Focus Groups: A Practical Guide for Applied Research* (SAGE Publications, 2014).

- **Session Replays**, which provide a video-like recording of a user's exact session, helping you diagnose frustration points like "rage clicks" and confusing UI interactions.

- **Heatmaps**, which offer an aggregate visual summary of user attention. You saw how to use different types, from click maps that show where users interact to move maps that reveal where their cursor pauses, indicating interest.

- **Targeted Surveys**, emphasizing the importance of using a mix of open-ended and closed-ended questions to gather direct feedback.

- **User Interviews**, which provide in-depth feedback using *The Mom Test* principles. You discovered how to focus on users' lives rather than your ideas, avoid hypotheticals, and separate problem discovery from solution validation.

- **Focus Groups**, which can reveal group dynamics and shared perceptions about concepts with strong subjective or social components, while understanding their limitations and risks like groupthink.

Next, we tackled the critical challenge of choosing the right product analytics platform. You explored the diverse landscape of tools, from feature-rich digital analytics platforms thoroughly evaluated in The Forrester Wave: Digital Analytics Solutions report like Amplitude to open source session replay specialists like OpenReplay. We categorized tools into Leaders, Strong Performers, and Contenders based on the Forrester research and examined open source alternatives for cost-conscious startups.

To cut through the overwhelming array of options, you learned a pragmatic four-step framework for tool selection: defining your audience and core use cases, assessing technical resources and existing infrastructure, evaluating true costs including startup programs, and determining feature and scalability requirements. We demonstrated this framework through the *Empathic* case study, where structured evaluation led to selecting Amplitude's startup program as the best fit for the startup's needs.

The chapter then explored the practicalities of building a robust event tracking plan. We discussed the importance of clean, reliable data and how to construct a single source of truth by focusing on critical user paths, consistent naming conventions, and enriching events with contextual properties. Common pitfalls were identified. *Empathic's* success story compellingly illustrated how a focused tracking plan can lead to significant improvements, such as their booking conversion rate through a data-driven A/B test.

This case study underscores a vital principle: an insight, however powerful, is only a starting point. To truly drive meaningful improvement and build a product that demonstrably gets better with every iteration, you must test your ideas systematically. Armed with a robust toolkit to diagnose problems and form educated hypotheses about user behavior, you are now ready for the next step. In Chapter 11, we will transition from observation to action, exploring the principles of experimentation and A/B testing, and learning how to design, run, and analyze controlled experiments to validate your hypotheses.

PART IV

Advanced Topics and the Future

CHAPTER 11

Experimentation and A/B Testing

To invent you have to experiment, and if you know in advance that it's going to work, it's not an experiment. Most large organizations embrace the idea of invention, but are not willing to suffer the string of failed experiments necessary to get there. Outsized returns often come from betting against conventional wisdom, and conventional wisdom is usually right.

—Jeff Bezos,[1]

Founder of Amazon

A startup is fundamentally one grand experiment—statistically unlikely to succeed, yet irresistibly appealing to those brave enough to challenge the odds. While larger companies experiment within established frameworks and safety nets, startups experiment with their very existence.

This chapter guides you through the complete journey of building an effective experimentation practice tailored specifically for startups. We'll begin by examining how experimentation differs fundamentally between early-stage companies and established enterprises, then dive into the practical mechanics of designing, running, and analyzing tests that drive meaningful business outcomes.

Here's what you'll master by the end of this chapter:

- **Understanding the Startup Experimentation Landscape:** How experimentation accelerates both success and failure, and why this dual outcome represents a strategic advantage rather than a limitation

---

[1] Amazon, "2015 Letter to Shareholders," SEC filing, 2016, https://www.sec.gov/Archives/edgar/data/1018724/000119312516530910/d168744dex991.htm.

P. Sidoruk, *From Data to Dollars*, https://doi.org/10.1007/979-8-8688-1898-1_11

- **Designing Effective Experiments:** Moving beyond simple A/B tests to understand when and how to employ different experimental approaches, from fake door testing to structured learning funnels

- **Implementing Your First Tests:** Selecting appropriate testing tools, validating your technical setup, and using frameworks like the Test Card to transform assumptions into rigorous hypotheses

- **Avoiding Common Pitfalls:** Recognizing and mitigating the biases and statistical traps that can derail even well-intentioned experiments

- **Building Sustainable Experimentation Capabilities:** Developing processes that generate continuous learning rather than one-off optimizations

Throughout this exploration, we'll emphasize practical application over theoretical perfection. You'll see real examples, learn from actual startup case studies, and gain the confidence to start experimenting immediately—even with limited traffic and resources.

Designing Experiments in the Startup Context

While experimentation is a vital tool for both startups and established corporations, its relative impact and strategic function differ dramatically. In a large corporation, experiments are often deployed to optimize products that are already operating at scale. Consequently, it is rare for anyone to point to a single experiment that proved existential for the company. Successful tests tend to produce incremental gains when measured against its vast overall revenue.

For a startup, however, the influence of an experiment is magnified. It is not uncommon for a founder to pinpoint a single, often lightweight, test that was truly existential for the business. The results can have a disproportionately large impact on performance and strategy. For example, a simple experiment that successfully optimizes the top of the marketing funnel might double the user conversion rate, allowing the company to reach breakeven and secure its survival.

This high-stakes environment demands a different approach. Large companies can dedicate substantial resources to comprehensive testing programs, complete with specialized teams and sophisticated infrastructure. At Microsoft, for example, their experimentation platform evolved over many years with dedicated developers, managers, and data scientists. Startups, by contrast, must embrace lean experimentation. Constrained by resources, they test critical hypotheses using smaller sample sizes and compressed timelines, forcing them to ruthlessly prioritize which questions truly require formal testing.[2]

To navigate this path from an initial idea to a mature, data-driven organization, startups can follow a clear, five-stage journey. Figure 11-1 illustrates this progression. It emphasizes a crucial philosophy: **don't run before you can walk**. The framework shows that sophisticated A/B/n testing is the destination, not the starting point, and that foundational work must come first. This journey can be broken down into five key phases:

1. **Qualitative Focus to Find Product-Market Fit:** In the earliest days, when traffic is low, the priority is not statistical testing but deep qualitative learning. The goal is to validate core ideas through user interviews, prototypes, and direct feedback to find product-market fit.

2. **Alternative Experiments:** With limited traffic and resources, startups can use clever, low-cost methods like fake door tests and structured learning funnels. These techniques help validate demand for new features by combining qualitative insights with early quantitative signals.

3. **Build the Data Foundation:** Before you can trust your data, you must build the systems to collect it reliably. This stage involves setting up the essential infrastructure, including a data warehouse, dashboards for key metrics, and robust event tracking and selecting the right experimentation solution to power your testing efforts.

4. **Validate and Run A/B Tests:** Once the data foundation is solid and traffic is sufficient, you can begin formal A/B testing. The first step is running an A/A test to validate the technical setup, followed by simple, high-impact tests on key user funnels.

5. **Mature Experimentation:** With significant traffic and established product-market fit, the experimentation practice can evolve. At this stage, startups can employ more advanced methods like multivariate testing and AI-driven optimization to refine the product at scale.

---

[2] Ron Kohavi, Thomas Crook, and Roger Longbotham, "Online Experimentation at Microsoft," Exp Platform, 2009, https://exp-platform.com/Documents/ExP_DMCaseStudies.pdf.

Startup Experimentation: From (No) Data To Dollars

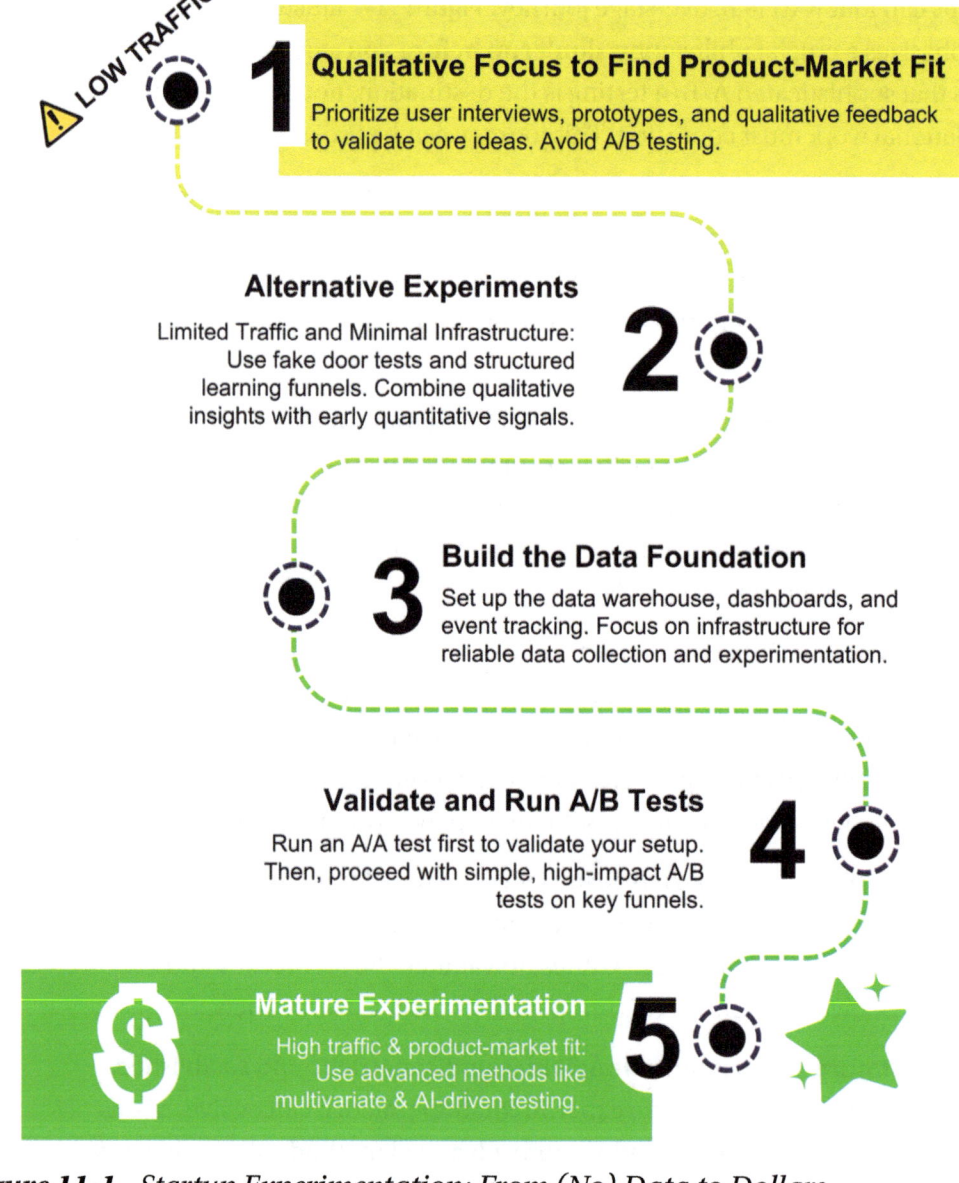

Figure 11-1. *Startup Experimentation: From (No) Data to Dollars*

This chapter will guide you through the key experimentation stages, from the practical mechanics of designing and running your first tests to avoiding the common pitfalls that can mislead even the most well-intentioned teams.

While the figure provides a simplified road map, the journey is not always linear, and nuances exist. For some startups, certain testing opportunities may arise earlier than the framework suggests, depending on the company's unique context. The goal of this model is to illustrate a general mindset: don't try to run complex A/B/n tests without sufficient traffic or product-market fit, and don't overcomplicate experimentation before the foundational elements are in place.

This lean, high-stakes environment indeed requires a specific mindset. As Jeffrey Bussgang, Co-founder and General Partner at Flybridge Capital, observes from his work with thousands of founders:

> *The most successful founders I've worked with—from teaching thousands at Harvard Business School to investing in hundreds through Flybridge Capital—approach their startups like scientists. They don't just build products; they run experiments. They don't just execute plans; they test hypotheses.*

> —Jeffrey Bussgang,[3]
> Co-founder and General Partner at Flybridge Capital

For startups, as Bussgang points out, experimentation serves a core strategic role. It is the primary process by which teams find and accelerate winning ideas, create entirely new products, or learn when to cut their losses. Unlike the marginal gains often seen in corporate settings, a single experiment in a startup can be a game-changer that redirects the entire future of the company.

I'm reminded of two friends from my university days whose careers took starkly different turns. The first accepted a coveted data science position at one of the tech giants. He imagined a dynamic, fast-paced environment but was soon disillusioned. His job consisted almost entirely of running A/B tests on trivial UI changes, like button colors, yielding work that felt unimportant and failed to challenge him.

---

[3] Jeffrey Bussgang, "Every startup founder needs a laboratory," The Experimentation Machine, February 18, 2025, `https://experimentationmachine.com/p/every-startup-founder-needs-a-laboratory`.

My other friend pursued the riskier path of entrepreneurship by joining a new startup. His role was that of a data generalist, a stark difference from the hyper-specialized position of his corporate counterpart. He did not run countless experiments on minor features; rather, he was responsible for a wide range of projects that directly shaped the business. His experiments targeted the top of the funnel, dealing with foundational pillars like the company's business model, core user experience, and user acquisition strategy, and their outcomes were directly tied to financial success. While one friend optimized pixels, the other optimized for the company's bottom line. Ultimately, his willingness to lead these high-stakes projects paid off, building a company that made him a millionaire and providing a dramatic contrast to the prestigious but limiting corporate role.

The success of this approach highlights a fundamental truth: a startup's ability to experiment on core business questions is precisely what determines its trajectory. In a comprehensive study, Rembrand Koning of Harvard Business School, along with Sharique Hasan and Aaron Chatterji, analyzed the performance of over 35,000 high-tech startups over a four-year period. Their work revealed that systematic experimentation does more than just fine-tune a business—it accelerates its ultimate fate.[4]

According to the research, A/B testing was found to push companies toward definitive outcomes much faster. As Koning emphasizes, for founders with flawed ideas, experimentation helps them "move onto their next opportunity faster."[5]

Key findings from their research include

- **Accelerated Scaling and Innovation:** Startups that consistently used A/B testing scaled their operations more quickly and launched more new products compared to their non-testing peers.

- **Efficient Failure:** The same startups were also more likely to fail faster when their fundamental hypotheses were incorrect. This outcome, while seemingly negative, is a strategic advantage, as it prevents founders from wasting valuable time and capital on unviable ideas.

---

[4] Rembrand Koning, Sharique Hasan, and Aaron Chatterji, "Experimentation and startup performance: Evidence from A/B testing," Harvard Business School Working Paper, September 10, 2021, https://www.hbs.edu/ris/Publication%20Files/AB_Testing_R_R_08b97538-ed3f-413e-bc38-c239b175d868.pdf.

[5] Kristen Senz, "Is A/B Testing Effective? Evidence from 35,000 Startups," Harvard Business School Working Knowledge, January 11, 2021, https://www.library.hbs.edu/working-knowledge/is-ab-testing-effective-evidence-from-35000-startups.

- **Decisive Results:** Experimentation led to more polarized outcomes. Companies were more likely to achieve significant success or shut down quickly, effectively helping them avoid the "zombie startup" scenario where a business languishes without real traction.

- **The Special Case of Early-Stage Startups:** The research reveals a distinct pattern among the youngest companies: those that adopt A/B testing often fail faster. This is typically attributed to two factors: the founders' inexperience in applying these methods correctly or, more fundamentally, that the core business idea itself was not viable. Industry experts note that A/B testing isn't universally beneficial at this stage; for it to be effective, certain criteria—such as sufficient traffic, product-market fit, and clarity of hypothesis—must be met. These prerequisites will be discussed later in this chapter.[6]

For startups, therefore, experimentation, when applied correctly, becomes a key competitive advantage. Its purpose is simple: to learn what works and, just as importantly, what doesn't.

What Are A/B Tests?

A/B testing, also known as split testing or online controlled experimentation, is a statistical method for determining which of two versions of a webpage, feature, or marketing message performs better. It allows you to replace intuition with data-driven evidence, making controlled changes that drive specific business goals, such as improving user retention, increasing conversions, or boosting customer lifetime value (LTV).[7]

Let's review a practical example drawn from my own experience that highlights a critical startup strategy. Encouraging users to choose an annual plan over a monthly one can significantly improve cash flow, increase LTV, and boost long-term retention.

---

[6] Kristen Senz, "Is A/B Testing Effective? Evidence from 35,000 Startups," Harvard Business School Working Knowledge, January 11, 2021, https://www.library.hbs.edu/working-knowledge/is-ab-testing-effective-evidence-from-35000-startups.

[7] Federico Quin, Danny Weyns, Matthias Galster, Camila Costa Silva, A/B testing: A systematic literature review, Journal of Systems and Software, Volume 211, 2024, 112011, ISSN 0164-1212, https://doi.org/10.1016/j.jss.2024.112011.

Optimizing how you present this offer on your landing page—before a visitor even signs up—can be one of the highest-leverage activities for a subscription business, often proving far more impactful than minor tweaks like changing button colors within your app. Figure 11-2 compares two designs for presenting an annual subscription plan.

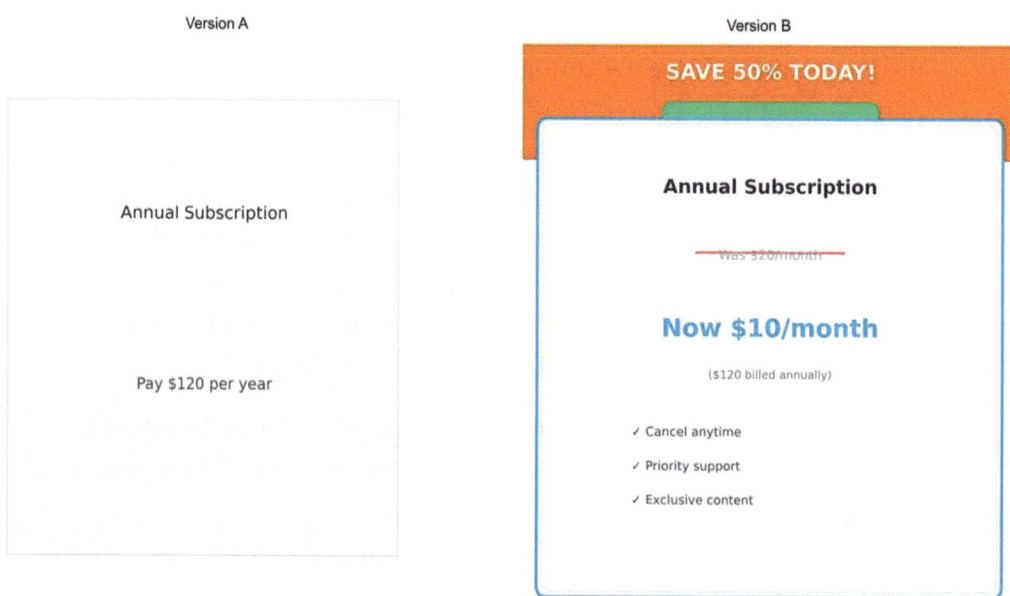

Figure 11-2. *A/B Test Example*

Version A is simple and straightforward, presenting the full annual cost of $120. Version B, however, employs several common marketing techniques. Instead of showing the total upfront cost, it frames the price in a more appealing monthly equivalent (*Now $10/month*), making it easier for customers to compare against a more expensive monthly subscription. It grabs attention with a prominent headline (*SAVE 50% TODAY!*), uses price anchoring (*Was $20/month*) to frame the new price as a great deal, and lists key benefits to reinforce the value.

On the surface, this sounds simple, right? Just launch the test and see which version gets more clicks. This is the moment, however, where discipline must overcome impatience. I've seen many well-intentioned teams sabotage their own work because they see statistical rules as an obstacle to speed. The arguments are always the same: *"Why wait two weeks when we can get a signal in two days?"* or *"We don't need a fancy tool, let's just switch the design on Monday and compare it to last week."* While born from a desire to move fast, this thinking confuses movement with progress and is the fastest way to be misled by your own data.

Statistical Methods Behind A/B Testing

For the results of an A/B test to be trustworthy, the process must adhere to strict statistical principles. A reliable test requires randomization and sufficient sample size. Randomization ensures that users are assigned to see Version A or Version B by pure chance, eliminating bias. A sufficient sample size—meaning enough users see each version—is crucial to ensure that the outcome isn't just random noise but a true reflection of user preference. Without this rigor, you risk making important business decisions based on flawed or misleading data.

At the heart of any A/B testing tool is a statistical engine that analyzes the results. The two most common statistical approaches are Frequentist and Bayesian. Complementing these engines are different testing protocols, like sequential methods, which offer more flexibility in how and when a test is concluded.

Sequential testing allows experimenters to monitor results in real time and potentially reach conclusions earlier than with traditional fixed-sample A/B testing. Unlike classic tests that require a pre-specified sample size and analysis only at the end, sequential testing enables valid, interim analyses at multiple points as data accumulates.[8]

Neither approach is universally superior. They involve important trade-offs between statistical power, flexibility, and ease of use.

Randomization Methods

Properly randomizing users into groups is the bedrock of a valid experiment. If specific types of users are more likely to see one version over another, the results will be biased. The most common randomization methods are

- **Simple Randomization:** This is the most straightforward method. Every user who enters the experiment has an equal chance of being assigned to the control (Version A) or the variant (Version B). Think of it as a coin flip for each user. It's easy to implement and works well for most tests.

---

[8] "The role of sequential testing in A/B testing," Kameleoon, July 24, 2024, https://www.kameleoon.com/blog/role-sequential-testing-ab-testing.

- **Stratified Randomization:** Sometimes, you know that specific user segments behave differently (e.g., new users vs. returning users, or users on iOS vs. Android). If one group is overrepresented in your test by random chance, it could skew the results. Stratified randomization helps prevent this by ensuring that the proportion of these segments is balanced across both the control and variant groups. For example, it ensures that both Version A and Version B are shown to a 50/50 split of new and returning users, giving you a more precise and reliable result.[9,10]

Sample Size and Test Duration

Before running a test, you must determine two critical parameters: the required sample size (how many users you need) and the test duration (how long it should run). These calculations are not a formality. They determine whether your experiment will have enough statistical power to detect a real change. For a startup with a small user base, this step is essential for assessing whether a proposed test is even feasible.

- **Sample Size:** The required sample size depends heavily on the expected difference between your control and variant. If you anticipate a small improvement (e.g., a 2% lift in conversions), you will need a much larger sample size to detect it confidently. Conversely, if you expect a massive change, you can get away with a smaller sample.

- **Test Duration:** A test must run long enough to capture a full user activity cycle. Depending on the industry, user behavior often differs between weekdays and weekends or even at different times of the day. Running a test for at least one, and preferably two, full weeks helps smooth out these fluctuations and prevents you from making a decision based on an anomaly.

Fortunately, you don't need to be a statistician to figure this out. There are numerous online calculators that can help you determine the necessary sample size and duration. A/B testing platforms like VWO offer calculators that work with both major statistical

[9] Anastasia Bogapova, "A/B Testing Essentials: Strategies, Metrics, and AI," DevToDev, November 7, 2024, https://www.devtodev.com/resources/articles/a-b-testing-essentials-strategies-metrics-and-ai.

[10] Allon Korem, Oryah Lancry-Dayan, "Randomization: The ABC's of A/B Testing," Statsig, June 30, 2025, https://www.statsig.com/blog/randomization-the-abcs-of-a-b-testing.

engines—the traditional Frequentist model and their enhanced SmartStats engine, which is based on Bayesian statistics. While the technical differences are complex, understanding the basics of each approach is helpful.[11]

Frequentist Statistics

The Frequentist approach is the more traditional method taught in most introductory statistics courses. The process begins by establishing a null hypothesis—the assumption that there is no real difference between your two versions and that any observed changes are due to random chance.[12]

- **Key Concepts:** The goal is to see if you can gather enough evidence to reject the null hypothesis. It uses concepts like the p-value and confidence level. A low p-value (typically under 0.05) suggests that your results are statistically significant. That allows you to reject the null hypothesis and declare a winner.

- **Pros:** As the established standard, its methods are widely published and understood in the scientific community. Many simple A/B testing tools use this approach, making it accessible without requiring deep statistical knowledge to set up a basic test.

- **Cons:** Its primary weakness is its rigidity. You determine your sample size in advance and should not stop the test early (a practice known as "peeking"). This can be a significant drawback for a startup that needs to make quick decisions with limited traffic. Furthermore, the focus on statistical significance can sometimes obscure practical significance. A result might be statistically significant but represent a lift so small that it provides no real business value and isn't worth the engineering effort to implement.[13]

---

[11] "A/B Test Sample Size & Duration Calculator," VWO, accessed August 29, 2025, `https://vwo.com/tools/ab-test-duration-calculator/`.

[12] Phil Burch, "Frequentist vs. Bayesian: Comparing Statistics Methods for A/B Testing," Amplitude, July 31, 2024, `https://amplitude.com/blog/frequentist-vs-bayesian-statistics-methods`.

[13] Sven Schmit, "Comparing Frequentist vs. Bayesian vs. Sequential Approaches to A/B Testing," Eppo, January 8, 2024, `https://www.geteppo.com/blog/comparing-frequentist-vs-bayesian-approaches`.

Bayesian Statistics

The Bayesian approach has become increasingly popular in the tech world because it is more aligned with business decision-making.[14]

- **Key Concepts:** Unlike the abstract outputs of Frequentist methods, the Bayesian approach is more intuitive because it directly answers the question startup leaders care about: *"What is the probability that this change is actually better?"*[15]

- **Pros:** Bayesian methods are highly flexible and provide more intuitive results. This allows you to make simple, powerful statements like, *"The new version has an 85% chance of beating the old one,"* which is far easier for non-technical stakeholders to understand. This flexibility also works seamlessly with sequential testing, allowing you to stop tests early the moment you have a confident answer—a major advantage for fast-moving startups.[16]

- **Cons:** The primary criticism of Bayesian methods is their reliance on a "prior"—an initial assumption about the likelihood of a version's success. The choice of this prior can be subjective, and if chosen poorly, it can influence the result. For this reason, the method can be perceived as less objective than its Frequentist counterpart.[17]

---

[14] Sven Schmit, "Comparing Frequentist vs. Bayesian vs. Sequential Approaches to A/B Testing," Eppo, January 8, 2024, https://www.geteppo.com/blog/comparing-frequentist-vs-bayesian-approaches.

[15] Deborah O'Malley, "Frequentist vs Bayesian AB Testing: Which Method Is Right for You?," Kameleoon, July 28, 2025, https://www.kameleoon.com/blog/ab-testing-bayesian-frequentist-statistics-method.

[16] Deborah O'Malley, "Frequentist vs Bayesian AB Testing: Which Method Is Right for You?," Kameleoon, July 28, 2025, https://www.kameleoon.com/blog/ab-testing-bayesian-frequentist-statistics-method.

[17] Sven Schmit, "Comparing Frequentist vs. Bayesian vs. Sequential Approaches to A/B Testing," Eppo, January 8, 2024, https://www.geteppo.com/blog/comparing-frequentist-vs-bayesian-approaches.

AI-Driven A/B Testing

AI-driven A/B testing platforms leverage machine learning to automate variant optimization and accelerate experiment cycles. Rather than evaluating variants in a fixed plan, AI-based tools can dynamically allocate more traffic to promising versions, potentially delivering reliable results sooner. They can also uncover non-obvious patterns in user behavior and automatically segment audiences, increasing statistical power and the ability to personalize at scale. For startups, AI-driven experimentation lowers the bar for running effective tests, especially as technical and analytical resources are limited.[18,19]

Multivariate Testing

While A/B testing focuses on changing one variable at a time, multivariate testing allows you to test multiple variables simultaneously to see which combination produces the best results. This technique is useful when you want to understand not just which element performs better but how different elements on a page interact with one another.[20]

For example, imagine you want to test two different headlines and three different hero images on your home page. A multivariate test automatically creates and tests all six possible combinations. In effect, you are no longer running a simple two-variant A/B test, but an A/B/C/D/E/F test, where each variant is a unique combination of a headline and an image, as shown in Table 11-1.

Table 11-1. *Multivariate Test Combinations*

	Image 1	Image 2	Image 3
Headline X	Variant A	Variant B	Variant C
Headline Y	Variant D	Variant E	Variant F

---

[18] "How to use AI for A/B testing," Kameleoon, accessed August 29, 2025, https://www.kameleoon.com/ai-ab-testing.

[19] Anastasia Bogapova, "A/B Testing Essentials: Strategies, Metrics, and AI," DevToDev, November 7, 2024, https://www.devtodev.com/resources/articles/a-b-testing-essentials-strategies-metrics-and-ai.

[20] "Multivariate Testing," Optimizely, accessed August 29, 2025, https://www.optimizely.com/optimization-glossary/multivariate-testing/.

The primary benefit of this approach is its efficiency in finding the most effective combination of changes, potentially saving you from running multiple, sequential A/B tests.

However, the main challenge with multivariate testing is the amount of traffic. Because the traffic has to be split among many different variations (six in the example above), it takes a much larger sample size and a longer time to reach statistically significant results compared to a simple A/B test. For this reason, multivariate testing is generally better suited for more mature companies with high-traffic websites, while early-stage startups are usually better off sticking to simpler A/B tests that test one significant change at a time.

When Not to A/B Test

The wisdom of experimentation isn't just in knowing how to A/B test but, more importantly, when. While a powerful tool for optimization, for an early-stage startup, it can become a counterproductive distraction from more critical activities. Knowing when to avoid testing is as important as knowing when to embrace it. Here are the key scenarios where you should pause and consider a different approach:

- **When You Don't Have Enough Traffic:** As discussed above, A/B tests rely on statistical significance to be trustworthy. Reaching significance requires data, which for most startups translates to a substantial amount of user traffic. If your website or app has a low number of users, an experiment might need to run for months to produce a reliable result. By that time, your product, market, or user base may have already changed, rendering the results obsolete. Making decisions based on tests with insufficient traffic is one of the most common ways to be misled by random noise.[21,22,23,24]

[21] Thomas Peham, "Startups, please stop talking about A/B testing!," Usersnap, accessed August 29, 2025, https://usersnap.com/blog/startups-stop-ab-testing/.

[22] "The Limitations of A/B Testing for Startups," RW Sugg, February 20, 2024, https://rwsugg.com/2024/02/20/the-limitations-of-a-b-testing-for-startups/.

[23] Dayana Marin, "B2B Startups Struggle to A/B Test Product Features: Strategies to Overcome Challenges," AdaSight, accessed August 29, 2025, https://www.adasight.com/blog/b2b-startups-struggle-to-a-b-test-product-features-strategies-to-overcome-challenges.

[24] Ian Vanagas, "10 things we've learned about A/B testing for startups," PostHog Newsletter, August 24, 2023, https://posthog.com/newsletter/what-we've-learned-about-ab-testing.

- **Before You Have Product-Market Fit:** A/B testing is purely a tool for optimization. Its purpose is to refine something that already shows signs of working. For instance, the pricing page experiment (Figure 11-2) is a classic optimization—it takes a functioning sign-up process and aims to improve it by encouraging more users to select an annual plan. However, A/B testing is not the right tool for making the big, strategic leaps required to find out if you have a viable business in the first place. If your core value proposition isn't resonating with an audience, then optimizing the headline on your landing page is like rearranging deck chairs on the Titanic. At this early stage, your focus must be on qualitative feedback—such as direct user interviews and rapid prototyping—to make bold, strategic bets.[25]

- **When Foundational Work Comes First:** For an early-stage startup, your most constrained resource is time. As the first data hire, you must be ruthless in your prioritization. A/B testing is just one of dozens of data-related activities you could be doing, and it is rarely the most important one at the start. Foundational projects—like setting up the data warehouse, building the first reliable dashboards, or establishing core ETL/ELT processes—often deliver far more leverage across the entire organization. Before you get drawn into optimizing a single page or a feature, ensure the foundational data infrastructure that enables company-wide intelligence is in place. Don't let the allure of tactical optimization distract you from the highest-impact strategic work.

- **When the Fix Is Obvious:** Some changes are simply common sense. If your checkout flow is broken or your sign-up form has a bug, you don't need to A/B test a version that fixes it. Running a test in this scenario means knowingly subjecting half of your users to a

---

[25] Robert Kaminski, "Early stage startups seeking product market fit should NOT do any AB testing," LinkedIn post, accessed August 29, 2025, https://www.linkedin.com/posts/heyrobk_early-stage-startups-seeking-product-activity-7226950981467394048-cEsh/.

broken experience for no valuable reason. The goal is to improve the product, and when a change is an undeniable improvement, the right move is to ship it immediately.[26]

— **When Making Radical Redesigns:** A/B testing is best suited for comparing specific, isolated variations. It is not the right framework for evaluating a complete product redesign or a major strategic pivot. When you change everything at once, you can't attribute an outcome to any single element, making the "why" behind the results impossible to understand. These transformative bets should be driven by a deep understanding of customer needs and a clear strategic vision, not a simple split test.[27]

For early-stage startups, the primary goals are learning and speed. Often, the fastest path to learning is not through a statistically rigorous A/B test but through decisive action, qualitative feedback, and a relentless focus on solving a real customer problem.

Other Types of Experiments

When formal A/B testing is impractical due to low traffic or a lack of product-market fit, it doesn't mean you should stop learning or iterating. You don't have to A/B test everything, especially as an early-stage startup.

A lot can be achieved by combining the experimentation methods described here with the qualitative and quantitative approaches detailed in Chapter 10. These include qualitative methods like user surveys and session replay analysis, and quantitative methods such as funnel and retention analysis based on event tracking. Instead of relying on strict statistical validation, you can adopt a more flexible approach focused on structured learning.

This flexibility is particularly powerful for evaluating multiple variants without splitting your limited traffic. For instance, you can conduct **usability tests**, where you observe a user interacting with different prototypes to see if they can complete key tasks

---

[26] Daniel Macák, "When NOT to use A/B tests," Dev.to, February 5, 2024, `https://dev.to/daelmaak/when-not-to-use-ab-tests-1ag7`.

[27] "The Limitations of A/B Testing for Startups," RW Sugg, February 20, 2024, `https://rwsugg.com/2024/02/20/the-limitations-of-a-b-testing-for-startups/`.

and where they encounter friction. Alternatively, you can run **preference tests**, which present multiple designs side by side and simply ask users which one they prefer and why. These qualitative methods allow you to compare several options at once to identify the most promising direction before committing development resources.[28,29]

Fake Door Testing

A fake door test is a simple yet powerful technique to gauge customer interest in a new feature or product before committing development resources. It works by placing a call to action, such as a button or link, for a feature that doesn't exist yet. When a user clicks it, they are typically shown a message explaining that the feature is coming soon and their interest has been noted.[30]

This method helps you validate demand with minimal effort. The number of clicks on the "fake door" serves as a strong indicator of how valuable the proposed feature would be to your users, allowing you to prioritize your product road map based on real user intent.[31]

The Structured Learning Funnel

Even when a startup's user base is too small for formal A/B testing, it is still crucial to adopt a structured approach to experimentation. Without a clear framework, you risk making decisions based on random noise rather than real insight.

Several articles from the startup world, like "Growth Experiments at SoundCloud" and "Startups, Please Stop Talking About A/B Testing!", offer a powerful alternative. They champion the idea of a "structured learning funnel"—a continuous learning cycle designed for environments where statistical significance is out of reach. This approach

[28] "Usability Testing: what it is, its benefits, and why it matters," Contentsquare, last modified November 7, 2024, `https://contentsquare.com/guides/usability-testing/`.

[29] Saffa Faisal, "Preference Testing: A Step-By-Step Guide," Userpilot, December 30, 2024, `https://userpilot.com/blog/preference-testing/`.

[30] Elena Mitsiou, "Fake Door Testing: Measuring User Interest," UXtweak, October 18, 2024, `https://blog.uxtweak.com/fake-door-testing/`.

[31] Sophie Grigoryan, "Fake Door Testing Explained: What Is It and How to Make An Effective Fake Door Test," Userpilot, August 9, 2025, `https://userpilot.com/blog/fake-door-testing/`.

moves away from a strict pass/fail mindset toward a "what can we learn?" perspective. This "experiment funnel" involves four key steps:[32,33]

1. **Brainstorming:** The process begins with generating ideas. Gather your team and brainstorm potential changes—big or small—that you believe could improve a key metric, whether it's sign-ups, user engagement, or conversion rates. All ideas are collected into a backlog, creating a repository of potential growth levers.

2. **Prioritizing:** With limited resources, you can't test everything at once. The next step is to prioritize your backlog. Evaluate each idea based on three simple criteria: potential impact, resources required, and probability of clear outcome. This third criterion is crucial: it prioritizes experiments that are likely to yield a conclusive result—successful or not—over those that may end with ambiguous data. Focus on the ideas that offer the best balance of high potential and low cost.

3. **Testing:** Once an experiment is prioritized, you execute it. This could be as simple as changing the call to action on a page or as complex as testing a new feature with a small group of users. The key is to document the process and track the relevant data, even if it's not enough for statistical significance.

4. **Analyzing:** This is the most critical step. After the experiment runs, analyze the outcome by asking a few key questions:

   a. Impact: What were the results? Did the metric move?

   b. Accuracy: How close were the results to your original hypothesis?

   c. Why: Why do you think you saw these results? This final question is what separates this method from a pure A/B test. It forces you to combine quantitative data with qualitative insights. The goal is not just to see what happened but to understand why it happened, often leading to deeper customer understanding and better ideas for future experiments.

---

[32] Thomas Peham, "Startups, please stop talking about A/B testing!," Usersnap, accessed August 29, 2025, https://usersnap.com/blog/startups-stop-ab-testing/.

[33] Stefanos Karakasis, "Growth Experiments at SoundCloud," UX Design, October 23, 2014, https://uxdesign.cc/growth-experiments-at-soundcloud-9a8579dc8b3a#.5ut8qj7d1.

This focus on structured learning over strict pass/fail outcomes highlights a crucial point: the philosophy and practice of experimentation are not one-size-fits-all. The approach a company takes is fundamentally shaped by its unique context—its resources, its market position, and the nature of the questions it needs to answer. For startups, these factors differ dramatically from those in established corporate environments, leading to distinct goals, methods, and mindsets when it comes to testing and validation.

What Startups Can Experiment With?

With a solid understanding of when and how to experiment, let's turn our attention to the practical question every startup faces: where should you focus your limited experimentation resources? The answer depends on your stage, your constraints, and your most critical business hypotheses. Here's a comprehensive look at the key areas where experimentation can drive the most significant impact.

Here's what you can A/B test:

- **Product Features and Design:** You can test different designs, layouts, and user flows within your product. For example, you might compare two different versions of a feature's user interface to see which one leads to higher engagement, better usability, or an increase in the completion of key actions.

- **Your Website:** Your website is a critical tool for converting visitors into customers, whether that means encouraging them to download an app, purchase a subscription, or sign up for a newsletter. You can A/B test nearly every element, including headlines, call-to-action buttons, images, and overall page layouts, to optimize your conversion rates.

- **Back-End Algorithms:** You can also test different versions of your back-end algorithms. For instance, if you have a recommendation system, you could test a new algorithm against the old one to see which leads to a higher click-through rate or more items being added to a cart.

- **Email Campaigns:** Email is a powerful channel for engaging with users. Many email marketing platforms, such as Mailchimp, have built-in A/B testing features that allow you to experiment with different subject lines, email copy, and calls to action to see what resonates with your audience. Crucially, you don't always need a finished product to start learning. If you have a waiting list, you can begin A/B testing your messaging on these early sign-ups to gather valuable insights before you even launch. This might seem to contradict the earlier advice to avoid A/B testing before achieving product-market fit, but the distinction lies in the goal of the experiment. While post-fit testing aims to optimize a working product, pre-fit testing on a waiting list is a form of strategic research. Its purpose is to validate your core value proposition and test your fundamental messaging. By A/B testing different headlines, benefits, and calls to action on a landing page, you are gathering quantitative data on what resonates with your target audience, a critical step in the search for product-market fit itself.

How to Approach Your First A/B Tests

The sheer number of opportunities—from back-end algorithms to email subject lines—can be exhilarating, but this breadth introduces the primary challenge for any startup: focus. Knowing what you can test is the easy part; knowing where you should actually start is what separates a successful experimentation program from a series of random activities.

When you're getting started, being strategic is paramount. A disciplined approach will save you time, conserve your limited traffic, and yield far more meaningful results. Here is the framework for developing that focus:

- **Prioritize High-Impact Hypotheses:** Think strategically about changes that could fundamentally impact your key business metrics. The pricing page example is a perfect illustration. A successful test here can dramatically increase revenue. Interestingly, it might even do so while decreasing average user engagement if you convert more users to an annual plan who then become inactive. This is a crucial trade-off to understand and monitor.

- **Keep It Simple:** Avoid the temptation to run numerous, complex experiments at once. Don't get bogged down with multivariate testing until you have both the experience and the significant user traffic required to get reliable results. Master the process with a small number of well-reasoned A/B tests first. Focus on what truly matters for your business and build your capabilities one step at a time.[34]

— **Focus on Top of Funnel:** For an early-stage startup, the highest-leverage experiments often happen at the top of the user funnel—the pages and flows that greet a user before they have even signed up. This is for three key reasons: it's where you have the most traffic, it's where you make your critical first impression, and it's where you can directly influence a customer's purchasing decision. Optimizing your pricing or sign-up page can be far more impactful than tweaking a feature deep inside your app. Think of a gym that aggressively markets an annual membership. It understands that a customer's motivation is highest at the moment of commitment. By persuading that customer to purchase the annual plan upfront, the gym secures immediate cash flow and increases that customer's lifetime value, even if they lose motivation and stop showing up soon after. The value was captured at the top of the funnel. This is precisely the logic behind the pricing page experiment in Figure 11-2. Successful companies apply this principle consistently. The online bank Monzo, for instance, concentrates its A/B testing on the top-of-funnel experience, running small, quick experiments where changes impact the largest number of people and can drive immediate business results.[35]

— **Test Meaningful Changes, Not Just Minor Tweaks:** When starting out, focus on significant changes that alter the user experience, not just minor cosmetic adjustments. While you want to avoid the complexity of a complete product redesign, you also shouldn't waste limited traffic on testing a different shade of blue on a button. The sweet spot lies in tests that are technically simple but psychologically profound. A simple change in messaging can test fundamentally

---

[34] "AB Testing Done Right - How Startups Grow EP7," How Startups Grow YouTube Channel, August 9, 2024, https://www.youtube.com/watch?v=XxLSlaJGagO.

[35] Ian Vanagas, "How YC's biggest startups run A/B tests (with examples)," PostHog, July 28, 2023, https://posthog.com/product-engineers/ab-testing-examples.

different user motivations to discover what truly drives your customers to act. This is where you can achieve a significant lift in conversions without a single line of new code. Consider a fitness app. Its users aren't a monolith; they have different goals. Some are motivated by long-term health, others by the specific goal of losing weight, and a third group might be most responsive to a financial incentive. Instead of using a generic call to action, you could run an A/B/C test to see which psychological trigger is most powerful:

— **Version A (Health Focused):** "Build a Lifelong Habit of Wellness"

— **Version B (Goal Focused):** "Achieve Your Weight Loss Goals Faster"

— **Version C (Savings Focused):** "Get 50% Off Your Annual Plan"

Each of these is a simple text change, but they are not minor tweaks. They are distinct value propositions aimed at different customer mindsets. This is how you move beyond optimizing pixels and start optimizing for motivation.[36]

With a backlog full of promising ideas sourced from competitive analysis, customer research, and your own data, the natural impulse is to start launching tests and hunting for big wins. But this is the critical juncture where an effective experimentation culture is either forged or broken. Before you can celebrate victories, you must first internalize the most difficult and important lesson in all of experimentation.

Managing Expectations: The Reality of Experimentation Outcomes

One of the most challenging truths for startup founders and data professionals to internalize is that most experiments will not yield the hoped-for positive results.

As Jeff Bezos points out, numerous big companies "are not willing to suffer the string of failed experiments"—for startups with way more limited budgets, time, and tools, it may be even more difficult.

---

[36] Anubhav Verma, "10 A/B test examples that work (From our analysis of 127,000 experiments)," Optimizely, April 15, 2025, https://www.optimizely.com/insights/blog/10-best-ab-testing-examples/.

It's crucial to celebrate learning over "winning." Experiments should be framed primarily as knowledge-generation tools rather than simple validation exercises. A "failed" experiment that conclusively disproves a faulty assumption is, in fact, a success because it prevents wasted resources and redirects entrepreneurial energy toward more promising avenues.

Teams should also learn to embrace small wins and understand their compound effects. Large, transformative A/B test victories are rare. More commonly, progress comes from modest, incremental improvements that, when stacked together over time, create significant cumulative impact.

Furthermore, recognizing the asymmetric value of negative results is key. Disproving a central, incorrect assumption early in a startup's life can save months or even years of misdirected effort and precious capital. The faster a startup can identify what doesn't work, the more quickly it can pivot to strategies that might.

Common Pitfalls and How to Avoid Them

Even well-designed experiments can lead to poor decisions if you fall into predictable traps. Before we dive into each one, Table 11-2 provides a quick cheatsheet of the most frequent pitfalls startup teams encounter and provides you with concrete strategies to avoid them.

Table 11-2. *Common Experimentation Pitfalls and How to Avoid Them*

Pitfall	Description	How to Avoid
The "Just Test It" Fallacy	Testing without a strategic hypothesis.	Articulate a clear, falsifiable hypothesis that addresses a meaningful question before launching any test.
The "Peeking" Problem	Calling a winner too early.	Commit to your pre-determined test duration or use a statistical engine designed for peeking.
Skipping Quality Assurance	A buggy experiment produces invalid data and harms your business.	Perform rigorous QA and monitoring.
The Novelty Effect	Users are often drawn to new things simply because they are new.	Run tests long enough for the novelty effect to wear off and for user behavior to stabilize.
Seasonality	User behavior changes with seasons, holidays, and business cycles, which can skew results.	Run experiments for at least one full business cycle.
Confounding Variables	Other factors, like different user cohorts (new vs. returning) or time of day, can influence and distort results.	When analyzing results, segment your data by these variables to uncover deeper insights and ensure your conclusions are accurate.
Confirmation Bias	Setting up tests in a way that is designed to confirm a pre-existing belief, which undermines objectivity.	Avoid designing tests where the alternative version is intentionally worse or framed to guarantee the success of your preferred option.
Sample Ratio Mismatch (SRM)	A significant deviation from your intended traffic split, which signals a technical bug.	Always check for SRM before analyzing results. If a mismatch is found, discard the results, find the bug, and rerun the experiment.
Ignoring Statistical Power	Running tests without calculating the required sample size.	Always use A/B test calculators.

Now, let's examine each of these pitfalls in greater detail to understand their underlying causes and consequences, beginning with the common temptation to test without a clear strategy:

- **The "Just Test It" Fallacy:** Testing without a strategic hypothesis. In a startup culture that prizes speed, the temptation to "just test it" can be overwhelming. This leads to a flurry of activity—testing button colors, minor copy changes, and small layout tweaks—that creates the illusion of progress. However, this "let's A/B test everything and see what happens" approach is a trap that wastes your most valuable resources: time, traffic, and focus. The goal is quality over quantity. One experiment designed to answer a fundamental strategic question is worth more than ten minor tests launched in a hurry. Before launching any test, you must articulate a clear, falsifiable hypothesis that tackles a meaningful uncertainty about your customer.[37,38]

  For example, compare these two hypotheses:

  - **Weak Hypothesis:** *We believe a new headline will increase signups.* (This is a guess, not a strategic question.)

  - **Strong Hypothesis:** *We believe a headline emphasizing social proof ("Join 10,000+ Happy Customers") will convert better than one emphasizing a key feature ("Powerful Analytics at Your Fingertips") because our user interviews suggest our target audience values community trust over technical specifications.*

  The second hypothesis is powerful because, win or lose, it generates a lasting insight into your customers' motivations that can inform your entire product and marketing strategy. The first tells you nothing beyond the performance of two random phrases.

---

[37] "10 Common A/B Testing Mistakes To Avoid," ContentSquare, October 18, 2024, `https://contentsquare.com/guides/ab-testing/mistakes/`.

[38] Jens-Fabian Goetzmann, "PM 101: Pitfalls of A/B Testing," Medium, June 9, 2019, `https://jefago.medium.com/pm-101-pitfalls-of-a-b-testing-d50919df6552`.

- **The "Peeking" Problem:** Calling the winner too early. It's the irresistible temptation to check your A/B test results every few hours. You see one version pulling slightly ahead and feel the urge to declare a winner and move on. This is called "peeking," and it is one of the most common ways that smart teams are misled by random noise. Every time you peek at the results, you're giving random chance another opportunity to fool you. Over the course of a test, it's almost guaranteed that each variant will be "winning" at some point due to normal fluctuations. If you stop the test during one of these temporary spikes, you will mistakenly conclude you have a winner when you don't. This practice dramatically increases the false-positive rate, rendering your results statistically invalid.

- **Solutions**

    - **Commit to Your Test Duration:** The most straightforward solution is to determine the required sample size and test duration before you begin. Then, commit to not analyzing the results or making a decision until the test has fully run its course. Hide the results if you have to.

    - **Use a Statistical Method Designed for Peeking:** Alternatively, use a testing platform that employs statistical methods—like sequential testing—that are specifically designed to mitigate the "peeking problem." These advanced engines allow you to monitor results continuously and stop a test the moment a variant reaches significance, without the statistical risks of traditional peeking. This approach can give you the best of both worlds: the speed of getting a fast result and the confidence that it is statistically sound. However, this is a sophisticated feature, and you must confirm that your tool explicitly supports it.[39]

- **Skipping Quality Assurance and Monitoring:** A buggy experiment is worse than no experiment at all. It doesn't just produce invalid data that leads to bad decisions; it actively harms your business

---

[39] Allon Korem, "Bayesian A/B Testing Falls Short," Medium, June 26, 2024, https://medium.com/data-science-collective/tldr-bayesian-a-b-testing-falls-short-f8646529a47a.

by frustrating users, damaging your brand, and potentially losing revenue. This critical mistake often happens in the rush to launch. Beyond simply validating your overall testing system, it is crucial to perform rigorous quality assurance (QA) on every single experiment before it goes live. You must verify that the specific changes you've implemented in your test variants are bug-free and render as expected for all users. Before launching, your team must confirm that both variants work flawlessly across key browsers, devices (desktop and mobile), and user states (e.g., new vs. logged-in users). Once the test is live, monitor its performance closely, especially during the initial hours. Look for any sharp, unexpected drops in your primary metrics or spikes in technical errors. This discipline is a cornerstone of product excellence. An experiment that hasn't been properly vetted isn't an experiment—it's a liability.[40,41]

- **The Novelty Effect:** Users are often drawn to new things simply because they are new. A new, bright red button might get more clicks initially, but this effect can wear off as users become accustomed to it. This initial spike can be misleading, so it's wise to run tests long enough for the novelty to fade and for user behavior to stabilize.[42]

- **Seasonality:** User behavior is not static; it changes with seasons, holidays, and business cycles. A test run during Black Friday will yield very different results than one run in a slow summer month. To mitigate this, run experiments for at least one full business cycle (e.g., one or two weeks) to capture variations between weekdays and weekends.

---

[40] David Kolodny, "The Startup Founder's Playbook for Effective A/B Testing," Inc., October 25, 2023, https://www.inc.com/david-kolodny/the-startup-founders-playbook-for-effective-a/b-testing.html.

[41] Usman Adepoju, "20 A/B Testing Mistakes To Avoid in 2025," Figpii, May 15, 2024, https://www.figpii.com/blog/ab-testing-mistakes-to-avoid/.

[42] "Ten common A/B testing pitfalls and how to avoid them," Adobe Experience League, September 1, 2023, https://experienceleague.adobe.com/en/docs/target/using/activities/abtest/common-ab-testing-pitfalls.

- **Confounding Variables:** Other factors can influence your results. For example, different user cohorts (e.g., new vs. returning users) may behave differently. The time of day or time zone of your users can also have an impact. When analyzing results, segmenting your data by these variables can help you uncover deeper insights and ensure your conclusions are accurate.

- **Confirmation Bias:** Avoid setting up tests in a way that is designed to confirm a pre-existing belief, for instance, by making the alternative version intentionally worse.

- **Sample Ratio Mismatch (SRM):** You set your experiment for a 50/50 split, but the results show a significant deviation, like 60/40. This isn't a statistical fluke; it's a red alert that your experiment's data is corrupt. SRM signals that a technical bug is preventing users from being properly assigned or tracked, which invalidates the entire test. Solution: Always check for SRM before analyzing conversion rates (most tools have this built in). If you find a mismatch, you must discard the results, find the underlying bug, and rerun the experiment.[43]

- **Ignoring Statistical Power:** Running tests without calculating required sample sizes, leading to underpowered tests that can't detect real improvements. Solution: Always use A/B test calculators before launching tests.

Steering clear of these common traps is the foundation of a trustworthy experimentation program. But a successful program is built on more than just avoiding mistakes—it requires a proactive, structured approach to turn ideas into reliable results.

Now that you're equipped with what to watch out for, let's shift our focus from theory to practice. This next section provides a step-by-step guide to get your first experiments up and running, from choosing the right tools to defining what success looks like.

---

[43] Emily Healy, "Sample Ratio Mismatch: What Is It and How Does It Happen?" AB Tasty, November 28, 2022, https://www.abtasty.com/blog/sample-ratio-mismatch/.

Getting Started with Experimentation

Moving from the theory of experimentation to the practice of running your first tests can feel like a significant hurdle for a resource-strapped startup. You understand the "why" and the "what," but the "how" can seem daunting. This section provides a practical, step-by-step guide to get you started. We'll cover the foundational actions: selecting a testing solution that aligns with your budget and technical capabilities, verifying your setup to ensure data integrity, and adopting a disciplined framework to transform your ideas into well-structured tests. By following these steps, you can build a robust experimentation practice from the ground up.

Choose Your A/B Testing Solution

Selecting the right A/B testing tool is a critical decision that balances cost, technical resources, and strategic goals. While enterprise-grade platforms like Optimizely offer powerful features, they are often out of reach for early-stage startups. Fortunately, the modern analytics ecosystem provides a range of options suitable for different needs and budgets. Your choice will generally fall into one of three categories: integrated platforms, dedicated experimentation tools, or a custom-built solution.

1. Integrated Platforms (Product and Marketing)

For many startups, the most efficient starting point is to leverage experimentation features already built into platforms you currently use. This approach consolidates your data and simplifies your tech stack.

- **Product Analytics Suites:** As discussed in Chapter 10, many product analytics platforms now offer robust A/B testing capabilities. This is often the ideal choice for experimenting with in-product features, user flows, and designs. Because the testing tool is unified with your core analytics, you can seamlessly analyze an experiment's impact on long-term retention, engagement, and other key product metrics.

  - **Examples:** Open source solutions like PostHog provide A/B testing as part of their core offering. Paid platforms like Amplitude also have powerful, integrated experimentation suites.

- **Marketing Automation Platforms:** For testing communications and top-of-funnel campaigns, look no further than your existing marketing tools. These platforms are designed to run experiments on specific channels, allowing you to optimize how you engage with users.

  - **Common Use Cases**

    - **Email Campaigns:** Test subject lines, email copy, calls to action (CTAs), and imagery using built-in features in tools like Mailchimp.

    - **Push Notifications and SMS:** Platforms like Braze allow you to rigorously test messaging copy, timing, and offers to improve open rates and conversions.[44]

2. Dedicated Experimentation Platforms

A dedicated experimentation platform is the right choice in two primary scenarios. First, you may simply not have A/B testing features in the product analytics or marketing tools you already use. In this case, a dedicated platform is the most direct way to add this capability to your stack.

Second, you might choose a dedicated platform when your needs become more complex and outgrow the capabilities of your integrated tools. If you require more advanced statistical models, sophisticated feature management, or deeper analytical power, a specialized tool is the logical next step. These platforms focus exclusively on testing, providing the robust features necessary to scale a mature experimentation program.

- **SaaS Solutions:** Paid tools like Optimizely, VWO, and LaunchDarkly offer sophisticated features, including visual editors for marketers and robust SDKs for engineers. Some of them provide startup-friendly pricing plans.[45]

---

[44] "Braze Customer Engagement Platform," Braze, accessed August 29, 2025, `https://www.braze.com/`.

[45] "LaunchDarkly," LaunchDarkly, accessed August 29, 2025, `https://launchdarkly.com/`.

- **Open Source Solutions:** For teams with engineering resources, an open source tool like GrowthBook can be a powerful and cost-effective option. It can be self-hosted—giving you full control over your data and infrastructure—and also offers paid self-hosted tiers that provide premium features and support under a commercial license.[46]

3. The Custom-Built Solution

For highly technical teams, the temptation to build a custom experimentation platform can be strong. This path, however, is typically only suited for later-stage startups that possess both a mature culture of experimentation and the specialized engineering resources to match. For these companies, the primary motivation is often to create a system perfectly tailored to specific or complex product needs.[47]

However, this path carries significant hidden costs. While it avoids subscription fees, the initial development and, more importantly, the ongoing maintenance require substantial engineering resources. Building a statistically sound, bug-free, and scalable platform is a massive undertaking. For most startups, this is a premature optimization. A custom solution should only be considered when off-the-shelf tools have been exhausted and experimentation has become a core, complex competitive advantage that justifies the significant investment.[48]

Run A/A Test

After you've selected and implemented an A/B testing tool, your first instinct will be to start experimenting with new ideas. However, there is a critical preliminary step that many teams overlook: running an A/A test.

---

[46] "GrowthBook Pricing," GrowthBook, accessed August 29, 2025, `https://www.growthbook.io/pricing`.

[47] Eric Metelka, "The Evolving Landscape of Experimentation Tooling: Build vs. Buy," Amplitude, August 28, 2025, `https://amplitude.com/blog/build-vs-buy-experimentation`.

[48] Graham McNicoll, "The Hidden Complexities of Building Your Own A/B Testing Platform," GrowthBook, February 18, 2025, `https://blog.growthbook.io/the-hidden-complexities-of-building-your-own-a-b-testing-platform/`.

An A/A test is a simple but powerful quality assurance check. It uses your A/B testing software to test two identical versions against each other. The goal is not to find a winning version—the results should be statistically inconclusive—but to validate that your technical setup is working correctly before you invest time and traffic in actual experiments.[49,50]

Here are the primary reasons why this is a crucial first step:

- **Verify Your Tool and Setup:** The main purpose of an A/A test is to confirm that your testing software is installed correctly and that its randomization and data collection are reliable. If you run a test on two identical pages and the tool reports a significant difference in performance, you've uncovered a problem. This could be anything from a faulty installation to a fundamental issue with the tool itself, and it's a problem you need to fix before running any real tests.[51]

- **Establish a Baseline and Understand Variance:** An A/A test helps you understand the natural variability in your conversion rates when no changes are made and provides a crucial baseline for future A/B tests.[52]

- **Build Confidence in Your Data:** By proving that your system produces an even split when there's no difference between variants, you build confidence in the entire experimentation process. When you later run an A/B test that shows a clear winner, you can trust that the result is due to your changes and not a flaw in the data collection.[53]

Think of an A/A test as a calibration for your experimentation engine. It might feel like a momentary delay, but it's a small upfront investment that prevents you from making poor decisions based on flawed data down the road.

---

[49] Mani Makkar, "What is an A/A Test? Why Should You Care? Learn More," VWO, May 5, 2025, https://vwo.com/blog/aa-test-before-ab-testing/.

[50] Pulkit Rastogi, "The Test of Tests - How To Do A/A Testing Like a Pro," Omniconvert, November 4, 2024, https://www.omniconvert.com/blog/a-a-testing/.

[51] Dionysia Kontotasiou, "All About A/A Testing: Why and When Should You Run A/A Tests?" Convert, February 8, 2023, https://www.convert.com/blog/a-b-testing/how-to-run-aa-tests-before-experimenting/.

[52] Pritam Roy, "AA Testing vs AB Testing: Differences Explained," Fibr, accessed August 29, 2025, https://fibr.ai/ab-testing/aa-testing-vs-ab-testing.

[53] Mani Makkar, "What is an A/A Test? Why Should You Care? Learn More," VWO, May 5, 2025, https://vwo.com/blog/aa-test-before-ab-testing/.

The Test Card

While the core idea of an A/B test is simple, running effective experiments requires discipline. It's easy to launch a test based on a vague idea, only to find yourself debating the results because you never defined what success would look like. To avoid this, you need a structured process for turning assumptions into testable hypotheses with clear success criteria.

A powerful tool for this is the Test Card, developed by Alex Osterwalder and the team at Strategyzer. The Test Card is a simple but effective framework designed to help you articulate and test your business ideas systematically. It forces you to move from vague notions to concrete, falsifiable hypotheses before you invest time and resources into building or changing anything.[54]

According to Osterwalder, every experiment should be structured around four core components:

1. **Hypothesis:** Clearly state the underlying belief you are about to test. The format is direct and powerful: "We believe that... for... will result in...."

2. **Test:** Describe the specific action you will take to test your hypothesis. This is the experiment itself.

3. **Metric:** Define exactly what you will measure and how you will measure it. This must be a quantifiable result.

4. **Criteria:** Set a clear success threshold. This is the target value for your metric that will prove your hypothesis right.

Let's apply this framework to our previous example of testing two different designs for an annual subscription offer (Figure 11-2).

Applying the Test Card to the Subscription Offer

Theory is one thing; practice is another. To make the concepts in this chapter concrete, this exercise walks you through a complete, end-to-end experiment at a hypothetical startup. Imagine you are the first data professional at a promising B2C subscription app.

---

[54] Alex Osterwalder, "Validate Your Ideas with the Test Card," Strategyzer, March 5, 2015, `https://www.strategyzer.com/library/validate-your-ideas-with-the-test-card`.

You've found product-market fit, but now the pressure is on to scale efficiently. Let's put you in the driver's seat.

Your startup is growing, but the CEO is concerned about cash flow and long-term user value. A key lever for both is the ratio of users choosing an annual plan over a monthly one. Your current annual-to-monthly sign-up ratio is troublingly low compared to industry benchmarks.

The catalyst for change comes at a tech conference. You strike up a conversation with the founder of a successful competitor who credits their high retention and strong cash position to a high uptake of their annual plan. While they don't share hard numbers, the message is clear: you're leaving money on the table.

Back at your desk, you conduct a swift competitive analysis. You scrutinize the pricing pages of your top three competitors and compare them to your own (Version A in Figure 11-2). A clear pattern emerges:

- **Your Page (Version A):** Presents the annual plan as a single, large sum ($120). Its muted design fails to highlight key benefits and causes the option to blend into the background, making the monthly plan the visually dominant and path-of-least-resistance choice.

- **Competitor Pages:** Aggressively market their annual plans. They frame the price in a lower monthly equivalent ("Just $10/month"), use bright colors and badges ("Best Value") to make the annual option pop, and clearly state the savings and benefits.

Your analysis concludes that your pricing page isn't just presenting options; it's actively—if unintentionally—suppressing the choice you want users to make. You have a clear, data-informed problem to solve. Now, you use the Test Card framework to transform your insights into a disciplined, measurable experiment.

Step 1: Hypothesis

You begin by clearly stating the belief you intend to test:

We believe that redesigning the pricing page to visually emphasize the annual plan and clearly articulate its value (savings, monthly equivalent cost, benefits) for new visitors arriving on the page will result in a significant increase in the percentage of users who choose the annual subscription.

Step 2: Test

Next, you define the exact experiment you will run to validate the hypothesis:

To verify this, we will run an A/B test.

- *Version A (Control): Our existing pricing page*

- *Version B (Challenger): The redesigned page that frames the price as "$10/month," adds a "SAVE 50%" headline, enumerates the benefits available with the annual plan ("cancel anytime," "priority support," "exclusive content"), and uses a more prominent design to highlight the annual option*

We will split all new, non-logged-in traffic 50/50 between the two versions.

Step 3: Metric

You must define precisely what data you will measure to determine the outcome:

We will measure the conversion rate to the annual plan for each variant. This is defined as (Number of Annual Subscriptions / Total Unique Visitors in the Variant).

As a secondary "guardrail" metric, we will monitor the overall signup conversion rate (monthly + annual) to ensure the new design doesn't decrease total conversions.

Step 4: Criteria

Finally, you set a clear, quantitative threshold for success before the test begins. This is arguably the most critical step. Let's imagine we have chosen VWO as our AB testing platform, and we can use the VWO calculator.[55]

We are right if Version B shows a statistically significant lift in the annual plan conversion rate.

- ***Baseline Conversion Rate:*** *Our current annual plan conversion rate is 5%.*

---

[55] "A/B Test Sample Size & Duration Calculator," VWO, accessed August 29, 2025, `https://vwo.com/tools/ab-test-duration-calculator/`.

- *Minimum Detectable Effect (MDE):* We decide a 30% relative lift (from 5% to 6.5%) is the minimum result that would be strategically meaningful.

- *Sample Size and Test Duration:* We assume the average number of daily visitors would be around 500. Based on these inputs, the online sample size calculator tells us we need approximately 14 days to run an A/B test using the Frequentist Statistics approach.

With your Test Card complete, the entire team has a shared and unambiguous definition of the experiment and what constitutes success.

Analyzing the Results and Making a Decision

After 14 days, your A/B testing tool provides the results presented in Table 11-3.

Table 11-3. *A/B Testing Results*

Variant	Visitors	Annual Subs	Conversion Rate
A	3,500	175	5%
B	3,501	281	8%

Version B's conversion rate of 8% represents a 61% relative lift over the control, exceeding your MDE and your expectations. The overall sign-up conversion rate (monthly + annual) stayed at the same level, so the new design doesn't decrease total conversions. The result is a clear, statistically significant win. This is where you translate the statistical win into business impact. You report to the CEO:

The pricing page experiment was successful. The new design drove a 61% increase in annual plan adoption. By shifting more users to an upfront payment, this single change will immediately improve our cash flow and increase Customer Lifetime Value (LTV), validating that our users respond to clear value communication. I recommend we roll this new design out to 100% of traffic immediately.

Additional Resources

Mastering experimentation is a continuous journey. While this chapter provides a startup-focused foundation, the resources below offer deeper dives into the statistical theory, technical implementation, and practical application of A/B testing and data-driven decision-making.

Foundational Books

- ***Trustworthy Online Controlled Experiments* by Ron Kohavi, Diane Tang, and Ya Xu:** This is the definitive guide to running reliable experiments at scale, written by leaders from Microsoft, Google, and LinkedIn. While its lessons are drawn from large tech companies, the principles of statistical power, experiment design, and avoiding common pitfalls are universally applicable.[56]

- ***Designing with Data: Improving the User Experience with A/B Testing* by Rochelle King, Elizabeth F. Churchill, and Caitlin Tan:** This book masterfully bridges the gap between data analytics and user experience (UX) design. It moves beyond simply finding a statistical winner and teaches you how to structure tests that generate deep insights into user behavior.[57]

- ***Hypothesis Testing: An Intuitive Guide for Making Data Driven Decisions* by Jim Frost:** If you need to strengthen your understanding of the statistical concepts behind hypothesis testing, this book is an excellent resource. Frost has a talent for explaining complex topics like p-values, confidence intervals, and statistical significance in a clear and intuitive way, without getting bogged down in dense mathematics.[58]

---

[56] Ron Kohavi, Diane Tang, and Ya Xu, *Trustworthy Online Controlled Experiments: A Practical Guide to A/B Testing* (Cambridge University Press, 2020).

[57] Rochelle King, Elizabeth F. Churchill, and Caitlin Tan, *Designing with Data: Improving the User Experience with A/B Testing* (O'Reilly Media, 2017).

[58] Jim Frost, *Hypothesis Testing: An Intuitive Guide for Making Data Driven Decisions* (StatisticsByJim.com, 2020).

– *Experimentation for Engineers: From A/B testing to Bayesian optimization* by **David Sweet:** This book is aimed at the technical-minded reader who is not just designing experiments, but also building the systems that run them. It covers the entire life cycle, from traffic splitting and data collection to statistical analysis and advanced topics like Bayesian methods.[59]

Online Learning

– **DataCamp:** For hands-on learning, DataCamp offers a wealth of courses and skill tracks directly relevant to experimentation. You can find courses covering the fundamentals of A/B testing in Python and R, customer analytics, and statistical simulation. Their interactive format, which combines video instruction with in-browser coding exercises, is ideal for quickly applying the concepts discussed in this chapter and developing the practical skills needed to analyze experiment results.

– **Testing Theory YouTube Channel:** Provides clear, concise videos on the statistical concepts and practical realities of A/B testing, making it an excellent resource for practitioners.[60]

– **Ron Kohavi's Curated List:** The lead author of *Trustworthy Online Controlled Experiments* shares his hand-picked list of the best A/B testing videos he has seen.[61,62]

---

[59] David Sweet, *Experimentation for Engineers: From A/B testing to Bayesian optimization* (Manning Publications, 2023).

[60] "Testing Theory," YouTube Channel, accessed August 29, 2025, https://www.youtube.com/@TestingTheory.

[61] Ron Kohavi, "What are the best AB Testing videos you've seen?" LinkedIn post, accessed August 29, 2025, https://www.linkedin.com/posts/ronnyk_best-ab-testing-talks-activity-71424 00073396592640-q8aB/.

[62] "Best A/B Testing Talks," Google Sheets, accessed August 29, 2025, https://docs.google.com/spreadsheets/d/1CLdCxXpmb2UhGOx-rCkj_8UsfSuNthasVuagseNVWrU/edit?gid=187719910#gid=187719910.

Summary

This chapter provided a comprehensive guide to building and running an experimentation practice tailored to the unique pressures and opportunities of a startup. You learned how to move from intuition-based decisions to a disciplined, data-driven process that accelerates both learning and growth.

Here are the key takeaways you can now apply:

- **Embrace Experimentation as a Strategic Tool:** For startups, A/B testing is not just for minor optimization; it's a strategic method to accelerate toward success or fail fast. This helps you avoid the "zombie startup" trap by quickly validating core business ideas.

- **Know When Not to Test:** A/B testing is counterproductive without sufficient user traffic, before achieving product-market fit, or for radical redesigns. Prioritize foundational data work first and ship obvious fixes without testing.

- **Focus on High-Impact, Top-of-Funnel Experiments:** Concentrate your limited resources on meaningful changes at the top of the user funnel—like your pricing and sign-up pages. This is where you have the most traffic and can directly influence revenue and customer lifetime value.

- **Adopt a Disciplined Framework:** Use tools like the Test Card to move from vague ideas to strong, falsifiable hypotheses. Clearly define your hypothesis, test, metric, and success criteria before you start to ensure you generate real insights, not just noise.

- **Understand That Most Experiments "Fail":** The goal of experimentation is learning, not "winning." A test that disproves a faulty assumption is a valuable success that saves time and resources. It is crucial to celebrate the knowledge gained, not just positive results.

- **Avoid Common Pitfalls:** Build a trustworthy experimentation practice by committing to your pre-determined test duration to avoid the "peeking" problem, rigorously performing quality assurance on every test, and always checking for Sample Ratio Mismatch (SRM) to ensure your data is valid.

- **Start with an A/A Test:** Before launching your first real experiment, run an A/A test—testing two identical versions against each other—to validate that your testing tool is implemented correctly and that your data collection is reliable.

The experimentation capabilities you've developed create a powerful feedback loop: better data leads to better decisions, which drive better business outcomes. But creating value is only half the battle; you must also be able to articulate it. This is the crucial bridge to our next topic.

In Chapter 12, we translate the impact of your data-driven work into the language of investors, acquirers, and stakeholders. Understanding valuation is not an abstract financial exercise—it is one of the most practical, high-impact skills a startup professional can possess:

- It empowers data leaders to justify budgets and prove ROI, reframing their work from a cost center to a core value driver.

- It gives employees with equity a clear understanding of how their contributions directly increase the financial worth of their shares.

- It equips founders to negotiate from a position of strength with investors and potential acquirers, translating product milestones and data trends into a clear, defensible valuation narrative.

The next chapter provides a clear guide to this complex world. You will learn how valuation evolves from a qualitative art to a quantitative science, mastering the pre-revenue and post-revenue methods used to value companies at every stage. We will also demystify the unique factors that drive the valuation of AI startups, equipping you to navigate one of the most dynamic sectors in the modern economy.

CHAPTER 12

Startup Valuation Methods

A good valuation is a bridge between stories and numbers.

—Aswath Damodaran,[1]

Valuation Expert and NYU Professor

Startup valuation represents one of the most challenging yet crucial aspects of the entrepreneurial ecosystem. It sits at the intersection of narrative and data, requiring a delicate balance between quantitative analysis and qualitative judgment. For data professionals working in startups, understanding these valuation mechanisms isn't merely academic—it's essential to connecting your daily work to the financial future of your organization. The numbers you generate, the insights you extract, and the systems you build all contribute to the story that determines your company's worth in the eyes of investors, acquirers, and stakeholders.

The journey of startup valuation differs fundamentally from traditional business valuation. While established companies rely on historical performance, predictable cash flows, and mature market dynamics, startups operate in a realm of uncertainty where potential often outweighs proven performance. This creates a unique challenge for data professionals who must navigate between the concrete world of metrics and the speculative nature of early-stage business development.

---

[1] Aswath Damodaran, "Laws of Valuation: Revealing the Myths and Misconceptions," Nordic Business Forum YouTube Channel, November 13, 2018, https://www.youtube.com/watch?v=c20_S-QgvsA.

P. Sidoruk, *From Data to Dollars*, https://doi.org/10.1007/979-8-8688-1898-1_12

This chapter provides a comprehensive guide to startup valuation, designed to show how your data capabilities can directly influence how your company is valued. We will systematically deconstruct the valuation process, moving from a qualitative art to a quantitative science. To do this, the chapter will guide you through

- **Foundational Principles:** We begin by establishing why startup valuation is a forward-looking exercise focused on future potential, not past performance, using the story of Snapchat's early, revenue-less multibillion-dollar valuation as a core example.

- **From Art to Science:** The chapter follows a single case study, *Viral Wardrobe*, to illustrate how valuation methods evolve as a company matures.

- **Pre-revenue Valuation (the Art):** We will first explore the qualitative methods used when a company has no revenue, where narrative and potential are key. These include

  - The Berkus Method

  - The Cost-to-Duplicate Method

  - The Scorecard Method

  - The Risk Factor Summation Method

  - The VC Method

- **Post-revenue Valuation (the Science):** As *Viral Wardrobe* begins to generate revenue, we will shift to quantitative models grounded in financial data. These include

  - The Revenue Multiple Method

  - The Discounted Cash Flow (DCF) Method

  - The First Chicago Method

- **The Special Case of AI Startups:** We will then cover the unique factors that drive the valuation of AI companies, such as proprietary data, elite talent, and defensible technology. A key focus will be a modern framework for quantifying "displacement risk"—the threat from larger foundation models.

—   **Valuation As a Diagnostic Tool:** Finally, the chapter concludes by showing that valuation is not a single, absolute number but a powerful tool that highlights a company's strengths and exposes its weaknesses, guiding you toward building a more valuable enterprise.

Introduction to Valuation

In November 2013, a scenario unfolded in the tech industry that left many observers stunned. Snapchat, then a two-year-old photo-messaging app led by 23-year-old Evan Spiegel, turned down Facebook's acquisition offer of reportedly $3 billion in cash. At the time, Snapchat had no revenue and was being courted by multiple investors. This decision seemed audacious—perhaps even reckless—to many industry watchers. Yet, it represented a profound confidence in the company's future value that transcended traditional valuation metrics.[2,3]

In a later interview, Spiegel elaborated on the context of this decision, revealing the intense initial skepticism the company faced. The doubt was so pervasive that he once overheard his own lawyers, unaware he was on a conference call, predicting the company was "basically going to zero" and dismissing it as a fad. Critics frequently advised him to sell while he could, emphasizing the fierce competition from established tech giants.[4]

However, Spiegel explained that they only saw a simple photo-sharing app and missed the broader vision for the future. He and his team were already developing new products that cemented their belief in Snapchat's massive long-term potential. This conviction, he noted, was a key reason for rejecting the offer.

---

[2] Scott Martin, "Snapchat turned down more than $3B from Facebook," USA Today, November 13, 2013, https://eu.usatoday.com/story/tech/2013/11/13/report-facebook-offered-snapchat-3-billion/3517929/.

[3] "Evan Spiegel explains why he didn't sell Snapchat to Mark Zuckerberg for $3 billion," Startup Archive, April 17, 2025, https://www.startuparchive.org/p/evan-spiegel-explains-why-he-didn-t-sell-snapchat-to-mark-zuckerberg-for-3-billion.

[4] Evan Spiegel, "Evan Spiegel explains why he didn't sell Snapchat to Mark Zuckerberg for $3 billion," Startup Archive YouTube Channel, April 17, 2025, https://www.youtube.com/watch?v=y3K6x523Xto.

Furthermore, a critical factor in their decision was a previous financing round that had allowed both Spiegel and co-founder Bobby Murphy to each sell $10 million of their stock. This financial security gave them the freedom to "swing for the fences" with the company's future, as Spiegel put it, without the personal pressure of needing a buyout to secure their own finances. This allowed them to pursue their much larger opportunity on their own terms.

As illustrated in Figure 12-1, which presents Snapchat's approximated valuation over time on its path to Wall Street, the decision to reject Facebook's offer proved to be remarkably prescient. While the $3 billion offer in 2013 may have seemed high to some at the time for a company with no revenue, Snapchat's value continued to soar. The company's valuation grew significantly over the following years, culminating in an IPO valuation of $24 billion in 2017. This chart's depiction of Snapchat's value evolution is based on approximations from various articles detailing its numerous funding rounds, from its initial seed investment through Series A, B, C, D, E, and F.[5,6,7,8,9,10]

---

[5] Sammy Abdullah, "All the Lessons from Snapchat's Founding," Blossom Street Ventures, accessed July 26, 2025, https://www.blossomstreetventures.com/post/all-the-lessons-from-snapchats-founding.

[6] "Snap Inc.'s Remarkable Journey in Fundraising: A Story of Innovation, Growth, and Numbers," FoundersToday, January 6, 2024, https://www.founderstoday.news/fundraising-story-snap-inc/.

[7] Dan Primack, "Snapchat confirms $50 million in new funding," Fortune, December 11, 2013, https://fortune.com/2013/12/11/snapchat-confirms-50-million-in-new-funding/.

[8] "Snapchat Raises Over $485 Million," Business Standard, January 2, 2015, https://www.business-standard.com/article/international/snapchat-raises-over-485-million-115010200011_1.html.

[9] Ari Levy, "Snap's Founders and Early Backers Stand to Make Billions," CNBC, October 12, 2016, https://www.cnbc.com/2016/10/12/snaps-founders-and-early-backers-stand-to-make-billions.html.

[10] Matthew Lynley, "Snap Values Itself at Nearly $24B with Its IPO Pricing," TechCrunch, March 1, 2017, https://techcrunch.com/2017/03/01/snap-values-itself-at-nearly-24b-with-its-ipo-pricing/.

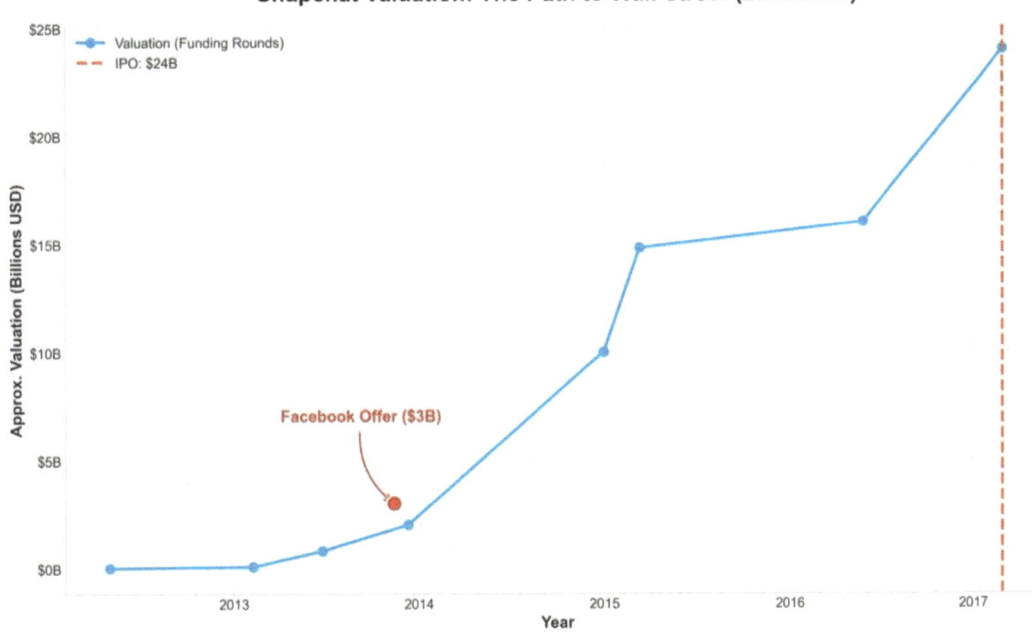

Figure 12-1. *Snapchat Valuation: The Path to Wall Street*

The Snapchat story exemplifies one of the most challenging aspects of the startup ecosystem: determining what a company with limited operating history, possibly no revenue, but enormous potential, is actually worth.

Traditional valuation approaches, like those famously employed by Warren Buffett, emphasize fundamental analysis, focusing on a company's intrinsic value based on its financial statements, competitive advantages, and management quality. However, startups often lack the established financial history that makes such analysis straightforward. Their value is predominantly tied to future growth potential rather than current performance, creating a valuation challenge that requires both art and science.[11]

The gap between traditional business valuation and startup valuation is particularly striking when considering how data factors into the equation. In established businesses, historical data provides the foundation for valuation. In contrast, startups often operate in a data-scarce environment, especially regarding their own performance metrics.

---

[11] "How to Value A Small Business - Warren Buffett Investment Strategy," The Investor's Podcast, accessed July 20, 2025, https://www.theinvestorspodcast.com/warren-buffett-investment-strategy/module-1/how-to-value-a-small-business/.

This is where data professionals can make an outsized impact by building systems that generate, collect, and analyze meaningful data that can substantiate valuation claims and growth projections.

Startup valuation serves multiple critical functions that extend far beyond fundraising conversations. While securing investment represents the most visible use case, valuation influences decisions throughout a startup's life cycle in ways that data professionals must understand and anticipate.

The Fundraising Foundation

At the heart of any equity exchange is valuation—the financial logic that determines what a piece of your company is worth. Two key terms anchor this calculation: pre-money and post-money valuation.

Pre-money valuation is what your company is deemed to be worth before an investor's capital comes in. **Post-money valuation** is the new value after the investment, combining the company's original worth with the fresh capital. This distinction is critical. For instance, a startup raising $2 million on a $10 million post-money valuation will sell 20% of its equity ($2 million is 20% of $10 million). This implies the company's pre-money valuation was $8 million.[12]

However, if that $10 million figure represents the pre-money valuation, the same $2 million investment creates a $12 million post-money valuation. In this scenario, the investors receive a smaller stake—approximately 16.7% ($2M ÷ $12M). While the math seems straightforward, the true value of that equity is far more complex. It's heavily influenced by factors like liquidation preferences, board control, and future dilution, which can dramatically alter an investor's actual return.

The relationship between valuation and equity dilution has profound implications for data teams. Higher valuations mean less dilution for existing shareholders, including employees with stock options. This mathematical relationship means that the metrics you track, the insights you generate, and the competitive advantages you help build through data can directly impact the financial outcomes for everyone in your organization.[13]

---

[12] "Pre-Money vs. Post-Money Valuation," Wall Street Prep, December 6, 2023, `https://www.wallstreetprep.com/knowledge/pre-post-money-valuation/`.

[13] "Understanding Equity Dilution" Morgan Stanley, November 25, 2024, `https://www.morganstanley.com/atwork/articles/what-is-equity-dilution`.

Employee equity compensation depends heavily on company valuation, creating a direct link between your data work and talent retention. Stock options become meaningful only when employees understand the underlying share value, and competitive recruiting often hinges on presenting compelling equity packages. Paul Graham's equity equation provides a useful framework here:

> *Whenever you're trading stock in your company for anything, whether it's money or an employee or a deal with another company, the test for whether to do it is the same. You should give up n% of your company if what you trade it for improves your average outcome enough that the (100 - n)% you have left is worth more than the whole company was before.*

> *For example, if an investor wants to buy half your company, how much does that investment have to improve your average outcome for you to break even? Obviously, it has to double: if you trade half your company for something that more than doubles the company's average outcome, you're net ahead. You have half as big a share of something worth more than twice as much.*[14]

For data professionals, this framework helps quantify your contribution to company outcomes. By demonstrating measurable improvements in customer acquisition, retention, or operational efficiency, data teams can use Graham's equation to justify equity compensation or budget allocations. If your data capabilities increase company outcomes by 20%, the calculation yields a significant equity justification that reflects the value you create.

Market Cycles and Timing

As we detailed in Chapters 4-6, the metrics a startup must prioritize are not static. The shift from focusing on aggressive growth in "good times" to capital efficiency in "bad times" is a direct response to the cyclical nature of the startup funding landscape. These cycles, characterized as "risk-on" and "risk-off" periods, profoundly impact valuations and create distinct operating environments that data professionals must understand and

---

[14] Paul Graham, "The Equity Equation," Paul Graham's Website, July 2007, https://www.paulgraham.com/equity.html.

anticipate. They are a reflection of broader economic conditions and investor sentiment, which collectively determine capital availability and the friendliness of investment terms.[15]

For founders and investors alike, navigating these shifts between "good times" and "bad times" is a fundamental challenge. These periods are defined by the collective mood of the market—investor sentiment—which dictates the flow of capital and shapes the fortunes of fledgling companies. During "risk-on" periods, optimism abounds, capital is plentiful, and the focus is squarely on rapid growth and capturing market share. Conversely, "risk-off" periods are marked by caution, scarce capital, and an investor pivot toward proven business models and profitability.

History provides a stark reminder of this cyclicality. The dot-com bubble of the late 1990s was a quintessential risk-on era, defined by irrational exuberance and soaring valuations, followed by a punishing crash that wiped out countless companies and fortunes. This pattern of boom and bust, driven by macroeconomic conditions, technological breakthroughs, and shifting investor psychology, is an enduring feature of the venture landscape.[16]

Figure 12-2, with data sourced directly from the Q2 2025 PitchBook-NVCA Venture Monitor report, offers a vivid illustration of the most recent venture capital cycle. It tracks the total deal value of US VC fundraising from 2020 through the first half of 2025, capturing a period of unprecedented heights followed by a sharp and sober correction.

---

[15] Ramana Nanda and Matthew Rhodes-Kropf, "Investment Cycles and Startup Innovation," Fung Institute for Engineering Leadership, UC Berkeley, January 30, 2012, https://funginstitute. berkeley.edu/wp-content/uploads/2013/12/investment-cycles-and-startup-innovation_20120130.pdf.

[16] "The Late 1990s Dot-Com Bubble Implodes in 2000," Goldman Sachs, accessed July 21, 2025, https://www.goldmansachs.com/our-firm/history/moments/2000-dot-com-bubble.

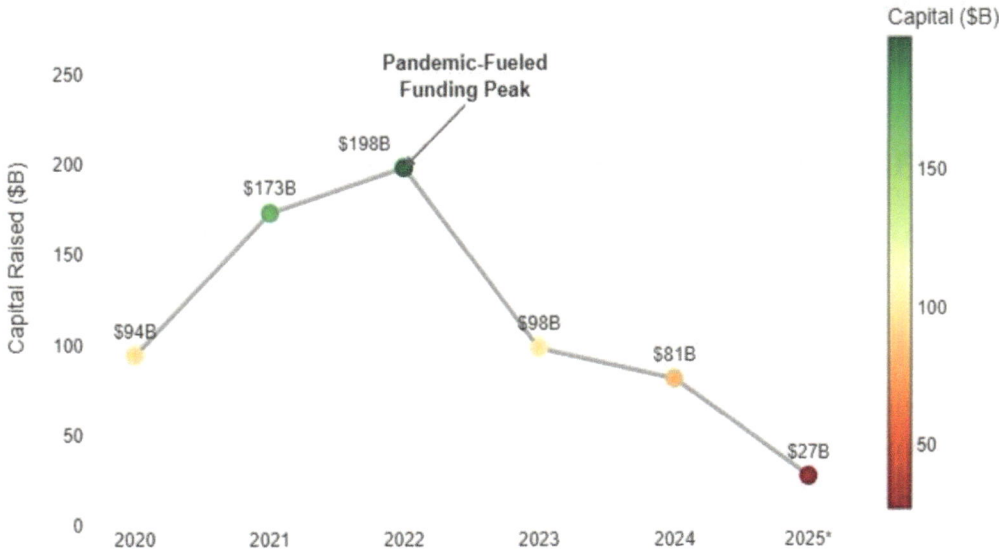

US VC Fundraising Deal Value, 2020 - mid-2025*

Figure 12-2. US VC Fundraising Deal Value, 2020–mid-2025, according to PitchBook-NVCA Venture Monitor, as of June 30, 2025[17]

The data shows a remarkable surge in fundraising, beginning with $94 billion in 2020 and climbing to $173 billion in 2021 before reaching what the chart labels the "Pandemic-Fueled Funding Peak" of $198 billion in 2022. This peak represented the apex of a historic risk-on environment. Following this zenith, investor sentiment shifted dramatically. Fundraising value fell to $98 billion in 2023 and continued its decline to $81 billion in 2024, reflecting a market grappling with a system-wide liquidity crunch and increased caution among investors.

Even as the venture ecosystem contracts from its recent highs, the market is not uniform. The decline in overall investor sentiment has created a bifurcated landscape where capital is highly selective. While many startups face a challenging fundraising environment, a select few continue to thrive. This is particularly true in the field

---

[17] PitchBook and National Venture Capital Association, "Q2 2025 PitchBook-NVCA Venture Monitor," July 14, 2025, https://pitchbook.com/news/reports/q2-2025-pitchbook-nvca-venture-monitor.

of artificial intelligence, where the generative AI boom has fueled an investment frenzy. Small, innovative AI startups are securing exceptionally large funding rounds, demonstrating that even in a risk-off market, a transformative technological shift can create its own powerful currents of investor enthusiasm and opportunity.

Popular Startup Valuation Methods

Valuing an early-stage company is a nuanced process, and the methods investors and founders employ depend greatly on the startup's stage of development. Many valuation techniques exist, but this section introduces some of the most recognized approaches, dividing them into pre-revenue and post-revenue methods. Each one will be discussed in greater detail later in this section.[18]

Pre-revenue valuation methods (see Table 12-1) are primarily used before a startup has meaningful revenue, making use of qualitative judgment and proxies for value:

- **Berkus Method:** Assigns a fixed monetary value to key success elements like the quality of the idea, prototype, management, relationships, and initial market traction—heavily reliant on expert assessment rather than hard data

- **Cost-to-Duplicate Method:** Estimates value based on what it would cost to replicate the company's assets

- **Scorecard Method:** Benchmarks the startup against comparable companies, adjusting for qualitative factors like management team, market size, and the strength of the business model

- **Risk Factor Summation Method:** Adjusts a base valuation up or down by systematically considering a set of risk factors (such as competition, technology, and funding), resulting in a composite qualitative score

- **VC Method:** Depends mainly on projections and qualitative insights

---

[18] Julian Stylianou, "Everything You Need to Know About Valuation Methods," Swiss Startup Association, January 29, 2021, https://swissstartupassociation.ch/2021/01/29/valuation/.

Table 12-1. Pre-revenue Valuation Methods

Method	Basis of Valuation	Key Takeaway
Berkus Method	Qualitative Milestones	Values progress in key risk areas (idea, team, prototype)
Cost-to-Duplicate Method	Replication Cost	Values what it would cost to build the company again from scratch
Scorecard Method	Peer Benchmarking	Compares the startup to similar pre-revenue companies
Risk Factor Summation Method	Risk Analysis	Adjusts a peer group valuation based on a checklist of 12 risks
VC Method	Future Exit & ROI	Works backward from a projected exit to find today's value

Once a company establishes real revenue, more data-driven techniques called **post-revenue valuation methods** (see Table 12-2) become appropriate:

- **Revenue Multiple Method:** Applies industry-standard multiples to current revenue to derive a market-based valuation; grounded in concrete financial performance

- **Discounted Cash Flow (DCF) Method:** Projects future cash flows and discounts them to present value using a risk-adjusted rate

- **First Chicago Method:** Combines scenario analysis (best, base, and worst case) with probability—a blend of projection and quantitative modeling

Table 12-2. Post-revenue Valuation Methods

Method	Basis of Valuation	Key Takeaway
Revenue Multiple Method	Current Revenue & Market Multiples	Applies a market-standard multiple to current revenue
Discounted Cash Flow (DCF) Method	Future Cash Flow Projections	Values the company based on its projected future cash flow
First Chicago Method	Scenario Analysis	Creates a probability-weighted valuation from best, base, and worst-case outcomes

While numerous other models exist, the methods outlined above are among the most widely used in the startup ecosystem and are the focus of this chapter.[19]

To visualize how these methods apply across a startup's life cycle, Figure 12-3 depicts this evolution graphically in a simplified way. The chart is divided into two distinct phases to illustrate an evolution in methodology. The green Pre-revenue Valuation section on the left represents the early stages, where the absence of financial data necessitates a reliance on the subjective and qualitative methods previously described, such as the Berkus and Scorecard methods.

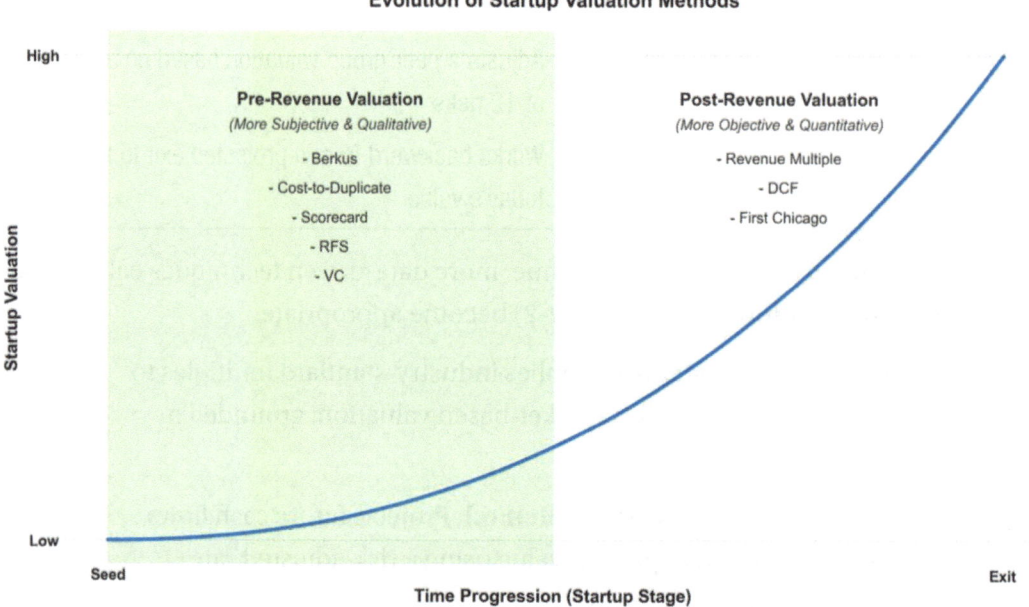

Figure 12-3. *Startup Valuation Methods: From Qualitative to Quantitative*

As the company matures and begins generating revenue, it moves into the Post-revenue Valuation phase on the right. Here, the approach becomes fundamentally more objective and quantitative. Valuation is no longer primarily an art of judgment but a science grounded in concrete financial performance, employing data-driven models like the Revenue Multiple and Discounted Cash Flow (DCF) methods. In essence, Figure 12-3 encapsulates the core narrative of startup valuation: a transition from assessing potential to analyzing performance.

---

[19] Adam Augusiak-Boro, Catherine Klinchuch, Andrew Zhan, and Jake Schwartz, "Startup Valuation Guide," Hangar8 Capital, September 2018, https://hangar8capital.com/wp-content/uploads/2019/01/Research-Report.pdf.

Pre-revenue Valuation Methods

When your startup has little to no revenue, traditional financial metrics become meaningless, requiring specialized approaches that focus on potential rather than performance. These methods have become particularly relevant for early-stage companies.

The Berkus Method

The Berkus Method is a simple valuation tool for the earliest stage of a startup, relying on qualitative assessment rather than financial projections. It operates on the premise that, at the pre-revenue stage, a company's worth is based on the progress it has made in reducing fundamental business risks. This method is based on subjective judgment across five key areas.

Developed by angel investor Dave Berkus, the method is designed for companies that have the potential to reach at least $20 million in revenue within five years. Once a company begins generating meaningful revenue, this model is no longer the right tool for the job.[20]

How the Berkus Method Works

Imagine you're running a startup and want to assess how much it's worth before talking to your first investors. The Berkus Method gives you a framework for this. An investor assigns a value of up to $500,000 for the progress your company has made in each of the following five areas. The sum of these values establishes a valuation of up to $2.5 million.

The five key factors are

- **Sound Idea:** How strong and unique is the core concept? Does it solve a real problem?

- **Prototype:** Do you have a working model or a minimum viable product (MVP)? This proves the technology or business model is feasible.

- **Quality Management Team:** How experienced and capable is your founding team? A strong team is more likely to navigate future challenges.

---

[20] Dave Berkus, "The Berkus Method: Valuing an Early-Stage Investment," Official Website of Dave Berkus, November 21, 2024, https://berkus.com/the-berkus-method-valuing-an-early-stage-investment-2/.

- **Strategic Relationships:** Have you secured any key partnerships that give you an advantage or open up market access?

- **Product Rollout or Sales:** Are you already in the market, or do you have a clear plan to launch and attract customers?

A Valuation Example

Let's say you've co-founded *Viral Wardrobe,* a fashion startup that partners with popular female social media influencers to design and sell limited-edition clothing collections to their followers. You haven't sold anything yet but need a valuation to start conversations with investors. Table 12-3 presents how you and a potential investor might assess the company using the Berkus Method.

Table 12-3. *Berkus Method Valuation Example*

Factor	Description	Assigned Value
Sound Idea	The concept capitalizes on the powerful creator economy and influencer marketing trends. It's a very strong, timely idea.	$450,000
Prototype	You have a functional ecommerce site and have produced a successful sample collection with a small test influencer.	$400,000
Quality Management Team	The founders are savvy marketers but lack deep experience in clothing manufacturing and supply chain logistics.	$300,000
Strategic Relationships	You have letters of intent from five mid-tier influencers to launch collections, which is a major asset.	$350,000
Product Rollout or Sales	There's a detailed launch plan and a pre-launch waitlist generated from the partner influencers' audiences.	$200,000
Total Pre-money Valuation		**$1,700,000**

This process gives *Viral Wardrobe* a pre-money valuation of $1.7 million. This figure isn't an absolute fact but a structured, defensible starting point for your negotiation with investors. It also clearly shows that while your concept and influencer partnerships are strong, you may need to address the team's experience gap in manufacturing to build more value.

While the Berkus Method provides a quick, risk-based assessment, other pre-revenue valuation methods offer different perspectives on a startup's worth. Each framework analyzes it through a unique lens, resulting in a range of potential valuations. Here is how other popular methods can be applied to our *Viral Wardrobe* example.

Cost-to-Duplicate Method

This approach provides a valuation based on a simple question: what would it cost to build this exact business from scratch today? It is a tangible, asset-based valuation that calculates the total investment required to replicate the company's progress. This includes all expenses for developing the product, purchasing physical assets, and other startup costs. While straightforward, its main limitation is that it ignores intangible assets that are difficult to duplicate like brand reputation, human capital, or future growth potential.[21]

To value *Viral Wardrobe* using this method, we would sum the costs of all tangible and developmental assets created to date. Table 12-4 details this calculation.

---

[21] Esteban Sastre, "What Is the Cost to Duplicate Startup Valuation Method?" Medium, September 5, 2022, https://medium.com/@fro_g/what-is-the-cost-to-duplicate-startup-valuation-method-7c888c96c4f0.

Table 12-4. *Cost-to-Duplicate Valuation for Viral Wardrobe*

Item	Description	Assigned Value
Ecommerce Platform Development	Cost for developers to build the website, back end, and payment integration	$45,000
Sample Collection Production	Cost of materials, manufacturing, and shipping for the initial test collection	$20,000
Legal and Administrative Fees	Company registration, trademark filing, and partnership agreements	$10,000
Founder "Salaries"/Sweat Equity	Estimated market-rate compensation for the time founders invested	$60,000
Marketing & Overhead	Costs for creating the pre-launch waitlist, office space, and utilities	$15,000
Total Cost to Duplicate		**$150,000**

As the table shows, this method yields a valuation of $150,000, representing only the hard costs invested so far. Perhaps this method could work for some businesses that strongly depend on experts' knowledge (e.g., specific deep-tech products). However, it does not seem to capture well the potential of *Viral Wardrobe*.

Scorecard Method

The Scorecard Method (also known as the Bill Payne Method) compares the target startup to similar pre-revenue companies in the same industry and region. It starts with the median pre-money valuation for these comparable startups and then adjusts it based on a weighted scorecard of qualitative factors. This provides a more nuanced valuation by factoring in key indicators of potential success.[22]

---

[22] Bill Payne, "Scorecard Valuation Methodology (Rev 2019): Establishing the Valuation of Pre-revenue, Start-up Companies," Angel Capital Association, October 21, 2019, https://angelcapitalassociation.org/blog/blog-scorecard-valuation-methodology-rev-2019-establishing-the-valuation-of-pre-revenue-start-up-companies/.

To apply this to *Viral Wardrobe*, we first establish a benchmark. Let's assume the median pre-money valuation for a pre-launch fashion-tech startup in our region is $1,500,000. We then score *Viral Wardrobe* against the median, as detailed in Table 12-5. Keep in mind that these factors and weights are adjustable. Every investor has their own preferences based on their unique experience. Some investors would treat marketing, sales, and relationships as one category and assign 10% to it. Some would also have categories like need for additional investment (5%) and other (5%). In our case, as we operate in the world of creator economy, we decided to differentiate between marketing/sales and strategic relationships where both categories are assigned 10%. However, this is a subjective decision every investor makes on their own. The goal is to be able to compare similar startups no matter how exactly you decide to define your factors and weights.[23]

Table 12-5. *Scorecard Valuation for Viral Wardrobe*

Factor	Weight	Score (vs. Median)	Calculation
Strength of Management Team	30%	0.9 (Slightly weaker due to lack of manufacturing experience)	30% x 0.9 = 0.27
Size of Opportunity	25%	1.3 (Strong, taps into major creator economy trend)	25% x 1.3 = 0.325
Product/Technology	15%	1.1 (Functional site and successful samples are a plus)	15% x 1.1 = 0.165
Competitive Environment	10%	1.0 (Average competition)	10% x 1.0 = 0.10
Strategic Relationships	10%	1.4 (Letters of intent from influencers are a major advantage)	10% x 1.4 = 0.14
Marketing/Sales	10%	1.2 (Pre-launch waitlist makes us stand out)	10% x 1.2 = 0.12
Total	**100%**	**Sum of Factors:**	**1.12**

---

[23] "Scorecard Valuation Method Explained," Eqvista, February 26, 2021, https://eqvista.com/scorecard-valuation-method-explained/.

The final valuation is the benchmark multiplied by the total factor from the table:

$1,500,000 (Benchmark Valuation) x 1.12 (Sum of Factors) = $1,680,000

Risk Factor Summation Method

This method also starts with a base valuation for a comparable startup and then adjusts it based on an analysis of 12 key risk areas. Each risk is rated on a scale from +2 (very positive) to -2 (very negative). The total score is then multiplied by $250,000 to get an adjustment value, which is added to or subtracted from the initial average valuation. This method provides a structured way to assess potential hurdles a startup may face.[24,25]

Using the same $1,500,000 average valuation for *Viral Wardrobe*, we assess the risks in Table 12-6.

Table 12-6. *Risk Factor Summation for Viral Wardrobe*

Risk Factor	Rating (-2 to +2)	Rationale
Management	-1	Strong marketing skills but a clear weakness in supply chain logistics and manufacturing.
Stage of the business	+1	Has a prototype, a waitlist, and clear letters of intent, which is advanced for pre-launch.
Legislation/Political risk	0	Standard ecommerce and fashion industry regulations.
Manufacturing risk	-1	The team's lack of experience in clothing production is a significant risk.
Sales and marketing risk	+2	Strong go-to-market plan leveraging influencer audiences.

(*continued*)

---

[24] "Risk Factor Summation Method: Everything you need to know," Eqvista, August 23, 2021, https://eqvista.com/risk-factor-summation-method/.

[25] Babu, A., Arikutaram, C., Mathews, A. (2023). Risk Factor Summation Method. In: Derindere Köseoğlu, S. (eds) A Practical Guide for Startup Valuation. Contributions to Finance and Accounting. Springer, Cham. https://doi.org/10.1007/978-3-031-35291-1_11

Table 12-6. (*continued*)

Risk Factor	Rating (-2 to +2)	Rationale
Funding/capital raising risk	0	Average risk for such a startup.
Competition risk	0	Average competition in our niche.
Technology risk	+1	The ecommerce platform is functional, reducing technical hurdles.
Litigation risk	0	No outstanding or unusual legal risks.
International risk	0	Currently focused on the domestic market.
Reputation risk	+1	Partnering with popular influencers provides an initial reputational boost.
Potential lucrative exit	+1	A successful exit in the creator economy space is highly plausible.
Total Score	**+4**	

Based on the total score from the table, the adjustment is calculated as +4 x $250,000 = +$1,000,000. That gives us the final valuation of

$1,500,000 (Base Valuation) + $1,000,000 (Risk Adjustment) = $2,500,000

VC Method

The Venture Capital (VC) Method is forward-looking and values a startup based on its potential exit value—the price it might be sold for or go public at in the future. An investor estimates this future value and then works backward, using their required return on investment (ROI), to determine a fair post-money valuation today.[26,27,28]

---

[26] Julian Stylianou, "Everything You Need to Know About Valuation Methods," Swiss Startup Association, January 29, 2021, https://swissstartupassociation.ch/2021/01/29/valuation/.

[27] "Early-Stage Valuation Methods: 5 Approaches for Evaluating Start-Up Worth," Business Angel Institute, July 25, 2024, https://businessangelinstitute.org/blog/2024/07/25/early-stage-start-up-valuation-methods/.

[28] Bill Payne, "Valuations 101: The Venture Capital Method," Gust Blog, November 1, 2011, https://gust.com/blog/startup-valuations-101-the-venture-capital-method/.

To see this in action with *Viral Wardrobe*, an investor would make the following assumptions:

- **Projected Exit Value:** Believing in the business model, they project *Viral Wardrobe* could be acquired for $40 million in 5 years.

- **Required ROI:** For a high-risk, early-stage investment, the VC targets a 20x return.

First, calculate the target post-money valuation today:

Post-Money Valuation = Exit Value / Required ROI

$40,000,000 / 20 = $2,000,000

This $2 million figure is the post-money valuation, meaning the value of the company after the investment is made. If the investor plans to invest $500,000, the pre-money valuation would be

Pre-Money Valuation = Post-Money Valuation - Investment Amount

$2,000,000 - $500,000 = $1,500,000

Under the VC Method, the pre-money valuation is $1,500,000.

Summary of Valuations: Pre-revenue Methods

No single method provides a definitive answer. Instead, they create a plausible range that informs negotiations between founders and investors. Each method emphasizes different aspects of the startup, from tangible assets to future potential. The various valuations for *Viral Wardrobe* are summarized in Table 12-7.

Table 12-7. *Summary of Pre-revenue Valuations for Viral Wardrobe*

Method	Valuation	Key Focus
Cost-to-Duplicate	$150,000	Tangible assets and sunk costs
VC Method	$1,500,000	Future exit potential and investor ROI
Scorecard Method	$1,680,000	Comparison to industry peers on key factors
Berkus Method	$1,700,000	Progress in de-risking the business model
Risk Factor Summation	$2,500,000	Adjusting an average valuation based on specific business risks

For *Viral Wardrobe*, the valuations range from a low of $150,000 (ignoring all future potential) to a high of $2.5 million (rewarding the company for its strengths in marketing and strategic partnerships). This range gives the founders a well-rounded and defensible set of figures to take into investor meetings.

This comparison clearly illustrates that valuing a pre-revenue startup is a subjective and qualitative art. The final figure is influenced by numerous factors. Different investors prefer different valuation methods and are guided by their unique personal experiences, which shape their perception of a startup's potential and its synergies with their own goals. The prevailing economic environment and overall investor sentiment also play a significant role in the process. Moreover, a startup's ability to attract interest from multiple potential investors can create a competitive bidding situation, often resulting in a better deal. While it might seem that a higher valuation is always preferable, this isn't necessarily the case. Theoretically, it's a win, but founders should be cautious. An inflated valuation can set unrealistic expectations that may be impossible to meet in the future.

While the methods discussed so far are essential for early-stage companies, the valuation landscape shifts dramatically once a startup begins to generate meaningful revenue. The conversation moves from assessing purely potential value to analyzing actual performance. This introduction of concrete financial data allows for more quantitative, data-driven valuation approaches. Although the "art" of valuation never fully disappears, it is now complemented by a "science" grounded in real-world metrics, opening the door to the post-revenue valuation methods we will explore next.

Post-revenue Valuation Methods

Once a startup begins to generate revenue, the focus of valuation shifts from pure potential to actual performance. While still forward-looking, post-revenue valuation methods ground their analysis in tangible financial data. These methods range from quick market comparisons to complex financial models.

Revenue Multiple Method

The Revenue Multiple is one of the most straightforward and widely used valuation methods. It determines a company's value by multiplying its annual revenue by an "industry multiple." This multiple is derived from what similar companies in the same sector have recently been valued at, either through acquisitions or private funding rounds. The method is fast and market driven, but it can be overly simplistic, as it doesn't account for profitability, growth rate, or unique company strengths.[29]

The Revenue Multiple is a flexible and widely used valuation method. While the core principle of applying a multiple to a financial metric remains constant, the specific metric used as the base can be tailored to the company's business model, industry, and stage of development. This adaptability has led to several specialized versions of the method:

- **Annual Recurring Revenue (ARR) Multiple:** Considered a standard for subscription-based businesses, this multiple focuses specifically on predictable, recurring revenue streams to better reflect the company's stable financial health.[30,31]

- **Gross Profit Multiple:** This version is used to evaluate a company's underlying profitability and margin structure, offering a clearer view of its core operational efficiency independent of sales and marketing expenses:

  Gross Profit = Revenue - Cost of Goods Sold (COGS).[32]

- **EBITDA Multiple:** For more mature, cash-flow-positive companies, the EBITDA multiple is common. EBITDA stands for Earnings Before Interest, Taxes, Depreciation, and Amortization. Its primary purpose is to provide a clean look at a company's core operational profitability

---

[29] Alejandro Cremades, "Fundraising Success: How the Revenue Multiple Factors in a Company Valuation," Alejandro Cremades Website, accessed July 24, 2025, https://alejandrocremades.com/fundraising-success-how-the-revenue-multiple-factors-in-a-company-valuation/.

[30] "ARR Multiple: A Key Metric for SaaS Valuation," Profit.co, accessed July 24, 2025, https://www.profit.co/blog/kpis-library/arr-multiple-a-key-metric-for-saas-valuation/.

[31] "ARR Multiple," Wall Street Prep, September 17, 2024, https://www.wallstreetprep.com/knowledge/arr-multiple/.

[32] "EV/Gross Profit Ratio," Corporate Finance Institute, accessed July 23, 2025, https://corporatefinanceinstitute.com/resources/valuation/ev-gross-profit-ratio/.

by removing the effects of taxes, financing costs, and non-cash expenses. Its calculation logic makes it useful for comparing businesses with different financial structures:

EBITDA = Operating Income + Depreciation & Amortization.[33]

- Here's a breakdown of the components:

    - *Operating Income: The profit generated from a company's core business operations. It excludes interest and taxes.*

    - *Depreciation and Amortization: A non-cash expense. The accounting method to spread the costs of tangible (e.g., laptops) and intangible (e.g., software) assets.*

- **User-Based Multiple:** In the case of pre-monetization startups, especially in the consumer tech space, a non-financial metric like Monthly Active Users (MAUs) may be used. Here, the "multiple" becomes a "per-user" value derived from what acquirers have paid for similar companies, shifting the focus to user engagement and growth potential.[34]

Ultimately, the choice of which metric to apply is critical, as it should align with the key drivers of value for that specific business at its particular stage of development. For the purposes of this exercise, we will stick to the basic revenue multiple.

Let's assume *Viral Wardrobe* is now one year post-launch and has generated $1 million revenue. To find the right multiple for *Viral Wardrobe*, your next step is to research the market. Let's say you identify five comparable companies in the fashion and creator economy space.

The first, a company called *Creators' Zone*, was recently valued at $9 million and has an annual revenue of $1.5 million. To find its multiple, you divide the valuation by the revenue:

$9,000,000 (Valuation) / $1,500,000 (Annual Revenue) = 6x

---

[33] Aswath Damodaran, "Value/EBITDA Multiple," NYU Stern School of Business, accessed July 24, 2025, `https://pages.stern.nyu.edu/~adamodar/New_Home_Page/lectures/vebitnote.html`.

[34] Christopher Haught, "Enterprise Value/Monthly Active Users: a Valid Sector Specific Multiple for the Valuation of Social Media Firms?" Miami University ETD, 2017, `https://etd.ohiolink.edu/acprod/odb_etd/ws/send_file/send?accession=miami1500280724256454&disposition=inline`.

After performing the same calculation for the other four companies, you find that the average multiple for this specific hypothetical group is 8x.

Valuation = Annual Revenue x Industry Multiple

$1,000,000 x 8 = $8,000,000

Based on this method, the valuation of *Viral Wardrobe* is $8 million.

Discounted Cash Flow (DCF) Method

The Discounted Cash Flow (DCF) method values a company based on the present value of its projected future cash flows. This is a highly quantitative, fundamentals-based approach. It involves forecasting a company's free cash flow over several years and then "discounting" it back to today's value using a discount rate that reflects the time value of money and the risk of the investment. The main challenge is that its accuracy depends entirely on the quality of its assumptions about future growth and risk, which can be highly speculative for a startup.[35]

The choice of discount rate is one of the most critical assumptions in a DCF model, as it directly quantifies the risk of an investment. A higher rate is used for riskier ventures to compensate investors for the significant uncertainty involved. While discount rates for other startups can often range from 30% to 50% or even higher, the specific context of *Viral Wardrobe* allows for a more nuanced approach.[36]

For this hypothetical valuation, we will assume the founders have consulted with several industry experts and mentors. These advisors, who highly value the work done so far—achieving $1 million in revenue and proving strong market traction—believe the company's progress has meaningfully reduced some of the typical startup risks. They have advised that a 25% discount rate is a justifiable figure. While acknowledging

---

[35] "Startup Valuation: Applying the Discounted Cash Flow Method in Six Easy Steps," EY Finance Navigator, March 20, 2019, https://www.ey.com/en_nl/services/finance-navigator/startup-valuation-applying-the-discounted-cash-flow-method-in-six-easy-step.

[36] "What is a Discount Rate and How to Calculate it?" Eqvista, accessed July 23, 2025, https://eqvista.com/company-valuation/discount-rate/.

that *Viral Wardrobe* remains a high-risk venture, their consensus is that its proven execution warrants a rate at the lower end of the early-stage spectrum. Therefore, for this example, we will use the expert-advised rate of 25%. This figure reflects a balanced view, recognizing both the inherent risks of a startup and the significant progress *Viral Wardrobe* has made to de-risk its future. Let's walk through the calculation with this rate.

Step 1: Calculate the Present Value of Forecasted Cash Flows

First, we discount the projected free cash flows (FCF) using our 25% rate. This projection assumes that in its first year of accelerated growth, *Viral Wardrobe* will generate approximately $1.7 million in revenue, and after covering all operating expenses and reinvesting heavily in technology and inventory to sustain its momentum, it will be left with $200,000 in free cash flow.

>   *Projected FCF for Year 1: $200,000*
>
>   *Projected FCF for Year 2: $450,000*
>
>   *Projected FCF for Year 3: $900,000*

Now, we find their present values:

>   *PV of Year 1 FCF: $200,000 / (1 + 0.25)^1 = $160,000*
>
>   *PV of Year 2 FCF: $450,000 / (1 + 0.25)^2 = $288,000*
>
>   *PV of Year 3 FCF: $900,000 / (1 + 0.25)^3 = $460,800*

The sum of these present values is $908,800.

Figure 12-4 uses a stacked bar chart to break down *Viral Wardrobe's* projected free cash flows over three years. In the chart, the total height of each bar corresponds to the total projected FCF for that year. Each bar is divided to illustrate the impact of discounting: the green area represents the present value of that cash flow after applying a 25% discount rate, while the yellow area on top shows the amount of the discount itself. This visualization makes it clear how the present value is a fraction of the future projection and how the impact of the discount grows larger in later years.

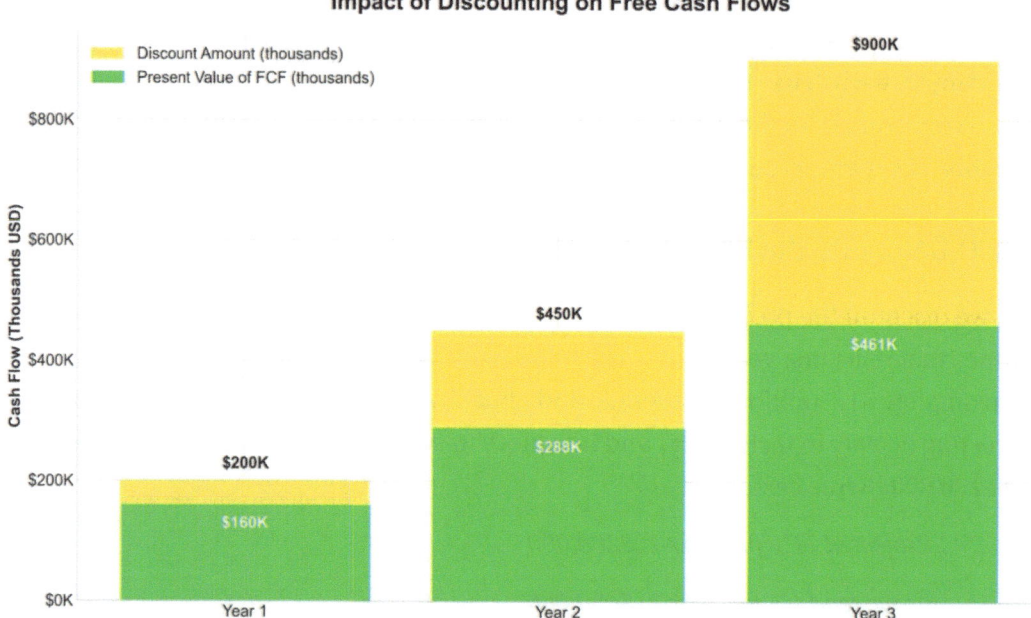

Figure 12-4. *DCF Method: Impact of Discounting on Free Cash Flows*

Step 2: Calculate the Terminal Value and Its Present Value

Next, we calculate the terminal value using the Exit Multiple Method based on the projected $4.5 million in revenue at the end of Year 3 and our 8x industry multiple.[37]

Terminal Value: *$4,500,000 x 8 = $36,000,000*

This $36 million is the company's estimated worth at the end of Year 3. We must now discount this figure back to its present value using the 25% rate:

Present Value of Terminal Value: $36,000,000 / (1 + 0.25)^3 = $18,432,000

---

[37] "The Pros and Cons of Using Exit Multiple vs. Perpetuity Growth Rate (PGR) in a Discounted Cash Flow," Apollo Financial Models, June 2, 2024, https://apollofinancialmodels.com/blogs/financial-modeling-insights-excel-blog/the-pros-and-cons-of-using-exit-multiple-vs-perpetuity-growth-rate-pgr-in-a-discounted-cash-flow.

Step 3: Determine the Total Company Valuation

Finally, the total valuation is the sum of the present values of the forecasted cash flows and the terminal value.[38]

Total Valuation = *PV of Forecasted FCFs + PV of Terminal Value*

$908,800 + $18,432,000 = $19,340,800

Under this DCF analysis, the valuation of *Viral Wardrobe* is approximately $19.3 million.

First Chicago Method

But how much can we be sure of the assumptions discussed above? What if something goes wrong—for instance, costs turn out to be higher or revenue turns out to be lower? What if it turns out we're too pessimistic about our assumptions—for example, we manage to acquire several top-tier creators and our revenue will be much higher than expected? This is exactly where we could analyze several scenarios, and the First Chicago Method is based on that.[39]

Developed by the venture capital arm of the First Chicago Bank, this method offers a structured way to handle the uncertainty of valuing early-stage companies. It works by creating a probability-weighted average from multiple valuation scenarios, which balances optimistic, pessimistic, and realistic outcomes to calculate a company's potential worth. This method combines elements of market-based multiples and DCF valuations.[40]

---

[38] "Terminal Value," Wall Street Prep, April 7, 2025, https://www.wallstreetprep.com/knowledge/terminal-value/.

[39] "First Chicago Method for Startup Valuation," Eqvista, April 7, 2021, https://eqvista.com/first-chicago-method-for-startup-valuation/.

[40] "Valuation Methods for Startups—The Easy Guide to Value a Startup," Valutico, October 7, 2024, https://valutico.com/business-valuation-methods-for-startup-companies/.

The approach involves creating three distinct financial forecasts for a company: a best-case, a worst-case, and a base-case scenario. A valuation is calculated for each, and a probability is assigned based on how likely each outcome is perceived to be. The final valuation is the weighted average of these scenarios, offering a more balanced view that accounts for risk and opportunity. Let's apply this to *Viral Wardrobe* using our previous calculations to inform the scenarios.

-   **Best-Case Scenario (Upside):** $40,000,000 Valuation (20% Probability). This optimistic outcome assumes *Viral Wardrobe* achieves exponential growth, perhaps by securing exclusive contracts with major creators or expanding into new markets more quickly than anticipated. Let's assume investors believe there is a 20% chance of this happening.

-   **Base-Case Scenario (Most Likely):** $19,340,800 Valuation (50% Probability). This scenario aligns with the detailed DCF analysis, representing a future where the company successfully executes its business plan and meets its projected growth targets. This is considered the most probable outcome, with a 50% likelihood.

-   **Worst-Case Scenario (Downside):** $8,000,000 Valuation (30% Probability). This more pessimistic forecast, aligning with the simple Revenue Multiple method, assumes the company's growth stalls due to unforeseen challenges or increased competition. In this case, its value is tied more closely to its current performance. Investors see a 30% chance of this scenario.

The next step is to calculate the probability-weighted valuation:

Valuation = (Best-Case Value x Probability) + (Base-Case Value x Probability) + (Worst-Case Value x Probability)

Valuation = ($40,000,000 x 0.20) + ($19,340,800 x 0.50) + ($8,000,000 x 0.30)

Valuation = $8,000,000 + $9,670,400 + $2,400,000 = $20,070,400

By weighing these potential futures, the First Chicago Method gives *Viral Wardrobe* a valuation of approximately $20.1 million. This comprehensive approach, which balances optimism with realism, is a favored tool among venture capital and private equity investors for its ability to quantify a range of outcomes.

Summary of Valuations: Post-revenue Methods

After a startup begins to generate meaningful revenue, valuation shifts from educated guesswork to more disciplined measurement. Investors now expect defensible numbers that spotlight operational traction without ignoring upside potential. While each technique approaches value from a different angle, they are best viewed as complementary lenses rather than mutually exclusive verdicts.

The simplest first pass is the Revenue Multiple Method, which multiplies current annual revenue by an industry multiple gleaned from comparable transactions. For our hypothetical *Viral Wardrobe's* $1 million annual revenue, applying an 8× industry multiple produced an $8 million figure. It is quick, market anchored, and easy to explain, yet it overlooks profitability, capital intensity, and growth durability.

To dig deeper, the Discounted Cash Flow (DCF) Method projects free cash flows over several years out and discounts them back at a rate that reflects startup risk. *Viral Wardrobe's* data—paired with a 25% discount rate—returned a $19 million present total value. DCF captures operating leverage and margin expansion but is notoriously sensitive to optimistic spreadsheets.

Finally, the First Chicago Method bridges narrative and numbers by blending three scenarios: worst case ($8 million valuation), base case ($19 million), and a $40 million best-case exit. Weighting them at 30%, 50%, and 20%, respectively, produced a probability-adjusted $20 million valuation. This hybrid view tempers blue-sky pitches with sober downside math. Table 12-8 summarizes these post-revenue valuations.

Table 12-8. *Summary of Post-revenue Valuations for Viral Wardrobe*

Method	Valuation	Key Focus
Revenue Multiple	$8 million	Anchoring value to current revenue and market comparables
Discounted Cash Flow	$19 million	Projecting intrinsic value via future cash flows
First Chicago	$20 million	Scenario-weighted outlook balancing downside and upside

Practical takeaway: when multiple methods diverge, treat the range as a decision space, not an error bar. Start conversations and align on assumptions—growth rate, discount rate, multiples—before debating the dollar output.

These results highlight the value in triangulating across valuation methods to inform both internal decision-making and external negotiations. Ultimately, the chosen valuation should reflect both hard data and the unique story of the startup.

AI Startups Valuation

Valuing an AI startup requires a different lens than a traditional non-AI company. While standard metrics like revenue and user growth are still important, they don't tell the whole story, especially for early-stage AI ventures that may not have significant revenue yet. The fundamental difference lies in what creates value and the unique risks AI companies face.[41]

For a non-AI startup (like a typical SaaS or ecommerce company), value is primarily measured by demonstrated market traction. Investors look at existing revenue, customer acquisition costs, and user engagement. The valuation is a reflection of current business performance and its projected growth.

For an AI startup, value is rooted in its foundational capabilities and proprietary assets. The valuation is less about current revenue and more about the potential locked within its core components. Investors focus on three key pillars that are not as central in non-AI valuations:[42]

- **Proprietary Data:** High-quality, unique datasets are the fuel for AI models. An AI startup with an exclusive data source has a powerful, defensible advantage—or "moat"—that is extremely difficult for competitors to replicate.

- **Talent and Team:** The expertise of the AI team is paramount. A company led by renowned researchers or engineers with a proven track record in building and scaling complex AI systems can command a significant valuation premium on its own. A fierce talent war has erupted in the sector, with tech giants like Meta, Google,

---

[41] Ryan E. Long, "3 Methods for Valuing Pre-Revenue Novel AI Startups," TechCrunch, December 13, 2022, https://techcrunch.com/2022/12/13/3-methods-for-valuing-pre-revenue-novel-ai-startups/.

[42] Lior Ronen, "Valuation Services for AI Startups," Finro Financial Consulting, February 13, 2025, https://www.finrofca.com/news/valuation-services-ai-startups.

OpenAI, and Apple engaging in an intense battle to recruit and retain the world's top AI researchers. This competition has driven compensation packages to unprecedented levels, drawing comparisons to the high-stakes world of professional sports.[43,44]

— **Technology and Models:** The sophistication, novelty, and defensibility of the AI algorithms and models are critical. Is the technology a genuine breakthrough or just an application of open source tools? The intellectual property (IP) behind the model is a core asset.

However, as the AI landscape matures, a fourth pillar has become critical: **the distribution moat**. This refers to a startup's ability to embed its technology deeply into customer workflows, build a strong brand, or leverage unique access to a large user base. Without effective distribution, even superior technology can fail to capture market value.[45]

In simple terms, you can think of it this way: valuing a non-AI startup is more like assessing a restaurant based on its current profits and number of customers. Valuing an early-stage AI startup is like assessing the value of a secret, revolutionary recipe held by a world-class chef—the potential is immense, even before the first meal has been sold.

While investors still lean on established pre-revenue valuation frameworks, they are adapted to reflect the unique drivers of AI.

— **Scorecard Method Re-weighted:** This method, which compares a target startup to typical angel-funded companies in a region, is adjusted to heavily favor AI-specific strengths. A common weighting for a seed-stage AI startup might be: Technology (35%), Data Defensibility (25%), Team (20%), and Market Size (20%).[46]

---

[43] Gerui Wang, "Meta And OpenAI's Talent Wars: How AI Mints Elites But Displaces Others," Forbes, July 12, 2025, `https://www.forbes.com/sites/geruiwang/2025/07/12/meta-and-openais-talent-wars-how-ai-mints-elites-but-displaces-others/`.

[44] Pragati Chougule, "AI Talent Wars: Meta vs. OpenAI and the Battle for Innovation," The Bridge Chronicle, June 18, 2025, `https://www.thebridgechronicle.com/tech/ai-talent-wars-meta-vs-openai-battle-for-innovation-2025`.

[45] Hamiz M. Awan, "FuturProof #239: Distribution is King," LinkedIn, October 21, 2024, `https://www.linkedin.com/pulse/futurproof-239-distribution-king-hamiz-m-awan-avpmc/`.

[46] Thomas Howard, "How to Make a Valuation Model for Your AI Startup: A Guide for Investors," Collateral Base, accessed July 24, 2025, `https://collateralbase.com/how-to-make-valuation-model-ai-startup/`.

- **Revenue Multiples with an AI Premium:** For post-revenue AI companies, investors apply significantly higher revenue multiples than for typical Software-as-a-Service (SaaS) businesses. This premium reflects the market's expectation of higher growth and margins.[47,48]

- **Discounted Cash Flow (DCF) Adjustments:** When using DCF models, analysts project higher long-term growth rates and terminal values for AI companies but also apply higher discount rates to account for the significant technical and market risks.

The Leonis Capital Framework: Quantifying AI Risk

A critical challenge in AI valuation is **displacement risk**—the threat that a startup's core technology could be rendered obsolete by a more powerful open source model or a new release from a major AI lab. To quantify this, some investors use the "Zero or Hero" framework developed by Leonis Capital, which introduces a "Displacement Factor," or D-Factor.[49,50]

The formula adjusts a standard valuation:

Valuation=ARR x Multiple x (1−D)

Here, D is a coefficient between 0 and 1 representing the probability of the startup's technology being displaced within a given timeframe. A higher D value signifies greater risk.

- D = 0.1–0.2: A true technical leader with a significant and defensible technological lead

[47] Tomasz Tunguz, "It Pays to be an AI Company," Tom Tunguz Website, February 3, 2025, https://tomtunguz.com/private_valuations_2025/.

[48] Lior Ronen, "M&A in AI: 2025 Valuation Multiples and Key Trends," Finro Financial Consulting, January 9, 2025, https://www.finrofca.com/news/ai-mna-valuation-2025.

[49] Jenny Xiao, LJW, and Jay Zhao, "Zero or Hero: A Technical Framework for Valuing AI Companies (Part II: AI Applications)," Leonis Newsletter, Substack, April 22, 2025, https://leonisnewsletter.substack.com/p/zero-or-hero-a-technical-framework-6e8.

[50] Jenny Xiao and Jay Zhao, "Zero or Hero: A Technical Framework for Valuing AI Companies (Part I: Foundation Models)," Leonis Newsletter, Substack, March 25, 2025, https://leonisnewsletter.substack.com/p/zero-or-hero-a-technical-framework.

- D = 0.3–0.5: A company with a moderate lead but facing credible competition

- D = 0.6–0.8: A startup at high risk of being matched or surpassed by competitors or open source alternatives

- D = 0.9–1.0: A company whose technology is already effectively matched or surpassed by available open source models, approaching what Leonis calls the "zero-value threshold"

This framework highlights a key dynamic: for foundation model vendors, the risk is that competitors or open source alternatives will match their performance at a lower cost. For application-layer AI companies, the risk is that a foundation model will absorb their feature set. This risk can have swift and devastating consequences on a startup's valuation, as illustrated in Figure 12-5.

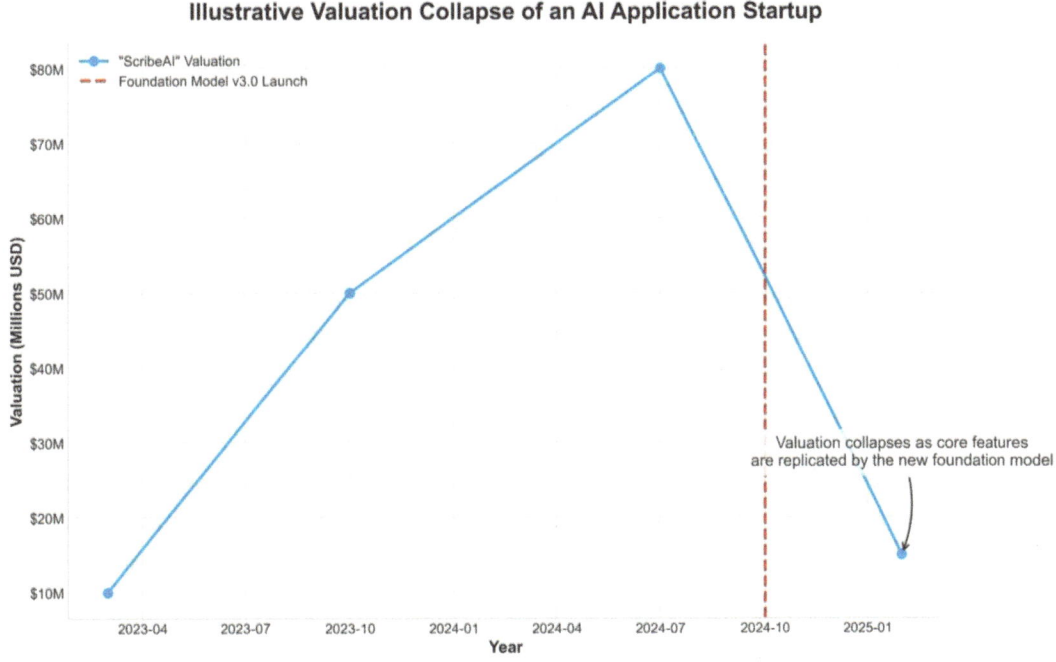

Figure 12-5. *Illustrative Valuation Collapse of an AI Application Startup*

The chart depicts a hypothetical AI application startup that enjoys initial valuation growth, only to see its value plummet after a major competitor—a large foundation model—integrates its core functionality, effectively rendering the startup's product obsolete overnight. This scenario represents a high "D" factor in action, where the "zero-value threshold" is crossed due to a lack of a durable moat.

The key to commanding a higher valuation and a lower D-Factor lies in building a "Reinforcing Moat System," where multiple layers of defensibility work together to create compounding value. The strongest defense, which can result in a D-Factor between 0.1 and 0.2, is typically built on the integration of three key pillars:[51]

- **Proprietary Data:** This goes beyond simply having a large dataset. The most valuable data is unique and improves with user interaction, creating a feedback loop where the product gets smarter with each use. This ever-widening competitive gap is difficult for generalized foundation models to close.

- **Domain Expertise:** Embedding specialized, industry-specific knowledge into the product creates a powerful moat. For example, a financial compliance tool that understands complex regulatory requirements or a healthcare platform built on clinical insights provides value that a generic model cannot replicate.

- **Workflow Integration:** When a tool becomes deeply embedded in a customer's critical daily operations, it creates high switching costs. The value comes not just from the AI's output but from its seamless integration into the user's process, increasing trust and dependency.

Together, these three elements create a synergistic effect where the whole is greater than the sum of its parts. The data improves the model's performance within a specific domain, which in turn makes the workflow integration more valuable, generating more proprietary data and starting the cycle anew.

The displacement risk factor is assessed based on

- **Foundation Model Capability Gap:** How much better is the application than a state-of-the-art foundation model? A slim lead means high risk.

---

[51] Jenny Xiao, LJW, and Jay Zhao, "Zero or Hero: A Technical Framework for Valuing AI Companies (Part II: AI Applications)," Leonis Newsletter, Substack, April 22, 2025, `https://leonisnewsletter.substack.com/p/zero-or-hero-a-technical-framework-6e8`.

- **Integration Depth:** How deeply is the product embedded in customer workflows? A tool that is easily swapped out is more vulnerable.

- **Data Network Effects:** Does the startup have a proprietary data loop where the product improves with more usage? Publicly available data does not constitute a strong moat.

- **Regulatory Barriers:** Operating in a regulated industry like healthcare (e.g., requiring FDA clearance) can create a strong, defensible moat against general-purpose models.

This framework highlights what Leonis calls the "Inverse Scale Law" for AI applications. In traditional SaaS, a larger Total Addressable Market (TAM) is always better. In AI, a broad, horizontal focus (e.g., generic marketing copy generation) is highly susceptible to being absorbed by foundation models. In contrast, a narrow, vertical focus (e.g., AI for legal contract analysis or for clinical trial documentation) often builds a stronger, more defensible moat and thus carries a lower "D" factor.

Ultimately, valuing an AI startup is a nuanced process that blends traditional financial modeling with a deep, technical understanding of the product's defensibility in a rapidly evolving technological landscape.

Market Evidence

The valuation premiums for AI startups are supported by overwhelming market evidence, though the precise figures vary across different data sources and analysis methods. The consistent theme is that AI companies command revenue multiples that are significantly higher than their non-AI software counterparts. However, it's crucial to understand that a single "AI multiple" is misleading; the devil is truly in the detail.

Recent analyses of M&A transactions and private funding rounds highlight this premium. For instance, studies of AI M&A deals in 2024 and 2025 show average revenue multiples in the range of 25x to 30x. In venture capital funding rounds, the multiples can be even more dramatic.[52,53]

---

[52] Lior Ronen, "M&A in AI: 2025 Valuation Multiples and Key Trends," Finro Financial Consulting, January 9, 2025, `https://www.finrofca.com/news/ai-mna-valuation-2025`.

[53] "AI Valuation Multiples: Funding rounds, valuations, and investor universe: 2010-2024," Aventis Advisors, March 2025, `https://aventis-advisors.com/wp-content/uploads/2025/03/AI-Valuation-Multiples-2025.pdf`.

Depending on the startup's maturity, multiples can range widely:[54]

- **Early-Stage AI Startups:** Multiples can span as broad as from 5x to 160x revenue, often reflecting investor confidence in future potential rather than current financial performance.

- **Growth and Mature-Stage AI Startups:** As companies scale and their market position becomes clearer, multiples tend to settle in the 5x to 20x revenue range. But still, a big discrepancy is seen between top companies and some of the less attractive ones.

This wide variation is driven by specific niches within the AI landscape. Foundational models and large language model (LLM) vendors command the highest valuations, with some analyses showing average multiples exceeding 50x revenue. In contrast, AI companies in more saturated markets, such as Marketing Tech or Computer Vision, may see multiples in the lower range of 12x to 15x.

According to Crunchbase, 2024 was a breakout year for investment in artificial intelligence. The sector became the leader in venture funding, attracting nearly a third of all global venture capital. Funding for AI-related companies exceeded $100 billion. This total surpassed funding levels for any single year in the past decade, including the pandemic-fueled funding peak. Of this capital, almost a third was directed toward foundation model companies, while the remaining two-thirds flowed into sectors impacted by these new models, including autonomous vehicles, healthcare, robotics, infrastructure, security, and defense applications.[55]

Summary

This chapter provided a comprehensive, step-by-step guide to startup valuation, demonstrating how it evolves from a qualitative art to a quantitative science as a company matures. We began by establishing that startup valuation is fundamentally

---

[54] "AI Valuation Multiples: Funding rounds, valuations, and investor universe: 2010-2024," Aventis Advisors, March 2025, `https://aventis-advisors.com/wp-content/uploads/2025/03/AI-Valuation-Multiples-2025.pdf`.

[55] Gené Teare, "Startup Funding Regained Its Footing In 2024 As AI Became The Star Of The Show," Crunchbase News, January 7, 2025, `https://news.crunchbase.com/venture/global-funding-data-analysis-ai-eoy-2024/`.

a forward-looking exercise focused on future potential, a principle exemplified by Snapchat's journey from a pre-revenue app to a multi-billion-dollar public company. From there, the chapter systematically explored the core methodologies and concepts that startup founders and data professionals must understand.

The key topics covered include

- **The Foundation of Valuation:** We established the core challenge: valuing companies with limited financial history, where narrative and data must intersect. We also examined how market cycles ("risk-on" vs. "risk-off" periods) influence investor sentiment and capital availability, directly impacting valuations.

- **Pre-revenue Valuation Methods:** For the earliest stages, we detailed qualitative approaches that assess potential rather than performance. Using the *Viral Wardrobe* case study, we demonstrated how to apply

  - **The Berkus Method:** Assigning value to core de-risking milestones

  - **The Cost-to-Duplicate Method:** Calculating the cost to replicate existing assets

  - **The Scorecard and Risk Factor Summation Methods:** Benchmarking against comparable startups while adjusting for qualitative strengths and weaknesses

  - **The VC Method:** Working backward from a potential future exit

- **Post-revenue Valuation Methods:** As *Viral Wardrobe* began generating revenue, we shifted to quantitative models grounded in financial performance, including

  - **The Revenue Multiple Method:** Applying an industry-standard multiple to current revenue

  - **The Discounted Cash Flow (DCF) Method:** Projecting future cash flows and discounting them to their present value

  - **The First Chicago Method:** Creating a probability-weighted valuation by analyzing best-case, base-case, and worst-case scenarios

- **The Special Case of AI Startups:** We explored the unique factors driving AI startups valuations, such as proprietary data, elite talent, and defensible technology. A key focus was on quantifying "displacement risk"—the threat from larger foundation models— using frameworks like Leonis Capital's "D-Factor" and the importance of building a "Reinforcing Moat System" through proprietary data, deep workflow integration, and domain expertise.

- **The Final Takeaway:** The chapter concluded by reinforcing that a startup's valuation is not a single, absolute number but a defensible range. The valuation process itself is a powerful diagnostic tool, highlighting a company's strengths and exposing critical weaknesses that must be addressed to build a more valuable and successful company.

Now that we have established how a startup's value is articulated and justified, we will turn our attention to the advanced techniques that help create that value in the first place. After all, a high valuation is the result, not the cause, of a successful product and a deep understanding of your market. In Chapter 13, we will dive into the world of advanced data science and artificial intelligence, exploring the practical machine learning techniques that can help you build personalized user experiences, predict customer behavior, and unlock the innovative potential that investors and employees will rally behind.

CHAPTER 13

Advanced Techniques in Data Science and AI

We have a lot of shared machine learning models that basically give you information that is very usable for many different use cases. Some examples are users' affinities to various things. How much do you like this artist, how much do you like this album, how much do you like this playlist? It could also be similarities. Are these two artists similar? Give me five artists that are similar to this artist, or give me five playlists that are similar to this playlist. It could also be clustering. Here are 20 artists, tell me how they belong together?

—Oskar Stål[1],

Vice President of Personalization at Spotify

Many startups today are built on data and the power of machine learning to ensure that data is used effectively to create personalized user experiences and keep customers satisfied with the product. Spotify's mission is to ensure each of their hundreds of millions of listeners has a personal and unique experience as they explore and enjoy the music they love. Netflix operates similarly, connecting users with a vast library of movies through sophisticated recommendation algorithms. Multiple startups across different sectors are working with various types of data, trying to make user experiences as personalized as possible by connecting their users with the right information at the right time.

---

[1] Oskar Stål, "Creating Personalized Listening Experiences with Spotify," Scale Events, accessed April 10, 2025, https://exchange.scale.com/public/videos/creating-personalized-listening-experiences-with-spotify.

To achieve this level of personalization, these companies need to build smooth and scalable algorithms based on people's characteristics and preferences, their past behavior, and clustering techniques that segment users using statistical methods. But how do you get there if you're just getting started, you don't have much data, you don't have many users, you don't have a big team of experts with technical expertise, and you don't have a big budget or a lot of time to build sophisticated systems?

Recommender systems represent just one area where machine learning can be leveraged by startups. Startups can use data science in multiple different areas, including reducing churn, building predictive models, automating customer support, and even using data science techniques when building data storytelling for startup investors through multidimensional, interactive, and colorful visualizations. However, according to Rand research, more than 80% of AI projects fail.[2]

This chapter explores three critical aspects of implementing advanced data science and AI in startups:

1. **Understanding the Technical Landscape:** We will demystify the buzzwords by clarifying the relationships between data science, AI, and machine learning and review the core ML techniques that are most relevant for startups.

2. **Selecting the Right Projects:** You will learn how to move from ideas to execution by using a prioritization framework to identify initiatives that deliver maximum business value with limited resources.

3. **Avoiding Common Pitfalls**: We will examine why most data projects fail and provide practical, real-world examples to help you navigate the challenges of running data science and AI projects in a startup environment.

With this road map in place, our goal is to provide a practical, field-tested guide for the startup environment rather than an exhaustive technical manual. For readers who wish to delve deeper into the foundational concepts we'll be discussing, the following section offers a curated list of essential learning resources.

---

[2] James Ryseff, Brandon De Bruhl, and Sydne Newberry, "The Root Causes of Failure for Artificial Intelligence Projects and How They Can Succeed," RAND, accessed June 20, 2025, https://www.rand.org/pubs/research_reports/RRA2680-1.html.

Learning Resources

The goal of this book and this chapter is not to explain all technical foundations behind particular data science and AI concepts. These topics are simply too broad for this book. Instead, the focus is on what's interesting and useful from the standpoint of a person getting started with data in a startup environment. The emphasis is on practical, field-tested knowledge that has proven valuable when working with startups, especially early-stage ventures.

For readers who want to dig deeper into the **general data science concepts**, there are several excellent resources providing comprehensive technical foundations:

- *Data Science for Business* by Foster Provost and Tom Fawcett offers a business-oriented approach to understanding data science applications.[3]

- *Hands-On Machine Learning with Scikit-Learn, Keras & TensorFlow* by Aurélien Géron provides practical implementation guidance.[4]

- *Introduction to Machine Learning with Python* by Andreas Müller and Sarah Guido focuses on Python-specific implementations.[5]

- For those preferring R programming, *R for Data Science* by Hadley Wickham, Mine Cetinkaya-Rundel, and Garrett Grolemund serves as an excellent foundation.[6]

- DataCamp's extensive course library, created by industry experts, delivers structured learning paths across topics like machine learning and data science.[7]

---

[3] Foster Provost and Tom Fawcett, *Data Science for Business: What You Need to Know About Data Mining and Data-Analytic Thinking* (O'Reilly Media, 2013).

[4] Aurélien Géron, *Hands-On Machine Learning with Scikit-Learn, Keras & TensorFlow*, 2nd ed. (O'Reilly Media, 2019).

[5] Andreas Müller and Sarah Guido, *Introduction to Machine Learning with Python: A Guide for Data Scientists* (O'Reilly Media, 2016).

[6] Hadley Wickham, Mine Cetinkaya-Rundel, and Garrett Grolemund, *R for Data Science: Import, Tidy, Transform, Visualize, and Model Data,* 2nd ed. (O'Reilly Media, 2023).

[7] "DataCamp: Learn Data Science and AI Online," DataCamp, accessed May 1, 2025, `https://www.datacamp.com/`.

MLOps has emerged as a crucial discipline for startups looking to move beyond experimentation to production-ready AI systems. For those looking to deepen their knowledge on how to operationalize machine learning models, these resources provide practical guidance on implementing MLOps practices:

- *Practical MLOps: Operationalizing Machine Learning Models* by Noah Gift and Alfredo Deza elaborates on MLOps tools, methods, and best practices.[8]

- *Introducing MLOps: How to Scale Machine Learning in the Enterprise* by Mark Treveil and the Dataiku Team provides a comprehensive overview of the machine learning life cycle and deployment strategies.[9]

Complementing these, several **AI ethics, policy, and governance** resources help navigate evolving legal and regulatory landscapes:

- The *AI and Ethics* journal (Springer) publishes peer-reviewed articles on ethical frameworks, policy implications, and emergent governance challenges across jurisdictions.[10]

- *The Oxford Handbook of Ethics of AI* offers interdisciplinary perspectives on privacy, fairness, accountability, and regulatory challenges.[11]

- The Electronic Frontier Foundation's website for ongoing analysis of AI policy developments in major markets.[12]

- Startups should engage legal counsel to stay current with jurisdictional differences—what's currently permitted in the United States or China may conflict with GDPR in the EU.

---

[8] Noah Gift and Alfredo Deza, *Practical MLOps: Operationalizing Machine Learning Models* (O'Reilly Media, 2021)

[9] Mark Treveil and the Dataiku Team, *Introducing MLOps: How to Scale Machine Learning in the Enterprise* (O'Reilly Media, 2021)

[10] "AI and Ethics," Springer, accessed June 17, 2025, `https://link.springer.com/journal/43681`.

[11] Markus D. Dubber, Frank Pasquale, and Sunit Das, eds., *The Oxford Handbook of Ethics of AI* (Oxford University Press, 2020).

[12] "Electronic Frontier Foundation," Electronic Frontier Foundation, accessed June 21, 2025, `https://www.eff.org/`.

If you want to get to know more about **AI agents and autonomous systems**, consider

- *Agentic Artificial Intelligence: Harnessing AI Agents to Reinvent Business, Work, and Life* presents a non-technical road map for deploying autonomous agents in enterprise workflows.[13]

- Matt Dancho's social media posts, GitHub repo, and courses offer hands-on tutorials for building AI agents that make data professionals' life easier.[14]

- *Human-Compatible: Artificial Intelligence and the Problem of Control* by Stuart Russell examines the design of provably beneficial agents under uncertainty.[15]

- *Architects of Intelligence* by Martin Ford features interviews with AI leaders on future agent capabilities and governance hurdles.[16]

This curated list balances practical startup needs, technical depth, ethical considerations, and legal compliance—essential for responsible AI adoption in a fast-changing world. With these foundational resources in mind, let's establish a clear understanding of the landscape itself. Before diving into specific applications and frameworks, it's crucial to clarify the relationships between the terms that are often used interchangeably but represent distinct approaches and capabilities.

Machine Learning, AI, and Data Science Methods

In today's technology landscape, terms like machine learning, artificial intelligence, and data science are frequently used interchangeably, often misused for marketing purposes as buzzwords to make products and services appear more sophisticated than they actually are. Understanding the precise differences between these interconnected fields

---

[13] Pascal Bornet et al., *Agentic Artificial Intelligence: Harnessing AI Agents to Reinvent Business, Work and Life* (independently published, 2025).

[14] Matt Dancho, LinkedIn profile, accessed June 3, 2025, `https://www.linkedin.com/in/mattdancho/`.

[15] Stuart Russell, *Human-Compatible: Artificial Intelligence and the Problem of Control* (Viking, 2019).

[16] Martin Ford, *Architects of Intelligence: The Truth about AI from the People Building It* (Packt Publishing, 2018).

is crucial for startup founders who need to navigate technical decisions, communicate effectively with investors, and avoid the trap of implementing trendy but inappropriate solutions. This clarification becomes even more important when evaluating vendor claims, hiring technical talent, or positioning your own product in a crowded marketplace where AI washing has become commonplace.[17]

Data science serves as the expansive foundation that encompasses the systematic extraction of insights from data. Think of it as the umbrella discipline that combines statistics, scientific computing, and domain expertise to transform raw information into actionable business intelligence. For startup founders, data science represents the practical toolkit for understanding customer behavior, optimizing operations, and making evidence-based decisions that can mean the difference between scaling successfully and burning through runway. What makes data science particularly valuable for startups is its interdisciplinary nature, bridging the gap between technical capabilities and business outcomes, incorporating everything from data collection and cleaning to visualization and strategic interpretation. This broad scope means that even early-stage companies generating modest amounts of data can benefit from data science approaches without necessarily requiring sophisticated AI infrastructure.[18]

Artificial intelligence represents the ambitious goal of creating systems that can perform tasks typically requiring human intelligence. While AI intersects significantly with data science, it extends beyond pure data analysis into areas like reasoning, perception, and decision-making. For startups, AI offers the potential to automate complex processes, enhance user experiences, and create entirely new product categories. The key distinction lies in purpose: while data science focuses on extracting insights to inform human decision-making, AI aims to enable machines to make those decisions autonomously. This difference has profound implications for startup strategy—data science initiatives typically enhance existing business processes, while AI implementations can fundamentally transform how your product or service operates.[19]

---

[17] "Artificial Intelligence (AI) vs. Machine Learning," Columbia Engineering, accessed June 20, 2025, https://ai.engineering.columbia.edu/ai-vs-machine-learning/.

[18] "What is Data Science?," IBM, accessed June 20, 2025, https://www.ibm.com/think/topics/data-science.

[19] "What Is Artificial Intelligence (AI)?," IBM, accessed June 20, 2025, https://www.ibm.com/think/topics/artificial-intelligence.

Generative artificial intelligence, commonly abbreviated as GenAI, represents a revolutionary subset of AI technology that focuses on creating new content rather than simply analyzing existing data. Unlike traditional AI systems that primarily process information to make predictions or automate specific tasks, generative AI can produce entirely original outputs including text, images, audio, code, and videos based on patterns learned from vast training datasets. What distinguishes GenAI from traditional AI is its creative capability—it doesn't just follow pre-programmed rules or make predictions based on historical data but actually generates novel content that mimics human creativity. For startups, GenAI offers transformative potential beyond traditional AI applications, enabling everything from automated content creation and personalized customer experiences to rapid prototyping and innovative product development.[20]

Large language models (LLMs) represent a specific type of foundation model within the generative AI ecosystem that specializes in understanding and generating human language. These sophisticated systems are trained on vast amounts of text data using deep learning techniques to develop a nuanced understanding of language patterns, context, and meaning. What makes LLMs particularly powerful is their scale and complexity, typically consisting of at least a billion parameters that determine how the model processes information. Popular examples include OpenAI's GPT models, Google's Gemini, Meta's Llama, and Anthropic's Claude, which serve as the foundation for many generative AI applications. For startups, LLMs offer powerful capabilities across various business functions, from automating customer service and generating marketing content to accelerating software development and providing personalized recommendations.[21]

AI agents represent a significant evolution within the artificial intelligence landscape, functioning as autonomous software systems designed to perceive their environment, make decisions, and take actions to achieve specific goals with minimal human intervention. Unlike basic AI applications or even sophisticated LLMs, these agents possess distinctive characteristics including autonomy, proactivity, reasoning capabilities, and the ability to learn from experiences. What distinguishes AI agents from AI assistants is their level of independence and initiative—while assistants primarily respond to user commands, agents can proactively identify problems, formulate

---

[20] "What is generative AI?," IBM Research, accessed June 21, 2025, `https://research.ibm.com/blog/what-is-generative-AI`.

[21] "What is LLM? - Large Language Models Explained," AWS, accessed June 20, 2025, `https://aws.amazon.com/what-is/large-language-model/`.

solutions, and execute actions based on their understanding of objectives. This autonomous capability allows startups to deploy agents for diverse applications ranging from customer service and data analysis to operational optimization and decision support.[22]

Machine learning sits at the intersection of data science and AI, representing the specific approach of enabling systems to automatically improve their performance through experience. Unlike traditional programming where you explicitly code every rule, machine learning algorithms identify patterns in data and make predictions or decisions based on those patterns. For startups, machine learning offers a scalable way to personalize user experiences, optimize operations, and automate decision-making processes that would be impossible to code manually. The practical advantage of machine learning for startups lies in its ability to continuously improve as your business grows—as you collect more customer data, user interactions, and operational metrics, your machine learning systems become more accurate and valuable, creating a compounding competitive advantage.[23]

Deep learning represents the most sophisticated subset within this hierarchy, using artificial neural networks to process information in ways that mimic human brain function. While requiring substantial computational resources and large datasets, deep learning excels at handling unstructured data like images, text, and audio. The advancement of deep learning techniques such as Generative Adversarial Networks (GANs) and Variational Autoencoders (VAEs) has been crucial in enabling the recent breakthroughs in generative AI, allowing models to develop rich understandings of large datasets and produce strikingly realistic synthetic content. For startups working with complex data types or seeking to build cutting-edge AI products, deep learning can provide capabilities that were impossible just a few years ago, though most startup applications don't require huge investments in deep learning's complexity—traditional machine learning approaches often provide better results with less computational overhead and faster development cycles.[24]

---

[22] "What are AI agents?," Google Cloud, accessed June 20, 2025, https://cloud.google.com/discover/what-are-ai-agents.

[23] Sara Brown, "Machine learning, explained," MIT Sloan, accessed June 19, 2025, https://mitsloan.mit.edu/ideas-made-to-matter/machine-learning-explained.

[24] C. Janiesch, P. Zschech, and K. Heinrich, "Machine learning and deep learning," Electronic Markets 31 (2021): 685-695, https://link.springer.com/article/10.1007/s12525-021-00475-2.

Understanding these relationships, as illustrated in the accompanying diagram showing the hierarchical and interconnected nature of data science, AI, machine learning, and deep learning, helps startups make informed decisions about their technical road map. You might begin with basic data science practices to understand your market, optimize operations, and build compelling data storytelling for your investors, gradually incorporating machine learning solutions as you scale, and only investing in sophisticated AI or deep learning capabilities when they directly address core business challenges. The most successful startups don't necessarily use the most advanced technology—they use the right technology to solve real problems efficiently. The simplified relationships shown in Figure 13-1 provide a road map for that strategic progression, helping you navigate from data-driven insights to intelligent automation as your startup matures and your data challenges evolve while avoiding the temptation to chase buzzwords at the expense of practical solutions.

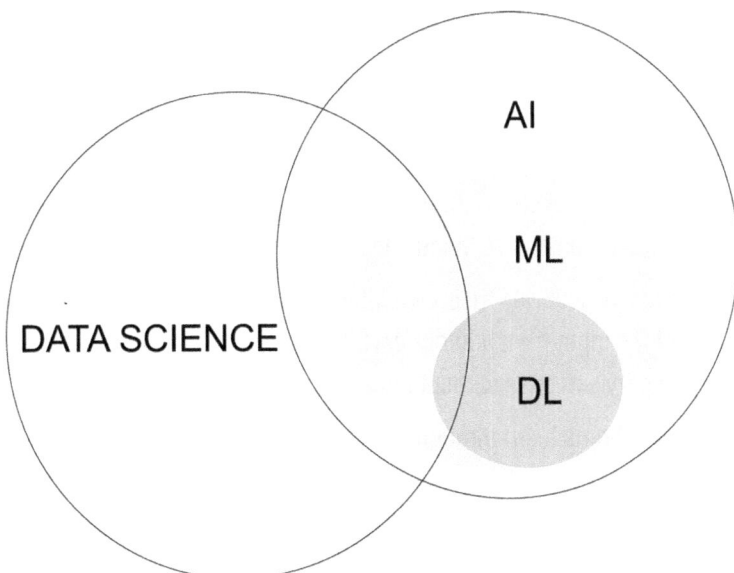

Figure 13-1. *The relationships between data science, AI, ML, and DL*

The Variety of Machine Learning Techniques

The machine learning ecosystem offers diverse approaches to extracting value from data, each with distinct characteristics and optimal applications. Understanding these methods enables startups to select the right tools for their specific challenges, whether predicting customer behavior, optimizing operations, or detecting fraudulent activities. Table 13-1 presents different types of machine learning.[25]

Table 13-1. *Machine Learning Types*

ML Type	Description
Supervised Learning	Training models on labeled datasets where input variables are mapped to known output variables. The model learns to predict outputs for new inputs based on the patterns observed during training.
Unsupervised Learning	Models learn from unlabeled data by identifying patterns and structures without predefined categories. Commonly used for clustering, association, and dimensionality reduction.
Semi-supervised Learning	Combines a small amount of labeled data with a large amount of unlabeled data. The labeled data guides learning for the unlabeled data, reducing the need for extensive manual labeling.
Self-supervised Learning	Models generate their own labels from unlabeled data by learning one part of the input from another part. This transforms unsupervised problems into supervised ones without human labeling.
Reinforcement Learning	Algorithms learn through a system of reward and punishment by interacting with an environment. Agents optimize their actions to maximize cumulative rewards over time.

---

[25] "Machine learning types," IBM, accessed June 18, 2025, https://www.ibm.com/think/topics/machine-learning-types.

Supervised Learning: The Foundation of Predictive Intelligence

Supervised learning represents the most mature and widely adopted branch of machine learning, utilizing labeled training data to build predictive models that can classify outcomes or estimate numerical values.[26]

This approach mirrors human learning patterns, where algorithms learn from examples with known correct answers to make accurate predictions on new, unseen data. The method's reliability and interpretability make it particularly valuable for startups seeking measurable business outcomes with clear performance metrics.

The power of supervised learning lies in its ability to solve two fundamental business problems: classification (predicting categories) and regression (predicting numbers). Classification applications include customer churn prediction, fraud detection, and market segmentation, while regression tackles challenges like sales forecasting, pricing optimization, and demand estimation. For startups, this versatility translates into immediate actionable insights that can drive revenue growth and operational efficiency. Table 13-2 shows how supervised ML methods may be used in startups.

Table 13-2. *Supervised ML in Startups*

Startup Type	Supervised ML Application
Fintech	Credit scoring, fraud detection
Healthcare	Disease diagnosis and treatment recommendation
Ecommerce	Customer churn prediction and personalized recommendations
Marketing	Lead scoring and conversion prediction
Manufacturing	Quality control and predictive maintenance

---

[26] Giovanni Cerulli, *Fundamentals of Supervised Machine Learning: With Applications in Python, R, and Stata* (Springer, 2023).

Unsupervised Learning: Discovering Hidden Patterns

Unsupervised learning operates without the safety net of labeled examples, instead discovering hidden patterns and structures within raw data. This exploratory approach proves invaluable for startups seeking to understand their customers, markets, or operational data in ways that weren't previously apparent. Unlike supervised learning's predictive focus, unsupervised methods excel at revealing insights that can fundamentally reshape business strategies and product development approaches.[27]

The primary applications of unsupervised learning center on clustering, dimensionality reduction, and association rule mining. Customer segmentation represents perhaps the most immediately valuable application for startups, enabling companies to identify distinct user groups based on behavior patterns, preferences, and engagement levels. This granular understanding facilitates targeted marketing campaigns, personalized product development, and optimized user experiences that drive retention and growth.[28]

Venture capital firms have embraced unsupervised learning to revolutionize startup evaluation processes. By clustering companies based on textual descriptions of their value propositions, investors can more efficiently identify promising opportunities and assess industry competition. This approach transforms traditionally subjective investment decisions into data-driven processes, improving both speed and accuracy of capital allocation decisions.

The business impact of unsupervised learning extends beyond immediate insights to long-term strategic advantages. Additionally, recommendation systems powered by unsupervised learning have become essential for ecommerce and content platforms, with companies like Netflix crediting such systems for significant portions of user engagement and retention. Table 13-3 shows how unsupervised ML methods may be used in startups.

---

[27] Ankur A. Patel, *Hands-On Unsupervised Learning Using Python: How to Build Applied Machine Learning Solutions from Unlabeled Data* (O'Reilly Media, 2019).

[28] Alok Malik and Bradford Tuckfield, *Applied Unsupervised Learning with R* (Packt Publishing, 2019).

Table 13-3. *Unsupervised ML in Startups*

Startup Type	Unsupervised ML Application
Retail	Customer segmentation for investor data storytelling and targeted marketing
Ecommerce	Recommendation systems
Cybersecurity	Anomaly detection
Marketing	Market basket analysis
Content Platforms	Content discovery

Semi-supervised Learning

Semi-supervised learning combines a small set of labeled examples with a large pool of unlabeled data to improve model accuracy when annotation is costly or scarce. This hybrid approach integrates supervised objectives—learning from labeled instances—with unsupervised objectives—capturing data structure—resulting in better generalization than training on labeled data alone. For startups, semi-supervised methods offer a cost-effective strategy to extract value from abundant raw data without the expense of exhaustive labeling.[29]

In practice, semi-supervised learning excels at tasks such as image and text classification, where only a fraction of samples are labeled and models propagate label information through data manifolds to unlabeled points. Startups can apply these techniques to build churn-prediction or sentiment-analysis systems by labeling a small subset of customer records and allowing the model to infer patterns across the rest of their logs. This accelerates insight generation and reduces reliance on domain experts for large-scale annotation.

Venture capital firms have begun adopting semi-supervised pipelines by combining limited historical exit labels (successful or failed startups) with extensive unlabeled features from pitch decks, market metrics, and team profiles to forecast investment outcomes. This data-driven framework refines deal sourcing and due diligence, bolstering risk assessment when explicit exit labels are incomplete or lag current market conditions.

---

[29] Olivier Chapelle, Bernhard Schölkopf, and Alexander Zien, eds., *Semi-Supervised Learning* (MIT Press, 2006).

Long-term, semi-supervised learning enables continuous model refinement as new unlabeled data arrives, keeping predictive systems up to date with minimal additional labeling effort. Startups employing these methods benefit from faster model development cycles, lower annotation costs, and enhanced performance, gaining a competitive edge over purely supervised approaches. Table 13-4 shows how semi-supervised ML methods may be used in startups.

Table 13-4. *Semi-supervised ML in Startups*

Startup Type	Semi-supervised ML Application
Fintech	Fraud detection with few labeled transactions propagated to large unlabeled sets
SaaS	Churn prediction combining sparse labeled events with broad usage logs
Healthtech	Diagnostic support from limited annotated scans plus bulk unlabeled images
Marketing	Sentiment analysis seeded by labeled reviews and extended to unlabeled text
Venture Analytics	Outcome forecasting blending exit labels with rich unlabeled startup data

Self-Supervised Learning

Self-supervised learning creates its own supervision by deriving pseudo-labels from the data—such as predicting masked tokens or reconstructing missing patches—eliminating the need for human annotations. This paradigm underpins modern foundation models (e.g., BERT), which pretrain on vast unlabeled corpora via pretext tasks like masked-language modeling or next-sentence prediction to learn generalizable embeddings.[30]

Startups can leverage self-supervised pretraining to obtain rich representations of text, images, or audio without manual labeling. These pretrained embeddings are then fine-tuned on downstream tasks—such as customer intent classification, image-based product search, or voice analytics—with minimal labeled examples. By reducing annotation overhead, self-supervised learning accelerates feature development and empowers lean teams to deliver robust AI capabilities rapidly.

---

[30] Robert Johnson, *Self-Supervised Learning* (HiTeX Press, 2024).

Some venture investors are experimenting with self-supervised embeddings to map startup profiles into learned feature spaces, enabling similarity searches against successful peers and revealing latent connections between companies. This novel use case aids deal sourcing by surfacing startups with analogous trajectories or complementary innovations.

The broader business impact of self-supervised learning is profound: it slashes data-annotation costs, produces versatile pretrained models, and democratizes access to high-quality representations. Startups harnessing this paradigm can iterate on AI-driven products swiftly and adapt to new data domains with only light supervision, establishing a formidable strategic advantage. Table 13-5 shows how self-supervised ML methods may be used in startups.

Table 13-5. *Self-Supervised ML in Startups*

Startup Type	Self-Supervised ML Application
Ecommerce	Product recommendations via pretrained image embeddings
Content Platforms	Personalized ranking using masked-language model embeddings
Autonomous Vehicles	Sensor fusion by predicting one sensor's data from another
Speech Tech	Voice intent modeling from raw audio without transcripts
Venture Analytics	Embedding-based similarity search for deal sourcing (experiments, not widely adopted yet)

Reinforcement Learning: Learning Through Experience

Reinforcement learning (RL) is a branch of machine learning in which an agent learns to make decisions by interacting with a dynamic environment and receiving feedback in the form of rewards or penalties that reflect the desirability of its actions. A prominent application of reinforcement learning is Spotify's method for personalizing music for its users, which involves learning their preferences to recommend relevant tracks, albums, and playlists.[31]

Unlike supervised learning, which depends on labeled input–output pairs, RL emphasizes the exploration–exploitation trade-off: the agent must balance trying new actions to discover their potential rewards with leveraging actions known to yield high

---

[31] Enes Bilgin, *Mastering Reinforcement Learning with Python: Build next-generation, self-learning models using reinforcement learning techniques and best practices* (Packt Publishing, 2020).

returns, with the ultimate goal of maximizing cumulative reward over time. An RL framework is commonly formalized as a Markov decision process (MDP) defined by a set of states, actions, transition probabilities, and reward functions; the agent's objective is to learn a policy—a mapping from states to actions—that optimizes expected future rewards under this MDP formulation. Take a look at this simplified diagram (Figure 13-2) to see how reinforcement learning works.

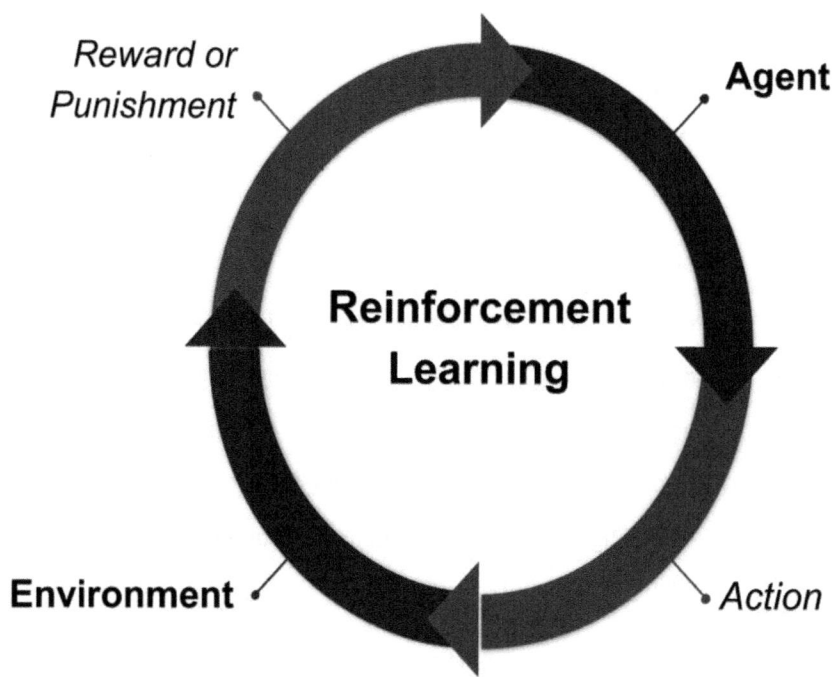

Figure 13-2. *Reinforcement Learning*

In startup contexts, RL enables the creation of adaptive, feedback-driven systems across domains such as autonomous navigation, robotic control, supply chain optimization, and dynamic pricing, allowing models to refine strategies continuously as new data arrives rather than relying on static training sets.

The strategic advantages of RL include robustness to changing environments, the capability to uncover novel decision pathways through trial-and-error learning, and the automation of complex sequential decision-making processes—properties that can yield significant operational efficiencies and competitive differentiation for startups. Table 13-6 shows how RL methods may be used in startups.

Table 13-6. *Reinforcement Learning in Startups*

Startup Type	RL Application
Autonomous Vehicles	End-to-end driving policies learned directly from raw sensor data
Robotics & Automation	Dynamic control for warehouse picking and assembly tasks through trial-and-error learning
Supply Chain Tech	Adaptive inventory optimization balancing holding costs and stock-out risks in real time
Energy Systems	Data-center cooling management agents that minimize energy consumption via continuous feedback
Fintech	Algorithmic trading strategies that adapt to market volatility to maximize risk-adjusted returns

Understanding these different ML approaches provides the foundation for making informed decisions about which techniques to pursue. However, knowing what's possible is only half the battle—the greater challenge for startups lies in choosing which projects will deliver the most impact with limited resources. This brings us to the critical question of prioritization.

Choosing the Right Projects

The most critical decision facing any startup data team isn't which algorithm to use or which tool to implement—it's deciding which projects deserve precious time and resources in the first place. Unlike enterprise environments where data teams might have the luxury of pursuing multiple parallel initiatives, startups operate in a world of brutal prioritization where the wrong choice can mean the difference between runway extension and running out of cash. The ultimate goal is to choose initiatives that directly advance the startup's business objectives. This requires evaluating not just the project itself but also its scope to focus only on elements that will impact your North Star Metric and other key goals. Stripping away "nice-to-have" features ensures you create maximum business value while minimizing effort and complexity. Most startups don't really need super sophisticated algorithms, trust me. The solution should be as cheap as possible and easy to scale and maintain. They can learn from those who have successfully implemented popular methods and adjusted them to their specific needs.

Aligning with Your Data Strategy

Before diving into any specific prioritization framework, it's crucial to remember that every data science and AI project must be well aligned with the comprehensive data strategy discussed in Chapter 2. Your data strategy serves as the North Star that guides all technical decisions, ensuring that individual projects contribute to broader business objectives rather than becoming isolated technical experiments. Without this alignment, even the most technically impressive AI implementation can become a costly distraction that drains resources without delivering measurable business value.

Stakeholder Alignment and Communication

Priorities need to be agreed upon with all key stakeholders inside the startup, and this requires establishing robust communication channels that go beyond ad hoc conversations. Schedule regular meetings to make it transparent for everyone to see what tasks the data team is working on, what new projects are being proposed, and what progress has been made on current initiatives. These meetings serve multiple purposes: they prevent scope creep, manage expectations, and ensure that the data team's work remains visible and valued across the organization.

As the startup data one-man army, you'll quickly discover that stakeholders often have unrealistic expectations about what's possible with AI and data science. Regular communication sessions help bridge the gap between business aspirations and technical reality, creating space for education while building trust through transparency. This becomes particularly crucial when leadership returns from conferences or networking events filled with ambitious ideas about AI capabilities.

The RICE Framework for AI Project Prioritization

While multiple prioritization frameworks exist for startups—including ICE (Impact, Confidence, Ease), MoSCoW (Must have, Should have, Could have, Won't have), and Value vs. Complexity model—the RICE framework has proven particularly effective for

AI and data science initiatives. RICE stands for Reach, Impact, Confidence, and Effort, providing a systematic approach to evaluate competing projects objectively.[32,33,34]

The framework calculates a priority score using the formula:

RICE Score = (Reach × Impact × Confidence) ÷ Effort.

This quantitative approach helps remove personal bias and political considerations from project selection, focusing instead on measurable business value. The original RICE framework evaluates projects based on how many people they'll reach, the impact they'll have, confidence in the estimates, and effort required:

- **Reach** quantifies the number of people who will be directly affected by this initiative within a specific period. It answers the question: "How many users will this touch in a month or a quarter?"

- **Impact** assesses the degree to which this initiative will affect each individual user. It measures the significance of the change, with a clear scoring system to represent its magnitude:

    - Massive: 3x

    - High: 2x

    - Medium: 1x

    - Low: 0.5x

    - Minimal: 0.25x

---

[32] Lenny Rachitsky, "The original growth hacker reveals his secrets | Sean Ellis (author of "Hacking Growth")," Lenny's Newsletter, accessed June 19, 2025, `https://www.lennysnewsletter.com/p/the-original-growth-hacker-sean-ellis`.

[33] Suchetha Vijayakumar, Krishna Prasad K., Raviraja Holla M., "Assessing the Effectiveness of MoSCoW Prioritization in Software Development: A Holistic Analysis across Methodologies," ResearchGate, October 2024, `https://www.researchgate.net/publication/385309437_Assessing_the_Effectiveness_of_MoSCoW_Prioritization_in_Software_Development_A_Holistic_Analysis_across_Methodologies`.

[34] "Understanding the RICE model and its framework," Microsoft, accessed June 20, 2025, `https://www.microsoft.com/en-us/microsoft-365-life-hacks/organization/understanding-the-rice-model-and-its-framework`.

- **Confidence** reflects your level of certainty about the estimates for Reach, Impact, and Effort. It's a percentage-based score that accounts for the strength of your data and the reliability of your assumptions:

    - High: 100% (Strong data and clear precedents)

    - Medium: 80% (Some data but with assumptions)

    - Low: 50% (A rough estimate with little data)

- **Effort** estimates the total time investment required from the entire team, measured in "person-months." This figure represents the work one person can complete in a single month. For simplicity, estimates should be made in whole or half-month increments to avoid getting lost in excessive detail.[35]

While these factors remain relevant, startups may need additional considerations that reflect their specific constraints and opportunities. Startups and particular teams within startups tend to adjust the framework to their specific needs. For AI projects specifically, the framework has been enhanced by some practitioners into frameworks like RICE-A, adding an "AI Complexity" factor to account for the unique challenges of machine learning implementations.[36,37]

A Day in the Life: When Your CEO Discovers AI

Picture this scenario: You're the founding team member and the only data professional at a 30-person SaaS startup. You've spent the last six months building basic analytics infrastructure and creating dashboards that the team actually uses. Then your CEO returns from a startup conference, eyes gleaming with excitement about AI possibilities.

---

[35] Sean McBride, "RICE: Simple prioritization for product managers," Intercom Blog, accessed June 22, 2025, https://www.intercom.com/blog/rice-simple-prioritization-for-product-managers/.

[36] Marily Nika, "Introducing RICE-A: a prioritization framework for AI products," LinkedIn, January 2025, https://www.linkedin.com/posts/marilynika_introducing-rice-a-a-prioritization-framework-activity-7287502969711869952-sPbg/.

[37] Sean Falconer, "Launching your first AI project with a grain of RICE: Weighing reach, impact, confidence and effort to create your roadmap," VentureBeat, accessed June 15, 2025, https://venturebeat.com/ai/launching-your-first-ai-project-with-a-grain-of-rice-weighing-reach-impact-confidence-and-effort-to-create-your-roadmap/.

"I just talked to three founders who are using AI for everything," she announces in the Monday all-hands meeting. "Customer support, sales forecasting, personalized marketing—they said it's all table stakes now. We need to get moving!"

By Tuesday, you have a list of "urgent" AI projects on your desk: an advanced customer churn prediction system that uses real-time behavioral analysis, a sophisticated personalization engine that adapts content in real time, and a customer support chatbot that can handle complex queries. Your CEO assures you that "the infrastructure is already there" and asks when these can be launched. Sound familiar?

This is where frameworks become your lifeline. Without a systematic approach to evaluate these requests, you'll either disappoint leadership by appearing negative or commit to projects that will consume months of effort without delivering promised results. The RICE framework provides a diplomatic and objective way to navigate these conversations while ensuring that your limited time produces maximum business impact.

Applying RICE to Real AI Projects

Let's walk through how to evaluate the three projects your enthusiastic CEO proposed. For each project, you'll need to estimate four key metrics: how many people it will reach, what impact it will have, how confident you are in the estimates, and how much effort it will require.

The Customer Support ChatBot emerges as the clear winner with a RICE score of 900 (see Table 13-7), despite having the lowest impact rating. Why? Because it leverages existing FAQ data, requires minimal custom development, and can be implemented with high confidence in just four weeks. The reach is moderate—automating 500 monthly support tickets—but the combination of low effort and high confidence makes it an excellent choice for demonstrating quick wins.

Table 13-7. *Rice Framework Example*

Project	Reach Score	Impact Score	Confidence Score	Effort Months	RICE Score
Customer Support ChatBot	500	2	90%	1	900
Personalization Engine	1000	3	40%	5	240
Advanced Customer Churn Prediction	200	3	60%	3	120

The Real-Time Personalization Engine, despite affecting 1,000 monthly users and promising high impact, scores only 240 due to low confidence and high effort requirements. As the solo data professional, you recognize that building sophisticated real-time infrastructure and A/B testing frameworks would consume 5 months—a significant portion of your startup's runway. The 40% confidence score reflects uncertainty about whether current data collection is sufficient for effective personalization.

The Advanced Customer Churn Prediction system falls last with a score of 120. While it promises high impact by potentially preventing revenue loss, the moderate confidence score reflects concerns about data quality and the 3-month development timeline. For a startup with limited historical customer data, building reliable churn models often proves more challenging than initial estimates suggest.

Making the Framework Work in Practice

When presenting these findings to your CEO, focus on the business rationale rather than technical complexity. Explain that the chatbot creates immediate value by reducing support load and improving response times, allowing the team to focus on higher-value customer interactions. Emphasize that starting with a high-confidence, low-effort project builds organizational trust in data initiatives while establishing foundation infrastructure that supports future AI projects.

The beauty of the RICE framework lies not just in project selection but in stakeholder education. By walking through the scoring process transparently, you help leadership understand the trade-offs inherent in AI development while establishing realistic expectations for future initiatives. **This approach transforms you from an order-taker into a strategic advisor** who guides data investments toward maximum business impact.

Remember that frameworks are tools, not rigid rules. Your startup's specific context—typical tasks, available data, technical infrastructure, team capabilities, and market pressures—should inform how you apply and adapt these prioritization methods. The goal isn't perfect scores but better decisions that align limited resources with business priorities while building sustainable momentum for your data program.

If you are interested in deepening your understanding of prioritization frameworks, several excellent resources are available. *Inspired: How to Create Tech Products Customers Love* by Marty Cagan elaborates on the principles behind effective prioritization applicable also to data projects. *The Lean Startup* by Eric Ries[38] offers essential insights into validated learning and iterative development that complement RICE methodology.[39]

Even with a solid prioritization framework in place, the path from concept to successful implementation is fraught with challenges. Understanding where projects typically fail can be just as valuable as knowing how to select them. Let's examine the most common pitfalls that derail AI initiatives, particularly those identified by recent research.

Common Implementation Mistakes That Kill AI Projects

The artificial intelligence gold rush has created both unprecedented opportunities and devastating casualties, particularly among startups attempting to compete with tech giants. According to the RAND research, more than 80% of AI projects fail—twice the rate of failure for information technology projects that do not involve AI. This sobering statistic represents billions in wasted investment and countless entrepreneurial dreams dashed against the rocks of implementation reality.[40,41]

---

[38] Eric Ries, *The Lean Startup: How Today's Entrepreneurs Use Continuous Innovation to Create Radically Successful Businesses* (Crown Currency, 2011).

[39] Marty Cagan, *Inspired: How to Create Tech Products Customers Love*, 2nd ed. (Wiley, 2017).

[40] James Ryseff, Brandon De Bruhl, and Sydne Newberry, "The Root Causes of Failure for Artificial Intelligence Projects and How They Can Succeed," RAND, accessed June 20, 2025, https://www.rand.org/pubs/research_reports/RRA2680-1.html.

[41] James Ryseff, Brandon De Bruhl, and Sydne Newberry, "The Root Causes of Failure for Artificial Intelligence Projects and How They Can Succeed," RAND, accessed June 20, 2025, https://www.rand.org/pubs/research_reports/RRA2680-1.html.

The RAND study, which interviewed a group of experienced data scientists and engineers across industry and academia, identified several leading root causes that systematically destroy AI initiatives.

Leadership-driven failures emerge as the primary killer, with 84% of practitioners citing misunderstanding or miscommunication about what problem needs to be solved using AI. For startups, this translates into founders who fall in love with the technology rather than the problem, building sophisticated solutions that solve no real customer pain point. **The "shiny object syndrome"** represents perhaps the most seductive trap for startup founders. As the RAND research reveals, organizations frequently focus more on using the latest and greatest technology than on solving real problems for their intended users. This pattern is amplified in startup environments where founders often begin with fascinating technological discoveries rather than clear market needs.

Lack of the necessary data constitutes another devastating category according to the RAND study. Startups face a cruel paradox here: they need massive amounts of high-quality data to train effective models, yet many of them lack the data collection infrastructure that large organizations have built over decades.

The **resource disparity between startups and big tech** companies creates fundamental asymmetries that go far beyond simple funding differences. As research conducted by Aspen Hopkins and Serena Booth demonstrates, professionals from smaller organizations face additional challenges stemming from deploying AI with limited resources, increased existential risk, and absent access to in-house research teams. Startups must navigate tensions between privacy and ubiquity, resource management and performance optimization, and access vs. monopolization that simply don't exist at the same scale for well-funded tech giants.[42]

Infrastructure inadequacies represent a silent killer that destroys AI projects long before they reach production. Organizations frequently lack adequate infrastructure to manage their data and deploy completed AI models, with data engineering consistently undervalued despite being the foundational requirement for success. Startups, already operating with constrained resources, often underestimate the computational and operational overhead required to maintain AI systems in production. The glamorous

---

[42] Aspen Hopkins and Serena Booth, "Machine Learning Practices Outside Big Tech: How Resource Constraints Challenge Responsible Development," ACM Digital Library, July 2021, https://dl.acm.org/doi/abs/10.1145/3461702.3462527.

work of model training gets attention and funding, while the unglamorous but critical work of data pipelines and model deployment gets neglected until projects collapse under their own operational weight.

Perhaps most critically for startups, the timing and complexity trap claims victims who attempt to solve problems that are simply too difficult for current AI capabilities. The RAND research emphasizes that AI is not a magic wand that can make any challenging problem disappear. Startups, pressured by investor expectations and competitive dynamics, often tackle problems that would challenge even the most sophisticated AI research labs. The key competitive advantage for startups lies not in matching big tech's computational resources but in identifying narrow, well-defined problems where AI can deliver clear, measurable value to specific customer segments.

Case studies from 2024 further illustrate how technological overreach and financial instability destroy AI startups regardless of technical merit. Ghost Autonomy exemplifies the dangers of technological overreach, having raised $238.8 million to revolutionize autonomous driving by integrating in-car AI with multimodal large language models (LLMs). Despite substantial funding, the company faced insurmountable industry skepticism about its core technical approach. Experts questioned the feasibility of using LLMs for self-driving applications in real-world scenarios, highlighting how even well-funded startups can fail when pursuing AI applications beyond current technological readiness.[43]

Similarly, Tally's collapse demonstrates that sophisticated AI implementation cannot overcome fundamental business model challenges. Despite raising over $200 million, Tally couldn't maintain financial stability in a deteriorating funding environment. Their pivot from direct-to-consumer loans to a B2B credit card debt management platform signaled desperation as global fintech investment plummeted from $210 billion in 2021 to just $15.9 billion by mid-2024. These examples underscore a sobering reality: neither AI sophistication nor impressive funding can save startups that either overestimate AI capabilities for their specific use case or fail to establish sustainable business models with clear paths to profitability.

The path forward for startups requires embracing constraints as competitive advantages rather than limitations. Startups can compete effectively by focusing on niche use cases and prioritizing agility over scale. Domain expertise becomes the

---

[43] "AI Startups That Failed in 2024 and Why," AIM Research, accessed June 17, 2025, https://aimresearch.co/ai-startups/ai-startups-that-failed-in-2024-and-why.

secret weapon: while big tech pursues general-purpose solutions, startups with deep understanding of specific industries can build AI applications that solve real problems with surgical precision. This approach allows smaller teams to move faster, iterate more rapidly, and build defensible positions in markets that may be too narrow for larger competitors to address profitably.

Successful startups consistently demonstrate certain characteristics. They start with validated customer problems rather than interesting technology. They build minimum viable solutions that deliver immediate value (regardless of how much AI is required). Moreover, the AI-first startups invest heavily in data quality and infrastructure from day one rather than treating these as afterthoughts.

The survivors in this space understand that competing with big corporations requires playing a fundamentally different game—one where deep domain knowledge, customer intimacy, and execution speed matter more than raw computational resources or the latest algorithmic breakthroughs. Table 13-8 summarizes the critical mistakes, their root causes, and practical mitigation strategies that can mean the difference between success and failure.

Table 13-8. *AI Implementation Mistakes*

Implementation Mistake	Root Cause	Suggested Mitigation
Poor product-market fit	Creating solutions without addressing pressing market needs.	Validate market demand before building complex AI solutions.
Financial instability	High development costs with limited funding opportunities.	Develop sustainable business models with a clear path to profitability.
Leadership-driven failures	Misunderstanding or miscommunication about what problem needs to be solved using AI. The "shiny object syndrome."	Focus on validated customer problems, not shiny technology. Ensure the team understands the problems well.

(*continued*)

Table 13-8. (*continued*)

Implementation Mistake	Root Cause	Suggested Mitigation
Resource disparity	Limited resources compared to big tech companies.	Leverage domain expertise to build niche solutions.
Infrastructure inadequacies	Underinvestment in data engineering and model deployment.	Prioritize data pipelines and deployment infrastructure.
Complexity and technological overreach	Attempting to solve problems too difficult for current AI capabilities.	Identify well-defined problems where AI can deliver clear value and conduct thorough technical feasibility assessment.
Lack of necessary data	Insufficient data collection infrastructure and quality issues.	Invest in data acquisition and quality from day one.

This framework provides a practical diagnostic tool for founders and founding team members to evaluate their AI initiatives against common failure patterns. By systematically addressing these implementation risks early in the development cycle, startups can significantly improve their odds of success in an environment where many AI ventures fail within their first year.

While these failure patterns paint a sobering picture, they also illuminate the path to success. The startups that navigate these challenges effectively share common characteristics: **they start with validated problems, build incrementally, and focus on practical value over technological sophistication**. Let's examine how these principles play out in practice across different applications.

Real-Life Projects in Startups

The following examples illustrate how startups across different stages and industries can successfully implement advanced data science and AI techniques. Rather than exhaustive technical tutorials, these cases focus on practical implementation strategies and business impact.

In this section, we move from theory to practice, exploring how core data science and AI concepts are brought to life within real startup environments. While earlier parts of the chapter established the foundational methods and frameworks, here we focus on concrete applications—showing how startups use these techniques to solve real business problems, drive growth, and create differentiated products. Each example highlights not only the technical approach but also the practical constraints and creative solutions that are unique to resource-constrained, fast-moving teams.

Whether it's leveraging large language models to automate customer support, deploying lightweight recommender systems to personalize user experiences from day one, or using segmentation and anomaly detection to optimize operations, these case studies illustrate the versatility and impact of data-driven thinking at every stage of the startup journey. The goal is to equip you with actionable insights and proven strategies that can be adapted to your own ventures, regardless of industry or scale.

LLMs

Large language models (LLMs) have fundamentally transformed how businesses interact with data and customers, creating unprecedented opportunities for startups to compete with established players. Unlike traditional AI models that excel at specific tasks like classification or prediction, LLMs offer remarkable flexibility through their ability to understand context, intent, and nuance in human language. This revolutionary technology has democratized access to sophisticated AI capabilities, allowing even resource-constrained startups to leverage powerful language processing without massive infrastructure investments.[44]

For startups specifically, LLMs represent a strategic advantage in addressing one of their primary challenges: operating efficiently with limited resources. By implementing LLMs, startups can automate content creation, enhance customer interactions, streamline operations, and make data-driven decisions without expanding their workforce. The global AI market has enormous growth potential for startups that effectively integrate these technologies. Open source LLMs like Meta's Llama series and DeepSeek's R1 have further accelerated adoption by offering performance comparable

---

[44] Jay Alammar, Maarten Grootendorst, *Hands-On Large Language Models: Language Understanding and Generation* (O'Reilly Media, 2024).

to proprietary models at a fraction of the cost, addressing concerns about vendor lock-in and operational expenses.

Startups across various sectors are already demonstrating the transformative impact of LLMs. In customer support, LLM-powered chatbots can engage in natural, empathetic conversations. For content creation and marketing, these models help startups draft compelling copy, summarize documents, and generate personalized communications that would otherwise require dedicated creative teams. In software development, tools powered by LLMs assist programmers in coding faster by autocompleting functions and suggesting fixes, enabling startups to accelerate product development despite limited engineering resources.

Integrating LLMs into core workflows not only amplifies a startup's productivity but also lays the groundwork for scalable innovation. While leveraging pre-trained language models via simple API calls allows for rapid prototyping and deployment of customer-facing features, some startups choose to invest in custom AI solutions to gain a distinct competitive advantage. Table 13-9 provides a concise overview of how startups in ecommerce, food services, analytics, and event management have implemented LLMs to reduce support workloads, personalize user experiences, automate report generation, and handle high-volume inquiries.

Table 13-9. *Examples of LLMs Implementations in Startups*

Startup Type	Use Case
Ecommerce	Reducing support workload and improving response times for product inquiries in the eyewear ecommerce industry (e.g., eye-oo[45])
Food	Streamlining online ordering, menu browsing, and personalized recommendations
Analytics	Handling queries and generating reports
Event Organizer	Handling high-volume attendee support requests during events

While LLMs provide powerful language capabilities, they represent just one component of more sophisticated AI systems. The next evolution involves combining these language models with autonomous decision-making capabilities through AI agents.

---

[45] Maryia Fokina, "11 Companies That Use AI-Generated Customer Support," Tidio Blog, accessed June 18, 2025, https://www.tidio.com/blog/companies-that-use-ai-generated-customer-support/.

AI Agents

While LLMs provide the foundation for understanding and generating human language, AI agents represent the next evolutionary step by adding autonomous action capabilities. Unlike passive AI tools that require constant human direction, agentic AI systems can independently perceive their environment, make decisions, and take actions to achieve specific outcomes. This shift from tools that merely suggest to systems that actually execute represents one of the most significant transformations in how businesses implement artificial intelligence.[46,47]

For startups, AI agents offer particularly compelling advantages in operational efficiency and customer experience. These autonomous systems can handle complex workflows without continuous human supervision, allowing lean startup teams to accomplish more with fewer resources.

The most forward-thinking startups are now exploring collaborative multi-agent systems to solve complex business problems. This approach involves multiple specialized AI agents working together seamlessly while appearing to the customer as a single coherent experience. For instance, one agent might understand customer style preferences, another could check inventory availability, and a third might handle logistics planning. The democratization of agent creation tools means startups can now build sophisticated agents with minimal technical expertise, allowing innovation to come from business units rather than just IT departments. As these technologies mature, startups that effectively implement agentic AI will gain significant competitive advantages through enhanced customer experiences, operational efficiencies, and the ability to anticipate customer needs in ways previously impossible for small companies.

Table 13-10 highlights how startups leverage autonomous AI agents to drive value across diverse operational domains, from end-to-end process orchestration and appointment scheduling to warehouse robotics and public-space sanitation. By showcasing these real-world implementations, Table 13-2 underscores the strategic impact of agentic AI in delivering efficiency gains and enhanced customer experiences for lean teams.

---

[46] Micheal Lanham, *AI Agents in Action* (Manning, 2025).

[47] Kapil Dabi, "The agentic AI revolution: Reshaping retail and consumer interaction," Google Cloud, accessed June 18, 2025, https://cloud.google.com/transform/the-agentic-ai-revolution-reshaping-retail-and-consumer-interaction.

Table 13-10. *Examples of AI Agents Implementations in Startups*

Startup	Use Case
ToothFairyAI	Orchestrating end-to-end business processes through intelligent workflow agents[48]
Penciled	Scheduling appointments automatically by interfacing with calendars and contacts (healthcare)[49]
Pixel Robotics	Warehouse automation through autonomous mobile robots[50]
HiveBotics	Performing autonomous sanitation and disinfection tasks in public spaces[51]

Matt Dancho's "AI Data Science Team" project exemplifies how specialized AI agents can transform data workflows, serving as a powerful case study in the startup application of agentic systems. His open source GitHub repository features an "army of agents" designed to help data professionals perform common tasks much faster than traditional methods. Each agent in this ecosystem specializes in specific functions—from data cleaning and SQL query generation to feature engineering and machine learning model creation—working either independently or as coordinated multi-agent systems.[52,53,54]

Moving from autonomous systems to personalization engines, one of the most proven applications of machine learning in startups remains recommendation systems. Unlike the cutting-edge nature of AI agents, recommender systems offer a mature, well-understood path to enhanced user engagement.[55]

---

[48] "ToothFairyAI," ToothFairyAI, accessed June 9, 2025, `https://toothfairyai.com/`.

[49] "Penciled," Penciled, accessed June 21, 2025, `https://penciled.com/`.

[50] "Pixel Robotics," Pixel Robotics, accessed June 19, 2025, `https://pixel-robotics.eu/`.

[51] "HiveBotics," HiveBotics, accessed June 19, 2025, `https://hivebotics.tech/`.

[52] "Business Science," YouTube channel, accessed June 21, 2025, `https://www.youtube.com/@BusinessScience`.

[53] Matt Dancho, "I Started Building an AI Data Science Team," LinkedIn, January 2025, `https://www.linkedin.com/posts/mattdancho_i-started-building-an-ai-data-science-team-activity-7280568332364263424-GFQ7/`.

[54] Matt Dancho, GitHub profile, accessed June 21, 2025, `https://github.com/mdancho84`.

[55] Charu C. Aggarwal, *Recommender Systems: The Textbook* (Springer, 2016).

Recommender Systems

Recommender systems have become fundamental to many successful technology companies, with Netflix serving as a prime example of their transformative power. The Netflix Prize competition demonstrated the significant impact that improved recommendation algorithms can have on user engagement and business success. Netflix's approach to personalization has evolved from simple collaborative filtering to sophisticated foundation models that analyze comprehensive user interaction histories.[56,57]

Beneath Spotify's seemingly simple interface lies a highly intricate, multi-stage recommendation pipeline that ingests hundreds of petabytes of content data and processes more than half a trillion user interactions each day to generate personalized listening experiences.[58]

This pipeline integrates diverse data sources—raw audio waveforms decoded into hundreds of acoustic features, rich metadata from artist biographies and track tags, and real-time user behavior logs—to feed an ensemble of machine learning models spanning collaborative filtering, high-dimensional embedding spaces, and reinforcement learning agents tuned for long-term satisfaction. These models operate in concert: embeddings create "fingerprints" for users and items to capture latent similarities, collaborative filters leverage these fingerprints at unprecedented scale, and reinforcement learners dynamically adjust recommendation policies to balance relevance, diversity, and serendipity over a user's journey.

Underpinning this is a robust data infrastructure of "golden datasets" that ensures consistency and reproducibility across teams, alongside rigorous A/B experimentation frameworks that validate every algorithmic tweak against engagement and retention metrics. Together, these layers form a sophisticated ecosystem that continually refines itself, exemplifying the state of the art in modern recommender system engineering.

---

[56] Ko-Jen Hsiao, Yesu Feng and Sudarshan Lamkhede, "Foundation Model for Personalized Recommendation," Netflix TechBlog, accessed June 7, 2025, https://netflixtechblog.com/foundation-model-for-personalized-recommendation-1a0bd8e02d39.

[57] Maureen Murtha, "History of Crowdsourcing: The Netflix Prize," HeroX blog, accessed June 7, 2025, https://www.herox.com/blog/839-history-of-crowdsourcing-the-netflix-prize.

[58] "Inside Spotify's Content Recommendation Engine," Scale Events, accessed April 10, 2025, https://exchange.scale.com/public/blogs/inside-the-content-recommendation-engine-at-the-heart-of-spotify.

OK, we've just explored colossal, intricate recommendation engines powering global platforms that serve millions every day. But what if you're a fledgling startup launching your very first product? Where should you begin when you need a recommendation system that's lightweight, cost-effective, and quick to build—yet still delivers a personalized experience?

The Cold Start Problem

For startups, one of the biggest hurdles in building an effective recommendation engine is overcoming **the cold start problem**—the challenge of making accurate suggestions when there's little or no historical data on new users or items. In a startup context, this issue is even more pronounced: as you onboard fresh cohorts of users—often with entirely different tastes and behaviors—you continuously face "cold" segments that lack interaction history. And if you aim to launch your recommendation system on day one, you have zero data to work with—no user profiles, no past behaviors, nothing to seed your models. From every angle, you're starting cold.[59,60]

Several proven strategies can help mitigate this problem, each varying in complexity and impact:

- **Onboarding Surveys and Preference Elicitation:** Prompting new users to rate a handful of items or select interests during sign-up captures initial signals that power more personalized recommendations from the outset.[61]

- **Popular-Item Recommendation**: The simplest tactic surfaces universally popular items, leveraging aggregate trends rather than personal history.

[59] Kuznetsov, S., Kordík, P. (2023). Overcoming the Cold-Start Problem in Recommendation Systems with Ontologies and Knowledge Graphs. In: Abelló, A., et al. New Trends in Database and Information Systems. ADBIS 2023. Communications in Computer and Information Science, vol 1850. Springer, Cham. https://doi.org/10.1007/978-3-031-42941-5_52

[60] Vatesh Pasrija and Supriya Pasrija, "The Cold-Start Problem in Recommender Systems: Challenges and Mitigation Techniques," International Research Journal of Modernization in Engineering Technology and Science 6, no. 5 (May 2024): 2579–88, https://www.irjmets.com/uploadedfiles/paper//issue_5_may_2024/55701/final/fin_irjmets1715656884.pdf.

[61] Charu C. Aggarwal, *Recommender Systems: The Textbook* (Springer, 2016).

- **Content-Based Filtering**: By matching item attributes (e.g., genre, tags, descriptions) to explicitly provided or inferred user preferences, you can generate reasonable suggestions without historical clicks or purchases.

- **Hybrid Models**: Combining collaborative and content-based methods often yields the best results, blending the strengths of both approaches to bootstrap recommendations even with sparse data.[62,63]

For brand-new ventures, starting with a lightweight onboarding questionnaire—asking users about their top genres, favorite topics, or must-see products—can supply the first critical data points your engine needs. As your user base grows and interaction logs accumulate, you can gradually introduce more sophisticated, hybrid algorithms to enhance relevance and drive engagement.

Understanding this challenge is crucial, but theory alone isn't sufficient. Let's work through a practical example to see how these concepts apply in a real startup scenario.

Real-Life Project: Building a Recommender System for a Fitness App Startup

Consider a fitness app startup that wants to recommend workout routines to users but has no historical data or existing user base. This example demonstrates how to implement recommendation functionality from day one using the techniques discussed in this chapter.

Step 1: Content-Based Foundation

Start by creating a comprehensive database of workout routines with detailed attributes including difficulty level, duration, equipment required, muscle groups targeted, and workout type (cardio, strength, flexibility). This content-based approach allows recommendations even for new users by matching workout characteristics to user preferences.

---

[62] Collaborative filtering is a method used to predict the preferences of a user by collecting preferences from a large group of users.

[63] "Hybrid Recommendation - An Overview," ScienceDirect, Accessed June 23, 2025. https://www.sciencedirect.com/topics/computer-science/hybrid-recommendation.

The content database should include rich metadata for each workout routine. Beyond basic attributes, consider factors like intensity level, calorie burn estimates, prerequisite skills, and progression pathways. This detailed content representation enables sophisticated matching algorithms that can recommend appropriate workouts even without user interaction history.

Step 2: User Onboarding Survey

Design a brief but comprehensive onboarding survey that captures user preferences including fitness goals, experience level, available time, preferred workout types, and equipment access. This survey data enables immediate personalized recommendations and achieves the highest effectiveness rate for cold start scenarios.

The survey design should balance comprehensiveness with user experience. Key questions should cover fitness goals (weight loss, muscle building, endurance), current fitness level, time availability, equipment access, and workout preferences. Advanced survey techniques can use adaptive questioning to gather more detailed preferences based on initial responses.

Step 3: Hybrid Recommendation Engine

Implement a hybrid system that combines content-based filtering with collaborative filtering as user data becomes available. Start with content-based recommendations for new users, then gradually incorporate behavioral data and user similarities as the platform grows.

The hybrid approach should seamlessly transition users from content-based to collaborative recommendations. As users interact with the system, track workout completion rates, ratings, and modifications to build behavioral profiles. Use this data to improve recommendations and identify users with similar preferences.

Step 4: Feedback Loop Integration

Build mechanisms to capture user feedback on recommended workouts through ratings, completion rates, and workout modifications. This feedback continuously improves recommendation accuracy and helps identify popular workout combinations.

Feedback collection should be unobtrusive but comprehensive. Track implicit feedback like workout completion, skip rates, and repeat selections alongside explicit feedback like ratings and reviews. Use this data to refine the recommendation algorithm and identify patterns in user preferences.

Step 5: Iterative Improvement

As user data accumulates, implement more sophisticated techniques like matrix factorization and deep learning models to improve recommendation quality. Monitor key metrics including user engagement, workout completion rates, and subscription retention to measure recommendation system effectiveness.

Advanced techniques can include temporal dynamics to account for changing fitness levels and goals, social features to leverage friend connections, and contextual recommendations based on time of day or location. Regular A/B testing helps optimize the recommendation algorithm and measure the impact of improvements.

This approach enables the fitness app to provide value from launch while building toward more sophisticated recommendation capabilities as the user base grows. The progressive enhancement strategy balances immediate functionality with long-term sophistication. This example demonstrates the fundamental approach, but successful implementation requires understanding the broader strategic context.

Conclusions

Launching a recommendation system as a startup is remarkably attainable—beginning with straightforward approaches such as popular-item and content-based suggestions can deliver immediate value without extensive historical data. Employing onboarding surveys or interactive preference-elicitation during user sign-up effectively mitigates the cold-start problem by capturing initial tastes. As users generate interaction logs, transitioning to a hybrid model that blends collaborative filtering with content signals enables more personalized and diverse recommendations.

Integrating continuous feedback loops—tracking explicit ratings and implicit behaviors like click-through and completion rates—ensures your engine evolves in response to real-world usage. Over time, iteratively enhancing your pipeline with matrix factorization, deep learning, and context-aware layers refines recommendation quality and drives long-term engagement. Finally, recent advances in large language and generative models provide new avenues for analyzing unstructured data—reviews, descriptions, and feedback—to craft even more nuanced user profiles and real-time recommendations.

Startup Segmentation Projects

By dividing users, products, or markets into meaningful groups, startups can tailor experiences, optimize marketing spend, and prioritize development efforts to maximize impact. Segmentation projects often begin with a simple question: "Which customers or items are similar, and how can we leverage these similarities?"[64]

For example, an ecommerce startup might segment shoppers by purchase behavior to personalize email campaigns, while a SaaS provider might group users by feature-usage patterns to improve onboarding. In each case, segmentation brings clarity to otherwise overwhelming datasets and helps teams focus on the most valuable opportunities. Startups that master customer segmentation can

- Develop personalized marketing for each group, increasing conversion rates and loyalty.

- Prioritize product enhancements by identifying underserved segments with high growth potential.

- Allocate limited resources more effectively, focusing on the segments most likely to drive revenue.

For instance, a meal-kit delivery startup might find that "Health-Conscious Millennials" respond best to plant-based recipes, while "Busy Families" prefer quick-prep comfort meals. By segmenting their user base, the startup can craft targeted offers and messaging for each group, boosting retention and word-of-mouth referrals.

Story: Building a Light Recommender in Your Travel App

Let's imagine you're working on your travel-app startup and tackling your first recommender system—a light, easy-to-implement version. You've collected data on dozens of destination attributes: culture scores, cost of living, climate indices, safety ratings, and more. As your dataset grows unwieldy, you turn to Principal Components Analysis (PCA) to reduce dimensionality, distilling most of the variation into just a few axes.[65]

---

[64] Tommy Blanchard, Debasish Behera, Pranshu Bhatnagar, *Data Science for Marketing Analytics*, (Packt Publishing, 2019).

[65] Gareth James, Daniela Witten, Trevor Hastie and Robert Tibshirani, *An Introduction to Statistical Learning*, 2nd ed. (Springer, 2021).

You then compute Euclidean distances between countries in this reduced space to quantify similarity: smaller distances mean more similar travel experiences. To add structure, you apply k-means clustering (with k=10) and color-code each country, identifying ten intuitive segments. Finally, you wrap it all up in an interactive 3D chart (Figure 13-1) that you can rotate, zoom, and explore, making the data instantly comprehensible to anyone.[66,67]

In Figure 13-3, each dot represents a country, positioned by its top three principal components (C1, C2, C3), colored by cluster membership. When a user picks "Italy" as a favorite, your recommender engine simply finds the closest dots to Italy in 3D space and suggests those destinations. Simple, yet powerful.

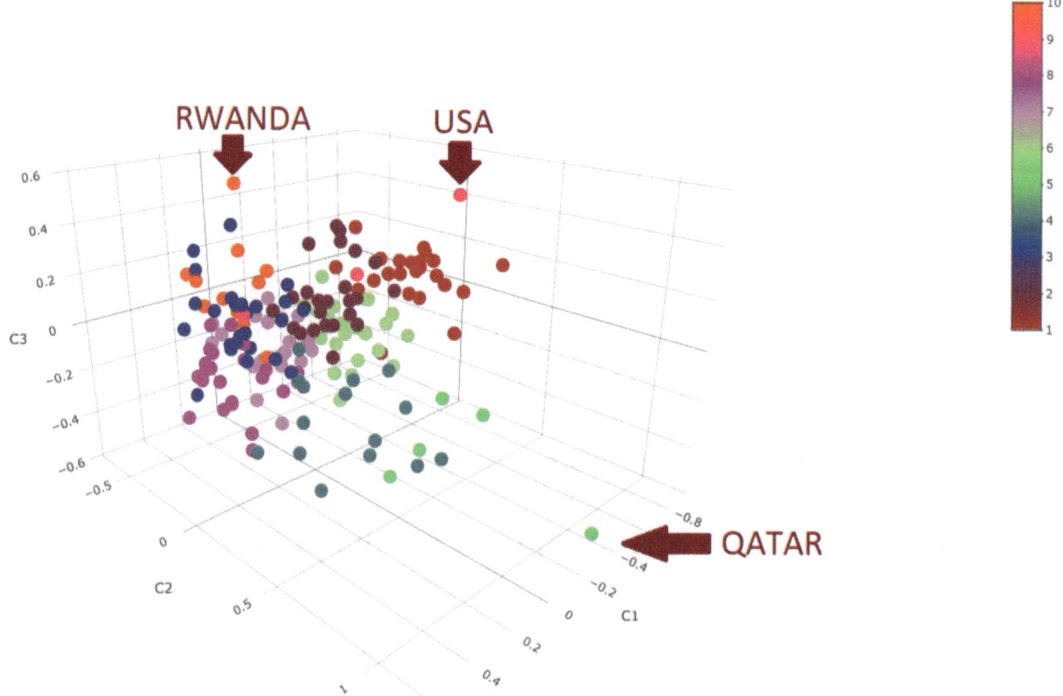

Figure 13-3. *Interactive 3D PCA-Based Chart of Global Travel Destinations, Colored by K-Means Segment*

---

[66] Vinod Chugani, "Understanding Euclidean Distance: From Theory to Practice," DataCamp Tutorials, accessed June 20, 2025, https://www.datacamp.com/tutorial/euclidean-distance.

[67] Eda Kavlakoglu and Vanna Winland, "What is k-means clustering?," IBM, accessed June 20, 2025, https://www.ibm.com/think/topics/k-means-clustering.

Notice three outliers in the chart: Rwanda, USA, and Qatar. Their positions tell a story:

- Rwanda stands apart because its data emphasize eco-tourism and wildlife indices far more than other countries.

- The USA stands out due to its vast geographic diversity and unique economic and cultural metrics.

- Qatar emerges separately, reflecting extreme values in per-capita income and infrastructure scores.

These outliers spark questions—and conversations with your first investor. Rather than explaining ten algorithms, you point to Figure 13-3 and say

Here's our world of travel, distilled into one chart. Each cluster shows destinations with similar vibes; those three dots stand out because they truly break the mold.

In a single glance, the investor sees your data pipeline (PCA → distances → clustering → visualization), understands how recommendations work, and appreciates the insight layered into the chart. That's the power of segmentation in a startup: turning complex data into simple, actionable stories. By weaving together dimensionality reduction, distance metrics, and clustering, you've built not just a recommender but a communication tool that bridges the gap between technical depth and business clarity. And it all starts with segmentation.

Building Predictive Models with No Historical Data

Another example of specific Data Science challenges when launching entirely new products, startups face the challenge of creating customer segments and predictive models without any historical data. A real-world example from the financial sector illustrates this challenge: a company developing a new fintech product needed to build a credit scoring model despite never having offered such services before.[68]

---

[68] "How Lean Startups Use Synthetic Data to Build Smarter AI Faster," thinslices, accessed June 20, 2025, `https://www.thinslices.com/insights/how-lean-startups-use-synthetic-data-to-build-smarter-ai-faster`.

In these situations, startups can generate synthetic data to simulate potential customer behavior and build initial models. Synthetic data generation involves creating artificial datasets that mimic the statistical properties of real data without exposing sensitive information. Recent advances in large language models have opened new possibilities for synthetic data generation, with LLMs able to generate task-specific training data that performs comparably to real-world data.[69,70]

Statistical modeling and simulation represent another fundamental approach to creating synthetic data. Data scientists can analyze the statistical properties of similar datasets from public sources, such as distributions, correlations, and frequencies, then generate new samples that mirror those properties. Monte Carlo simulations can create realistic financial market data by sampling from probability distributions, while epidemiological simulators can produce synthetic patient health records by modeling disease progression.[71,72]

For fintech startups specifically, this process involves several critical steps including market research to understand typical customer segments, analysis of publicly available financial data, creation of synthetic customer profiles based on demographic and behavioral patterns, and validation of synthetic data against industry benchmarks. The key is ensuring that synthetic data maintains the essential characteristics needed for model training while providing sufficient diversity to handle real-world scenarios.

Churn Prediction

Imagine you're the only data person at a subscription startup that just launched its mobile app. After four weeks of excitement and user sign-ups, you notice a worrying pattern: a growing number of new users drop off within days of joining. You realize that

---

[69] Mostafa Ibrahim, "Generating Synthetic Data with LLMs," Packt How-To Tutorials, accessed June 20, 2025, https://www.packtpub.com/en-us/learning/how-to-tutorials/generating-synthetic-data-with-llms.

[70] Khaled El Emam, Lucy Mosquera, Richard Hoptroff, *Practical Synthetic Data Generation: Balancing Privacy and the Broad Availability of Data* (O'Reilly Media, 2020).

[71] Huang, CY., Tsai, YS., Wen, TH. (2016). Simulations for Epidemiology and Public Health Education. In: Mustafee, N. (eds) Operational Research for Emergency Planning in Healthcare: Volume 2. The OR Essentials series. Palgrave Macmillan, London. https://doi.org/10.1007/978-1-137-57328-5_9

[72] Nick T. Thomopoulos, *Essentials of Monte Carlo Simulation: Statistical Methods for Building Simulation Models* (Springer, 2013).

predicting and preventing churn could be the key to the company's survival and growth. This is a familiar story for countless startups.

Startups embarking on churn prediction should begin by clearly defining what constitutes churn for their business model, whether it be subscription cancellations, prolonged inactivity, or contract non-renewal. This definition anchors all subsequent steps, ensuring that data collection and modeling efforts target the right outcomes and timeframes.[73]

Once churn is defined, the next focus is on data collection and preparation. Early-stage companies often rely on limited datasets—typically basic customer profiles, activity logs, and transaction history. Even these minimal data can be sufficient to build a minimum viable model, provided features are carefully selected and cleaned. Regular data audits and imputation techniques help maintain data quality and address missing values, which is critical for reliable predictions.

With a clean dataset in place, startups should start with simple, interpretable models such as logistic regression or decision trees. These models require fewer resources, are easy to explain to stakeholders, and can yield useful insights quickly. As confidence in the process grows, more advanced techniques like random forests or gradient boosting can be introduced to improve accuracy without sacrificing transparency.

Building a proof-of-concept typically involves splitting data into training and validation sets, tuning basic hyperparameters, and evaluating performance using recall or F1-score to balance the cost of false negatives and false positives. A lean approach emphasizes rapid iteration—launching a basic model within weeks and refining it based on real-world feedback rather than striving for perfection from the outset.

Once a model demonstrates acceptable predictive power, integrating predictions into workflows is essential. Predictions can be surfaced through dashboards or simple scoring scripts that flag at-risk customers to sales and support teams. You can funnel your predictions to marketing tools like Mailchimp as well. This operationalization step transforms analytics into action by enabling targeted retention campaigns and automated alerts.

---

[73] Carl Gold, *Fighting Churn with Data: The science and strategy of customer retention* (Manning Publications, 2020).

Startups should monitor both model performance and business metrics post-deployment, tracking how interventions based on predictions affect churn rate and customer lifetime value. Periodic retraining—weekly or monthly—keeps models aligned with evolving customer behavior and new data sources. Transparent reporting on these metrics builds trust and secures ongoing investment in data science initiatives.

Finally, resource-constrained startups may leverage open source tools (e.g., Python's scikit-learn, pandas) or affordable no-code platforms to accelerate development while minimizing infrastructure overhead. As data volumes and team expertise grow, startups can transition to more comprehensive solutions without disrupting existing processes. Table 13-11 compares the different approaches to churn prediction for startups vs. large corporations.

Table 13-11. *Churn Prediction Approach for Early-Stage Startups*

Aspect	Lightweight Startup Approach	Enterprise Approach
Data Volume	Low	High
Model Complexity	Simple (e.g., logistic regression)	Complex (sophisticated models)
Time to First Model	Weeks	Months
Infrastructure	Simple (e.g., simple scripts or no-code tools)	Complex
Team Size	1–2 generalists	Dedicated teams of specialized experts

By following this lean, iterative road map—defining churn, preparing data, starting simple, iterating rapidly, operationalizing insights, and leveraging accessible tools—startups can build effective churn prediction capabilities that scale with their growth.

Anomaly Detection

Anomaly detection harnesses the power of data science and artificial intelligence to uncover unexpected behaviors, irregularities, or outliers within streaming datasets, empowering startup teams to spot emerging issues before they escalate. By embedding these detection algorithms into their workflows, founders and data professionals gain forward-leaning insights that bolster operational efficiency, fortify security measures, and accelerate growth.[74]

Startups, often constrained by tight budgets and lean teams, reap particular advantage from real-time anomaly detection, which catches small glitches—like budding system errors—before they spiral into revenue-draining crises. It also underpins fraud prevention by continuously monitoring transactions and user activity to flag suspicious patterns early. Likewise, spotting performance dips—such as server slowdowns or logistics bottlenecks—enables rapid root-cause analysis, dramatically reducing downtime. Predictive maintenance powered by anomaly detection further trims unplanned repairs, cutting costs and preserving productivity, while positive anomalies—such as a sudden spike in user engagement—can point the way to new market opportunities. Table 13-12 presents simple, startup-relevant use cases of anomaly detection, illustrating data sources, detection goals, and business impacts.

Table 13-12. *Anomaly Detection Examples for Startups*

Use Case	Data Source	Detection Goal	Business Impact
Fraud Monitoring	Payment transactions, login logs	Identify unauthorized transactions or logins	Prevent revenue loss and protect brand trust
Marketing	Web analytics	Uncover unexpected dips or spikes in conversions	Allocate budget effectively
Predictive Supply	Sensor readings, delivery times	Predict equipment failure or logistic delays	Minimize delays and maintenance costs

Social Network Analysis

Social network analysis (SNA) is critical for startups that rely on network effects and referral programs to achieve rapid growth. By modeling users and their connections as graphs, SNA uncovers key influencers, community structures, and information flow pathways, enabling more effective targeting and engagement strategies for

---

[74] Venkata Krishna Parimala, *Anomaly Detection - Recent Advances, AI and ML Perspectives and Applications (Artificial Intelligence)* (IntechOpen, 2024).

referral-based customer acquisition. In Creator Economy startups, SNA reveals the relationships between creators and their audiences, highlighting content consumption patterns and collaboration networks that drive platform value.

Network effects arise when each additional user increases the overall value of a product or service, creating a virtuous cycle of user acquisition and retention. Applying SNA techniques—such as centrality measures, community detection, and influence maximization—helps identify the most impactful users and referral pathways, guiding the design of features and incentives that amplify this cycle.[75]

Creator Economy often follows a power-law distribution in which a small number of creators capture the majority of attention, making platforms vulnerable to overreliance on "superstars." SNA assists in diagnosing this imbalance by quantifying creator centrality and detecting emerging niche communities where rising talents could be nurtured, thus fostering a more resilient ecosystem.

Startups operate within broader digital entrepreneurial ecosystems that include partners, competitors, investors, and regulators. SNA maps these actors and their relationships—such as co-investment ties or partnership networks—allowing startups to pinpoint strategic allies, potential threats, and collaborative opportunities for innovation.

Table 13-13 summarizes the primary data science techniques for analyzing relationships among users and between users and creators across various startup contexts. It provides a concise reference to help practitioners select appropriate methods based on their specific growth and analytical needs.

Table 13-13. *Data Science Techniques for Relationship Analysis*

Startup Context	Key Techniques	Purpose
General Network-Effect Startups	Centrality measures (degree, betweenness), community detection (Louvain method)	Identify influencers and core communities to drive referrals
Referral-Based Growth Programs	Influence maximization, cascade modeling	Optimize seeding strategies to maximize referral spread
Creator Economy Platforms	Link prediction, bipartite graph analysis, node embeddings (node2vec)	Predict emerging collaborations and recommend creator–audience connections

---

[75] Stephen P. Borgatti, Martin G. Everett, Jeffrey C. Johnson, and Filip Agneessens, *Analyzing Social Networks*, 3rd ed. (SAGE Publications Ltd, 2024).

For readers eager to explore social network analysis in the context of startups more deeply, the book *Social Network Analysis for Startups: Finding Connections on the Social Web* by Maksim Tsvetovat and Alexander Kouznetsov offers an in-depth, practical guide to the subject. It covers both the theoretical foundations of graph-based methods and hands-on techniques using open source Python libraries. Whether you are interested in uncovering hidden community structures, mapping influence pathways, or implementing network-driven growth strategies, Tsvetovat and Kouznetsov provide a clear framework and real-world examples to help practitioners apply social network analysis effectively to startup challenges.[76]

Summary

The central message of this chapter is that there is no universal data science and AI playbook for startups. Instead of a single recipe for success, it is your unique business needs and your data-driven strategy that must guide your path. This exploration is designed to inspire you to take the right actions for your specific context—actions that are simple, cost-effective, and reasonable.

This chapter provided the strategic frameworks and practical knowledge to navigate this complex landscape. Key takeaways include

- **No Universal Recipe**: Your startup's unique needs and data strategy must dictate your AI approach, not a generic playbook. The goal is to find the simplest, most reasonable solution that addresses a concrete business problem.

- **Strategic Prioritization Is Paramount**: The chapter demonstrated how to use the RICE framework (Reach, Impact, Confidence, Effort) to objectively evaluate and select high-value projects, ensuring limited resources are invested for maximum impact.

- **Learn from Common Failures**: We analyzed why over 80% of AI projects fail, covering critical mistakes in leadership, technology, data infrastructure, and business model alignment so you can avoid them.

---

[76] Maksim Tsvetovat and Alexander Kouznetsov, *Social Network Analysis for Startups: Finding connections on the social web* (O'Reilly Media, 2011).

- **Master the Startup ML Toolkit**: The chapter provided a hands-on guide to essential machine learning methods and seven key real-life projects, including leveraging LLMs for automation, predicting customer churn, and building recommender systems from day one.

- **Focus on Value, Not Hype**: The most successful startups build a culture of experimentation focused on solving real customer problems. This means treating data as a core strategic asset rather than simply chasing the latest AI buzzwords.

By internalizing these principles, you can build a strong, data-driven foundation for your venture. This chapter provides the ideas and frameworks to do just that, setting the stage as we look toward the next chapter—the emerging trends that will shape the future of data-driven ventures.

CHAPTER 14

The Future of Data in Startups

What if the equivalent of a Stanford education cost a penny? What if print-ing a house cost a penny? What if curing prostate cancer cost a penny? That's the world we're entering—one where prices crash toward zero through exponential productivity gains.

—Marc Andreessen,[1]
Co-founder of a16z

Is this the blueprint for a utopia of abundance, or a road map to mass displacement? This question lies at the heart of the current AI revolution. Andreessen's vision extends beyond mere cost reduction; he argues that for artificial intelligence to usher in an economic utopia, it must first cause a dramatic fall in human wages, leading to a future defined by a massive surge in productivity and a collapse in the prices of goods and services.[2]

This perspective, while controversial, highlights a critical tension. Critics suggest these forecasts stem from a harsh economic logic that treats technological progress as inevitable and accepts widespread disruption as a necessary means to an end. Yet regardless of one's position on these debates, the underlying reality remains clear: AI and data technologies are fundamentally reshaping how startups operate, and those who fail to adapt risk being left behind.

---

[1] Andreessen, Marc. Interview by Martin Casado. a16z Podcast, June 10, 2023. YouTube video, 0:41:06. https://www.youtube.com/watch?v=OwIUKOnsyUg.

[2] Marc Andreessen (@pmarca), "A world in which human wages crash from AI …," X (formerly Twitter), January 25, 2025, https://x.com/pmarca/status/1882993091784880557.

517

© Piotr Sidoruk 2026
P. Sidoruk, *From Data to Dollars*, https://doi.org/10.1007/979-8-8688-1898-1_14

The startup environment presents unique advantages in navigating this transformation. Unlike large corporations constrained by legacy systems and bureaucratic processes, startups possess the agility to make bold decisions quickly and implement cutting-edge technologies without the burden of existing infrastructure. This inherent flexibility positions startups as the primary drivers of innovation in the data and AI space, making them both beneficiaries and catalysts of the technological shifts ahead.

This chapter examines how artificial intelligence and data technologies are transforming the startup landscape, empowering individual founders to achieve unprecedented scale with minimal teams. We begin by exploring current trends—AI as a "force multiplier" for market research, strategy, and product development—and spotlight insights from industry leaders like Gian Segato and Mark Cuban on how agency now outweighs credentials. Next, we explain why the traditional growth-and-distribution playbook is giving way to deep technical risk, where "full-stack AI" ventures generate proprietary data flywheels to secure defensible moats. We then trace the evolution of the modern data stack—toward AI-native, composable architectures and lakehouse foundations—and highlight cost, observability, and unstructured-data trends shaping its future. Finally, we explore the most promising frontiers for investment, including quantum computing, autonomous vehicles, biotech, fintech, agriculture, space, and defense. The analysis shows how specialized AI applications in the physical world, not just in software, are launching the next generation of billion-dollar startups.

Future Trends in Startups, Data, and AI

One of the most significant developments is the emergence of AI as what Stanford professor Steve Blank calls a "force multiplier" for startup activities. Rather than simply automating existing processes, AI is enabling entirely new approaches to market research, business planning, and strategic decision-making. Blank, who has guided eight different startups and currently teaches innovation at Stanford, advocates for founders to begin their entrepreneurial journey by engaging with AI chatbots to rapidly validate ideas, analyze market opportunities, and develop comprehensive business strategies. This represents a fundamental departure from traditional startup methodologies, where such strategic analysis might require expensive consultants or extensive manual research.[3]

---

[3] Tom Huddleston Jr., "Stanford professor who co-founded 4 startups: How to use AI as a 'force multiplier' to start a business," CNBC Make It, April 1, 2025, https://www.cnbc.com/2025/03/17/stanford-professor-steve-blank-how-to-use-ai-to-start-a-business.html.

The implications for data professionals in startups are profound. As AI tools become more sophisticated and accessible, the role of data teams is shifting from basic reporting and analysis toward more strategic functions like AI system design, data infrastructure optimization, and cross-functional collaboration. However, the quality of a startup's team is still crucial.[4]

This profound shift in technological capability does more than just optimize existing business models; it fundamentally redefines the scale of what a single, determined individual can achieve. As artificial intelligence handles tasks that once required entire teams—from market analysis to code generation—some industry leaders predict we are moving toward a future where the primary bottleneck for innovation will no longer be capital or specialized knowledge, but the vision and drive of the founder. This empowerment of the individual entrepreneur marks the beginning of a new era, one where personal agency becomes the most valuable asset in creating economic value. We are entering a time when the traditional structures of a company are being questioned, and the prospect that a lone innovator could build a billion-dollar enterprise is rapidly becoming reality.

The Billion-Dollar Mindset: Agency in the Age of AI

This vision of the future comes from a founder operating at the very heart of this transformation: Gian Segato. As a founding data scientist and engineer at Replit, Segato is building the tools he predicts will define the next economy. Replit, an AI coding agent, allows anyone—even those with no development experience—to create bespoke apps instantly through simple natural language prompts. It has become one of the fastest-growing products and developer communities in history, proving the immense demand for democratized creation. But Segato's insights aren't just theoretical; they are forged from experience. Before Replit, he founded Uniwhere, the first mobile app to manage college life end to end in Italy, and as an occasional angel investor, he now puts his capital behind the very principles he champions.[5]

---

[4] Erik Wikander, "How Generative AI Empowers Startup Founders," Startups Magazine, November, 2023, `https://startupsmagazine.co.uk/article-how-generative-ai-empowers-startup-founders`.

[5] Gian Segato, "Agency Is Eating the World," April 2025, `https://giansegato.com/essays/agency-is-eating-the-world`.

Segato posits that we are witnessing a fundamental economic shift where individual "agency" is supplanting traditional qualifications as the most critical asset for success. In his view, the rise of powerful and accessible artificial intelligence has made the prospect of a one-person, billion-dollar company not just a theoretical possibility but an emerging reality. The core of his argument is that AI acts as a great equalizer, democratizing skills that were once the exclusive domain of specialists. For decades, deep expertise in a specific field was a defensible advantage, but now, complex tasks like drafting legal documents or creating financial models can be accomplished in minutes with AI tools. This tectonic change means the critical dividing line is no longer what you know, but rather your will to act and orchestrate these technologies to solve problems and create value.[6]

This new era, as described by Segato, represents the unraveling of credentialism, where the bias is toward action rather than accumulated knowledge. He points to companies generating hundreds of millions of dollars in revenue with just a handful of employees as proof that this is a structural shift, not an outlier. While this empowers a new class of "high-agency" entrepreneurs who are curious, defiant, and driven to build, Segato also cautions that the transition will be messy. He anticipates a wave of new businesses built on brittle ground by founders who, in their rush to market, may neglect the technical foundations necessary for security and scale. In high-stakes industries where imperfect AI presents an unacceptable risk of costly and even fatal mistakes, he expects a growing demand for specialized human experts to ensure accountability and oversight. Success in this new landscape will hinge on an individual's drive to innovate, effectively transforming the very definition of what it takes to build an empire. This profound empowerment of the individual, as described by Segato, is the very reason why seasoned investors like Mark Cuban believe the next wave of wealth creation will look radically different.

This sentiment is echoed by Mark Cuban. As a billionaire entrepreneur, *Shark Tank* investor, and the former owner of the Dallas Mavericks, Cuban has built his career on identifying and capitalizing on technological shifts. He sees the current AI revolution not just as an incremental change but as a force with the power to rewrite the rules of wealth creation entirely, placing unprecedented power in the hands of individual innovators.

Cuban predicts that the artificial intelligence industry will enable an individual to break the trillion-dollar wealth threshold. He argues that the rapid pace of technological

---

[6] Gian Segato, "Agency Is Eating the World," Pirate Wires, April 21, 2025, https://www.piratewires.com/p/agency-is-eating-the-world.

change means this individual could be an unknown innovator rather than an established figure. He compares the current AI landscape to the early days of personal computers and the internet but believes AI's potential "dwarfs all that." This belief underscores the idea that the next great breakthrough may not come from a corporate lab but from an unexpected source, armed with little more than curiosity and access to AI. As Cuban puts it:

> *We haven't seen the best or the craziest of what it's going to be able to do. And not only do I think it'll create a trillionaire but it could be just one dude in a basement. That's how crazy it could be.*[7]

The End of the Old Growth Playbook

Gian Segato's essay, "The Dawn of a New Startup Era," reframes the landscape of software startups in the age of AI. He moves beyond the now-exhausted growth-and-distribution playbook, instead focusing on three distinct categories shaping the next wave of innovation: AI Frontends, AI Infrastructure, and AI Full-Stack Products.[8]

The "growth-and-distribution playbook" refers to the strategy that dominated the past two decades, where startups achieved success by rapidly scaling internet-based products through viral growth tactics, aggressive user acquisition, and leveraging network effects—an approach that has become saturated and less effective as digital markets matured and competition intensified.

The first category, **AI Frontends**, consists of products that layer user-friendly interfaces atop existing large language models. These startups thrive on distribution, often acting as "wrappers" around third-party AI. While they can reach users quickly, their core technologies are easily replicated, making defensibility a challenge. The low technical barriers mean competition is fierce, and only those targeting difficult-to-access markets such as the military stand much chance of building really big companies.

AI Infrastructure startups, by contrast, provide the essential tools and platforms that power the broader AI ecosystem. These companies build the scaffolding: model training environments, data pipelines, compute resources, and developer tools. Their value lies in technical depth and integration, requiring significant expertise and capital. Because

---

[7] Mark Cuban, interview, High Performance, YouTube video, June 30, 2025, `https://www.youtube.com/watch?v=dyRlhORLG6s`.

[8] Gian Segato, "The Dawn of a New Startup Era," September 2024, `https://giansegato.com/essays/dawn-new-startup-era`.

of this complexity, infrastructure startups can achieve defensibility through ecosystem effects and platform lock-in, but they also demand patience and substantial investment.

The most ambitious are **AI Full-Stack Products**. These startups control the entire feedback loop, designing products that generate proprietary datasets unavailable to competitors. By owning both the product and the data it collects, they create a powerful flywheel: user interactions feed unique data back into the system, enabling continuous improvement of their AI models.

Segato's analysis suggests that while AI front ends are easy to build and launch, the most enduring and valuable companies will emerge from those willing to take on the technical and product risk of building infrastructure or full-stack solutions that generate and leverage proprietary data. Table 14-1 summarizes the key distinctions among these three categories.

Table 14-1. *Defining Features of AI Startup Types*

Feature	AI Front Ends	AI Infrastructure	AI Full-Stack Products
Main Value Driver	Distribution, UX	Tools, Platforms	Proprietary Data & Model
Defensibility	Low	Medium-High	Very High
Risk	Low	High	Very High
Investor Appeal	Low	High	Very High

Applying the AI Full-Stack approach is especially potent in domains where data is scarce or specialized—robotics, biotech, hardware, and defense. While the risks and capital requirements are high, so too is the potential for building generational companies with deep competitive moats. As we will explore later in this chapter, the application of full-stack AI is poised to drive foundational shifts in sectors as diverse as defense, biotech, space, fintech, and agriculture.

Small Data Teams, Niche Startups, High Impact

This paradigm shift toward technical depth has profound implications for how startups organize their data capabilities. As AI democratizes complex analytical tasks once requiring entire departments, we are witnessing the emergence of a new breed of venture: small data teams tackling highly specialized markets that were previously economically inaccessible.[9]

The mathematics are compelling. AI automation has transformed tasks that once required extensive manual effort, significantly reducing the need for large data teams. Startups can now leverage AI to automate repetitive tasks, freeing up valuable time for team members to focus on strategic initiatives. This fundamental shift enables data teams as small as one person to achieve what previously demanded entire analytics departments, fundamentally altering the economic calculus for niche market exploration.[10]

A data flywheel (a term introduced in Chapter 2) in the context of AI is a self-reinforcing loop where data, AI models, and product usage continuously feed and improve each other. Consider a hypothetical startup, "Agri-Sense." It deploys proprietary drones to capture unique thermal data on crop hydration. This data, which no competitor has, is used to train a specialized AI model that predicts early signs of drought stress with 99% accuracy. Agri-Sense then sells these precise, actionable alerts to high-value vineyards. With each flight and each customer interaction, the dataset grows, the model becomes more accurate, and the product becomes more indispensable—a perfect illustration of the data flywheel in action.

For startups, this means that products designed to generate proprietary data that no one else can access create a powerful competitive moat. Small teams can now build AI solutions that aggregate or collect user behavior data, which becomes the AI-equivalent of network effects. This dynamic allows lean teams to create defensible businesses by focusing on specialized datasets and unique AI models rather than competing on traditional growth metrics.

The real opportunity lies in applying this data-driven approach to physical world applications that demand deep technical expertise but offer substantial barriers to entry. Unlike Software-as-a-Service models that have become commoditized, ventures in robotics, biotech, defense, and hardware create proprietary datasets from real-world interactions that cannot be replicated by competitors. Small teams can now validate and iterate on these technically complex ideas using AI-powered tools for market research, competitor analysis, and MVP development, dramatically reducing the upfront

---

[9] Ruben Domínguez Ibar, "The Billion-Dollar Startup Formula: Why AI-Driven Small Teams Are Beating Giants," The VC Corner, March 21, 2025, https://www.thevccorner.com/p/the-billion-dollar-startup-formula.

[10] AI-First Business Solutions, "AI & Startups: How Small Teams Achieve Big Results," LinkedIn post, May 7, 2025, https://www.linkedin.com/posts/aifirstbusiness_ai-startups-how-small-teams-achieve-big-activity-7325995418176147456-EiVH/.

investment traditionally required to explore such specialized markets. What once required expensive consultants or extensive manual research can now be accomplished by lean teams using AI chatbots and validation tools to rapidly assess market opportunities and develop comprehensive business strategies.

The Evolution of the Modern Data Stack

The modern data stack is entering a new, transformative phase. The era of simply moving data to the cloud has matured, and the new frontier is defined by the deep integration of artificial intelligence, a strategic shift toward open and composable architectures, and a renewed focus on economic efficiency. We are moving from a rigid pipeline to a dynamic, intelligent ecosystem.[11]

A View from dbt Labs

In a conversation with Andreessen Horowitz, dbt Labs co-founder and CEO Tristan Handy provides a nuanced perspective on the future of the data ecosystem, arguing that the "modern data stack" as we've known it (discussed in Chapter 3) has reached a plateau and is entering a new phase of evolution. While the last decade was defined by solving data transformation and orchestration in the cloud, Handy suggests the next chapter will be shaped by AI and a re-evaluation of how analytics creates value.[12]

Handy challenges the notion that AI's primary role is to simply write code or replace data analysts. While he acknowledges AI's proficiency at generating well-formed SQL, he argues this is not the most difficult part of analytics. Instead, he posits that the core function of a data analyst is to "socially construct truth" within an organization. Metrics like revenue are not abstract concepts; they are the result of a shared agreement on how they are defined and measured.[13]

---

[11] "Modern Data Stack," Domo Glossary, accessed July 14, 2025, `https://www.domo.com/glossary/modern-data-stack`.

[12] Tristan Handy, Jennifer Li, and Matt Bornstein, "AI, Data Engineering, and the Modern Data Stack," a16z Podcast, June 23, 2025, `https://a16z.com/podcast/ai-data-engineering-and-the-modern-data-stack/`.

[13] Tristan Handy, Jennifer Li, and Matt Bornstein, "AI, Data Engineering, and the Modern Data Stack," a16z Podcast, June 23, 2025, `https://a16z.com/podcast/ai-data-engineering-and-the-modern-data-stack/`.

From this perspective, the true impact of AI is not the automation of query writing but the empowerment of "human-in-the-loop" self-service analytics across a company. The goal is to enable broader access to data, but Handy cautions that doing so without a human to verify the results is a "very scary thing." This highlights a critical challenge for the next generation of data tools: balancing the democratization of data with the need for accuracy and trust.

The AI-Native and Composable Future

The next evolution of the data stack is being built around AI as a core component, not an afterthought. Every layer of the stack, from ingestion to analytics, is gaining AI-powered assistants or "copilots." These agents are designed to accelerate work by automating complex tasks like generating data pipelines, suggesting code transformations, and proactively flagging data quality issues before they impact business decisions. The result is a more agile and intelligent data culture where data teams are freed from repetitive work to focus on higher-value strategic initiatives.[14,15]

This intelligence is being built on a foundation of composable architecture. We continue to move further away from the monolithic, one-size-fits-all approach, trending instead toward a modular design where best-in-class tools can be snapped together like LEGO blocks. This allows startups to build flexible, customized stacks that can evolve with their needs, avoiding vendor lock-in and maintaining a consistent governance model across different tools. As a company evolves from a nimble startup to a larger organization, this shift allows it to adopt a more mature data strategy. Many are moving to decentralized "data mesh" models, where individual domain teams take ownership of their own data products to foster the accountability and scalability that sustained growth demands.[16,17]

---

[14] Srinivas Lakkireddy, "Distributed Data Engineering: The Backbone of Modern Data Ecosystems," World Journal of Advanced Research and Reviews, 2025, 26(02), 3288-3295, https://journalwjarr.com/node/1806.

[15] Gopinath Govindarajan, "Data Quality and Automation in Modern Data Ecosystems," International Journal of Advances in Engineering and Management, Volume 7, Issue 04 April 2025, pp: 228-239, https://ijaem.net/issue_dcp/Data%20Quality%20and%20Automation%20in%20Modern%20Data%20Ecosystems.pdf.

[16] Fivetran, "Data and AI Predictions for 2025," blog post, January 3, 2025, https://www.fivetran.com/blog/data-and-ai-predictions-for-2025.

[17] Airbyte, "The Essential Modern Data Stack Tools," April 9, 2025, https://airbyte.com/top-etl-tools-for-sources/the-essential-modern-data-stack-tools.

The AI-Optimized Lakehouse Architecture

A logical next step beyond the foundations outlined in Chapter 3 for many AI-centric startups is the AI-optimized lakehouse. This fundamental architectural shift represents the rise of data lakehouse solutions built on open table formats like Apache Iceberg and Delta Lake, which experts predict will become the backbone of modern data infrastructure for more and more companies.

This model combines the scalability and flexibility of a data lake with the reliability and performance of a data warehouse. This consolidation streamlines the entire data science life cycle, enabling startups to accelerate innovation, reduce costs, and satisfy stringent governance requirements without the complexity of separate systems.[18]

Key Trends Defining the Next-Generation Stack

Several key trends are emerging as data stacks evolve:

- **Cost Containment As a Priority:** Building a modern data stack has become remarkably easy, but managing its cost is an increasingly complex challenge. As more vendors move to consumption-based pricing models, infrastructure costs can spiral. For startups, which must operate with financial discipline, cost containment and proving the ROI of their data stack have become a top priority.[19]

- **Real-Time Observability and Governance:** As data pipelines draw from an ever-increasing number of sources, ensuring trust and transparency is critical. Real-time data observability provides teams with full visibility into how data is transformed at every stage. This is essential for tracing data lineage, understanding its journey, and building the trust required for confident, data-driven decision-making.

---

[18] S. Al-Amin, "AI/ML-Optimized Lakehouse Architecture," World Journal of Applied Engineering & Technology 3, no. 1 (2025): 32–46, https://journalwjaets.com/node/840.

[19] Etai Mizrahi, "Here's What Trends To Expect for the Modern Data Stack This Year," Secoda Blog, April 9, 2025, https://www.secoda.co/blog/predictions-for-the-modern-data-stack.

- **Unlocking Unstructured Data:** The focus is expanding beyond traditional structured data. Technologies like knowledge graphs are becoming crucial for organizing and analyzing unstructured data from sources like emails, chat logs, and documents. These graphs represent information in a way that mimics human thought—connecting entities and relationships—to uncover hidden patterns and valuable insights.[20,21]

- **Centralization for Smarter AI:** To fully leverage AI for complex operations like supply chain management or personalization, data fragmentation is a major obstacle. Forward-thinking companies are focusing on building centralized data ecosystems that break down silos. This unified data foundation is a prerequisite for AI to deliver meaningful insights and scale automation across the business.

This evolution of the data stack is not merely a technical upgrade; it is the foundational platform upon which the next generation of industry-defining companies will be built. An intelligent, composable, and economically efficient data ecosystem is the prerequisite for tackling the world's most complex and valuable problems.

With these powerful new tools in hand, startups are now equipped to move beyond the digital-only landscape and apply the power of data and AI to the physical world, creating breakthroughs in sectors that have long been resistant to technological disruption. The following industries represent the new frontier where this advanced data infrastructure will unlock unprecedented innovation and investment opportunities.

Top Emerging Investment Industries

The true breakthrough in artificial intelligence isn't a chatbot that helps you plan a vacation or finish your homework. The real revolution is in specialized AI, tailored with unmatched precision to solve our greatest challenges—powering self-driving cars, discovering cures for diseases once thought incurable, and helping to end starvation.

---

[20] Thomas H. Davenport and Randy Bean, "Five Trends in AI and Data Science for 2025," MIT Sloan Management Review, January 8, 2025, https://sloanreview.mit.edu/article/five-trends-in-ai-and-data-science-for-2025/.

[21] Joseph D. Stec, "Knowledge Graphs: The AI Engine Powering Modern Business Intelligence," Strategy Software blog, July 31, 2024, https://www.strategysoftware.com/blog/knowledge-graphs-the-ai-engine-powering-modern-business-intelligence.

In this section, we spotlight a handful of industries—quantum computing, autonomous vehicles, healthcare and biotechnology, fintech and blockchain, agriculture, space technology, and defense—where AI and data are already rewriting the rules.

However, these are just illustrative examples: AI's transformative power will ripple across countless other fields, from law and entertainment to education, logistics, marketing, manufacturing, and beyond. As governments and tech giants pour billions into foundational models and infrastructure, the real revolution will be driven by startups that harness these investments to build industry-focused solutions, creating verticalized innovations that redefine each sector's economics and open up entirely new markets.

Agriculture

The agricultural industry, long perceived as a realm of intuition and tradition, is undergoing a profound and disruptive transformation, making it one of the most compelling sectors for data-centric investment. The days of farming as a gamble against the elements are fading, replaced by a new paradigm of precision agriculture powered by artificial intelligence and data analytics.

AI algorithms now process vast datasets from satellite imagery, IoT sensors in the soil, and real-time weather feeds to create what is essentially a "digital twin" of the physical farm. This allows for unprecedented accuracy in decision-making. Instead of relying on experience alone, farmers can use AI-generated prescription maps to apply water and fertilizers with surgical precision, optimizing resource use and boosting crop yields. This data-driven approach transforms farms into highly efficient, productive, and profitable operations, rewriting the fundamental economics of growing food.[22,23,24,25]

---

[22] Jon Stojan, "Disrupting Agriculture: How AI and Data Are Powering the 2025 AgTech Revolution," Old National Bank Insights, April 30, 2025, https://www.oldnational.com/resources/insights/disrupting-agriculture-how-ai-and-data-are-powering-the-2025-agtech-revolution/.

[23] Mark Gildersleeve, "From Seed to Server: The Evolution of Modern Agriculture," IBM Watson Newsroom, accessed July 14, 2025, https://newsroom.ibm.com/IBM-watson?item=30660.

[24] Patil, Dimple, Artificial Intelligence Innovations In Precision Farming: Enhancing Climate-Resilient Crop Management (November 23, 2024). Available at SSRN: https://ssrn.com/abstract=5057424 or https://doi.org/10.2139/ssrn.5057424

[25] Tushar, P. Abhinash, and P.P.N. Srija, "Data Analytics and AI-driven Innovations Across Agriculture Industry," International Research Journal of Modernization in Engineering Technology & Science 7, no. 4 (April 2025): 533–539, https://www.irjmets.com/uploadedfiles/paper/issue_4_april_2025/71764/final/fin_irjmets1743968823.pdf.

Beyond pure efficiency, innovative data solutions are fortifying the agricultural sector against its most significant systemic threats: climate change and global food insecurity. Predictive analytics, driven by machine learning models, can now forecast extreme weather events, pest infestations, and crop diseases, enabling farmers to shift from reactive to proactive management. AI is also revolutionizing agricultural research, accelerating the development of climate-resilient crop varieties that can withstand drought or require less time to grow. At a global level, organizations like the Food and Agriculture Organization of the United Nations (FAO) are championing the development of ethical and inclusive AI, ensuring these transformative tools are accessible to smallholder farmers and do not exacerbate the digital divide. This focus on sustainability and equity adds a layer of social impact to the investment thesis, promising not just financial returns but a more resilient global food system.[26,27]

The integration of AI extends across the entire value chain, from field-level operations to the end consumer, signaling a systemic shift in how the world produces and manages food. AI-powered platforms are streamlining complex supply chains and helping food companies meet growing consumer demands for transparency and sustainability. Startups in this space are not merely creating apps; they are building the foundational infrastructure for the future of food. Initiatives like the global AI for Good Summit underscore the international commitment to harnessing these technologies to solve fundamental human challenges. For investors and entrepreneurs, agriculture is no longer just about land and labor; it has become a frontier of innovation where data and AI are cultivating a more efficient, sustainable, and secure future for everyone.[28,29]

---

[26] "FAO and ITU launch Robotics for Good – Youth Challenge 2025-2026 to tackle global food insecurity," News release, July 8, 2025, `https://agenparl.eu/2025/07/08/fao-and-itu-launch-robotics-for-good-youth-challenge-2025-2026-to-tackle-global-food-insecurity-fao-news-release/`.

[27] Antonia Davison, "AI and the future of agriculture," IBM Think Blog, September 24, 2024, `https://www.ibm.com/think/topics/ai-in-agriculture`.

[28] Safeer, Sajid and Gallo, Pierluigi and Pulvento, Cataldo, Agri-Farming with Computer Vision, Iot and Blockchain Towards Climate Smart Cultivation. Available at SSRN: `https://ssrn.com/abstract=5217467` or https://doi.org/10.2139/ssrn.5217467.

[29] Food and Agriculture Organization of the United Nations, "Shaping the Future of Agriculture: FAO AI for Good Summit 2025," July 4, 2025, `https://www.fao.org/e-agriculture/news/shaping-future-agriculture-fao-ai-good-summit-2025`.

Defense

Visionary investors like Peter Thiel are betting heavily on this sector, arguing that amid rising geopolitical instability, the integration of AI into defense technology is not just an option but a necessity. This creates a fertile ground for startups that can develop solutions for autonomous systems, enhanced cyber warfare capabilities, and sophisticated intelligence and surveillance platforms. For investors, the logic is compelling: as governments worldwide prepare to invest trillions in these technologies, the companies that dominate even a single, critical niche within defense are positioned for tremendous growth and could become the giants of the future.[30]

The immense potential of the defense market is rooted in the tangible, transformative impact of data and AI on modern warfare and national security. The integration of AI moves beyond theoretical advantages to offer concrete solutions that enhance operational superiority and strategic readiness. AI-powered systems are revolutionizing defense through capabilities like predictive threat detection, which analyzes vast datasets to identify and neutralize dangers before they escalate. Furthermore, these technologies provide enhanced situational awareness, processing real-time information from multiple domains—land, air, cyber, and space—to give commanders a clear and immediate tactical advantage. This data-driven approach optimizes everything from resource allocation to predictive maintenance on critical assets, ensuring forces are better equipped and positioned while also driving significant long-term cost savings. For the startups building this future, the defense industry offers a unique opportunity: a market with clear, high-stakes challenges where innovative data solutions can provide a decisive edge.[31,32]

---

[30] Scott Nover, "Palantir Delivers Two Key AI Systems to the US Army," Gzero, March 11, 2025, https://www.gzeromedia.com/gzero-ai/palantir-delivers-two-key-ai-systems-to-the-us-army.

[31] Justin Orcutt, "The Strategic Advantage of AI for the Defense Industrial Base," Microsoft Public Sector Blog, October 23, 2024, https://techcommunity.microsoft.com/blog/publicsectorblog/the-strategic-advantage-of-ai-for-the-defense-industrial-base/4276817.

[32] Fabio Macrì, "The Future of Defense: How AI Is Transforming the Industry," May 19, 2025, https://www.eurodev.com/blog/defense-industry-ai-transformation.

Healthcare and Biotechnology

The convergence of data science and artificial intelligence is set to redefine the landscape of healthcare and biotechnology, presenting an unprecedented opportunity for innovation and investment. In the realm of digital health and AI-powered healthcare, companies are moving beyond the traditional, linear models of research and development, which have long been plagued by high costs and a high failure rate for drug candidates. The new paradigm leverages AI and machine learning to analyze immense and complex datasets, uncovering insights into human biology that were previously unattainable. A key strategy in this transformation is the "lab in a loop" approach, an iterative cycle where AI models are trained on laboratory and clinical data to predict potential drug targets and design new medicines. These AI-generated predictions are then tested in the lab, and the results are fed back into the models, creating a self-improving system that accelerates the discovery process and drastically reduces the reliance on conventional trial-and-error methods. This not only speeds up the timeline for developing novel therapies but also enhances the success rate of research programs across the board. Of course, realizing this potential requires overcoming significant hurdles, such as ensuring patient privacy and mitigating the risk of bias in the datasets used to train these powerful AI models.[33,34]

The integration of AI extends deeply into biotechnology and life sciences, fueling significant growth and creating new frontiers. AI-driven platforms are revolutionizing the core processes of drug discovery by rapidly generating and testing virtual structures for thousands of new molecules and simulating their interactions with therapeutic targets. This capability is particularly impactful in the field of personalized medicine, where AI can analyze genomic data to identify the most promising neoantigens for individualized cancer vaccines. By integrating complex genomic, proteomic, and clinical data, generative AI can recognize intricate patterns, formulate novel hypotheses, and even generate designs for new molecules and proteins, accelerating breakthroughs in drug development and synthetic biology. The strategic application of AI is already showing a massive return, with some industry leaders investing billions to integrate these

---

[33] Roche, "Harnessing the Power of AI," January 2022, https://www.roche.com/stories/harnessing-the-power-of-ai.

[34] Roche, "AI and machine learning: Revolutionising drug discovery and transforming patient care," January 30, 2025, https://www.roche.com/stories/ai-revolutionising-drug-discovery-and-transforming-patient-care.

technologies into their development processes, signaling a market ripe with potential. This fusion of advanced computation and biological science is not just an incremental improvement; it represents a fundamental shift in how we discover, design, and deliver treatments, opening doors to previously intractable challenges in medicine.[35,36,37]

Financial Technology and Blockchain

The new frontier of finance is not being built on concrete and steel but on data and algorithms. At the heart of this transformation lies the powerful convergence of artificial intelligence and blockchain technology, a pairing that is fundamentally reshaping our concepts of trust and security in the digital age. Blockchain provides a decentralized and immutable ledger, creating a tamper-proof foundation for transactions that historically required validation from central authorities like banks or governments. Layered on top, AI acts as an intelligent guardian, automating fraud detection and performing real-time risk analysis with a speed and accuracy far beyond human capability. This synergy creates a system of intelligent, self-verifying security, where the integrity of data is guaranteed not by a single entity but by the very architecture of the network itself. For startups navigating this new landscape, this means the opportunity to build financial products that are not only innovative but are secure by design, fostering a new level of consumer confidence in an increasingly digital world.[38,39]

Beyond fortifying security, the fusion of AI and blockchain is a catalyst for unprecedented operational efficiency and the creation of entirely new markets. The technology facilitates frictionless and automated value exchange, eliminating costly intermediaries and streamlining processes from cross-border payments to complex

---

[35] Feed the AI, "How AI Is Revolutionizing Drug Discovery," January 3, 2025, `https://www.feedtheai.com/how-ai-is-revolutionizing-drug-discovery/`.

[36] Ashfaq Ur Rehman, Mingyu Li, Binjian Wu, Yasir Ali, Salman Rasheed, Sana Shaheen, Xinyi Liu, Ray Luo, Jian Zhang, Role of artificial intelligence in revolutionizing drug discovery, Fundamental Research, Volume 5, Issue 3, 2025, Pages 1273-1287, ISSN 2667-3258, `https://doi.org/10.1016/j.fmre.2024.04.021`.

[37] Bulashevska, A., Nacsa, Z., Lang, F., Braun, M., Machyna, M., Diken, M., Childs, L., & König, R. (2024). Artificial intelligence and neoantigens: paving the path for precision cancer immunotherapy. Frontiers in Immunology, 15, 1394003. doi:10.3389/fimmu.2024.1394003.

[38] Geert Theys, "Fintech Thought Leadership Series - Ep. 3: AI & Block chain in Fintech," YouTube video, April 17, 2024, `https://www.youtube.com/watch?v=crmubD7OUiI`.

[39] Jerome Knyszewski, "The Future of Payments: Jed Morley on AI, Blockchain, and Fintech Disruption," interview, March 18, 2025, `https://valiantceo.com/jed-morley-plat-pay/`.

contract execution. Imagine smart contracts, powered by AI, that can autonomously adapt to real-time market data, or global payment systems that settle in seconds rather than days, all at a fraction of the cost of traditional methods. This is not a distant vision; it is the reality that innovative fintech companies are building today. For investors and entrepreneurs, this translates into a compelling proposition: an industry poised to unlock immense value by rewiring the foundational infrastructure of commerce. By leveraging data-driven insights and decentralized automation, the financial technology and blockchain sector offers a rare opportunity to invest in the very architecture of the future economy.[40,41,42,43]

Autonomous Vehicles and Self-Driving Technology

The world of autonomous vehicles is undergoing a seismic shift, driven not by mechanics but by the very essence of modern innovation: data and artificial intelligence. The traditional, rigid, rule-based systems that once governed automated driving are being replaced by something far more dynamic and intelligent. We are witnessing the birth of vehicles with brains rewired by end-to-end AI models. These sophisticated models merge perception, prediction, and planning into a single, cohesive neural network, granting vehicles the ability to learn and adapt to the complexities of real-world driving with unprecedented speed. The fuel for this revolutionary learning process is data, but not just any data. Generative AI is a key enabler, creating vast and realistic synthetic datasets that allow these systems to train and be tested in a multitude of virtual

[40] Chantal Fouad, "Fintech revolution: How blockchain, AI is still making waves in financial services," Gulf News, November 8, 2024, https://gulfnews.com/business/markets/fintech-revolution-how-blockchain-ai-is-is-still-making-waves-in-financial-services-1.1730791251663.

[41] Michael Treacy, "Innovation for All: How AI and Blockchain Revolutionise Embedded Finance," The Paypers, accessed July 13, 2025, https://thepaypers.com/fintech/interviews/innovation-for-all-how-ai-and-blockchain-revolutionise-embedded-finance.

[42] Oyewole, Adedoyin and Adegbite, Michael, The impact of Artificial Intelligence(AI), Blockchain, Cloud Computing and Data Analytics on the future of the Fintech Industry in the US. (June 22, 2023). Available at SSRN: https://ssrn.com/abstract=4487815 or https://doi.org/10.2139/ssrn.4487815.

[43] Columbia University, "The Future of AI in FinTech," Columbia Engineering, accessed July 13, 2025, https://ai.engineering.columbia.edu/ai-applications/fintech/.

scenarios. This approach overcomes one of the most significant hurdles in autonomous development: safely and efficiently exposing the AI to the rare "edge cases" that are critical for building robust and reliable self-driving capabilities.[44,45]

This technological leap is paving the way for a disruption that will redefine urban mobility itself: the rise of the robotaxi. These driverless vehicles are more than just a futuristic novelty; they represent the flagship application of this new data-driven ecosystem and the cornerstone of a burgeoning Mobility-as-a-Service (MaaS) framework. As an investment, the robotaxi market is projected to expand into a multi-billion-dollar industry, fundamentally altering transportation economics. By offering a potentially cheaper, safer, and more convenient alternative to private car ownership, robotaxis promise to reshape our cities by reducing traffic congestion and improving overall road safety. However, the path to widespread adoption is not without obstacles, as the industry must still navigate significant regulatory hurdles and earn broad public trust to become a ubiquitous feature of urban life. Businesses in this space are not just building cars; they are creating intelligent networks that continuously learn and improve from the immense volumes of data collected during operation, creating a powerful feedback loop that perpetually refines performance and decision-making.[46,47,48]

Quantum Computing

As we venture further into the age of data, a new computational paradigm is quietly emerging from research labs and stepping into the commercial arena, promising to redefine the boundaries of what is possible. This is the world of quantum computing,

---

[44] World Economic Forum, "How GenAI Is Helping Drive Vehicle Autonomy," April 3, 2025, https://www.weforum.org/stories/2025/04/how-genai-is-helping-drive-vehicle-autonomy/.

[45] World Economic Forum, "Autonomous Vehicles: Timeline and Roadmap Ahead," April 24, 2025, https://www.weforum.org/publications/autonomous-vehicles-timeline-and-roadmap-ahead/.

[46] Alexander S. Gillis, "The Rise of Robotaxis," TechTarget, May 20, 2025, https://www.techtarget.com/whatis/feature/The-rise-of-robotaxis.

[47] Grand View Research, "Robotaxi Market Size, Share & Trends Analysis Report By Propulsion Type, By Component Type, By Level Of Autonomy, By Vehicle Type, By Service Type, By Application, By Region, And Segment Forecasts, 2025 - 2030," accessed July 13, 2025, https://www.grandviewresearch.com/industry-analysis/robotaxi-market-report.

[48] Goldman Sachs, "The Autonomous Vehicle Market Is Forecast to Grow and Boost Ridesharing Presence," Insights, July 3, 2025, https://www.goldmansachs.com/insights/articles/autonomous-vehicle-market-forecast-to-grow-ridesharing-presence.

a technology poised to be as revolutionary as the internet or artificial intelligence. For startups and venture capitalists, it represents not just another trend but a fundamental shift in how we will process information and solve the world's most complex problems.[49]

The core of quantum computing's promise lies in its departure from classical bits. Instead, it uses qubits, which leverage the principles of superposition and entanglement to exist in multiple states at once and process information at speeds exponentially faster than today's most powerful supercomputers. While the technology is still in its early stages, the race to build a fault-tolerant quantum computer is well underway. Tech giants like Google, IBM, Microsoft, and Amazon are investing billions, while governments globally have committed over $50 billion to quantum research and development, signaling immense confidence in its transformative potential. This robust, global investment surge is creating a vibrant ecosystem where startups can thrive.[50,51,52]

The investment appeal of quantum computing is rooted in its vast and disruptive applications across numerous sectors. In medicine and materials science, quantum computers can simulate molecular interactions with incredible precision, dramatically accelerating drug discovery and the creation of novel materials. Pharmaceutical and automotive companies are already partnering with quantum startups to explore these possibilities. In the financial services industry, quantum algorithms are being developed to optimize investment portfolios, enhance fraud detection, and manage risk with a level of analysis previously unattainable.[53,54]

---

[49] Jonathan Selby, "Quantum Computing Startups: The Future of Technology," Founder Shield, May 29, 2025, https://foundershield.com/blog/quantum-computing-startups-the-future-of-technology/.

[50] Maya Netser, "Quantum Computing: The Next Frontier in Venture Capital," Qbeat VC, March 23, 2025, https://www.qbeat.vc/post/quantum-computing-the-next-frontier-in-venture-capital.

[51] Pinnacle Digest, "Quantum Computing 2025: Milestones, Hype, and Investment Realities," May 18, 2025, https://pinnacledigest.com/blog/quantum-computing-2025-milestones-hype-investment-realities.

[52] Sandhya Michu, "Quantum Computing in 2025: From Promise to Reality," CIO Inc., December 31, 2024, https://www.cio.inc/quantum-computing-in-2025-from-promise-to-reality-a-27184.

[53] Ryan Morrison, "Emerging Quantum Computing Start-ups," Tech Monitor, August 15 2022, https://www.techmonitor.ai/technology/emerging-technology/quantum-computing-start-ups.

[54] Jonathan Selby, "Quantum Computing Startups: The Future of Technology," Founder Shield, May 29, 2025, https://foundershield.com/blog/quantum-computing-startups-the-future-of-technology/.

Furthermore, quantum computing is set to fundamentally reshape cybersecurity. While a powerful quantum computer could theoretically break many of today's encryption standards, the field is simultaneously creating new, quantum-resistant cryptographic methods to ensure secure communications for the future. This dual role places quantum technology at the heart of the next generation of digital security. The technology also stands to unlock powerful synergies with other fields, accelerating AI model training and optimizing complex systems in logistics and climate science.[55]

For startups, the barrier to entry is lowering significantly thanks to the rise of Quantum-as-a-Service (QaaS) platforms. Cloud services like Amazon Braket, Microsoft Azure Quantum, and IBM Quantum Platform are democratizing access to quantum hardware, allowing businesses and researchers to experiment with quantum algorithms without the prohibitive cost of building their own machines. This accessibility is fostering a new wave of innovation focused on developing quantum software and applications. The global innovation landscape is also rapidly expanding, with nations like Israel making significant strides in building national quantum computers and fostering a dynamic startup culture through government and industry collaboration. Events like MIT's iQuHACK are bringing together numerous students and professionals, expanding the talent pool and accelerating the development of practical applications.[56,57,58,59,60]

[55] McKinsey & Company, "The Year of Quantum: From Concept to Reality in 2025," June 23, 2025, https://www.mckinsey.com/capabilities/mckinsey-digital/our-insights/the-year-of-quantum-from-concept-to-reality-in-2025.

[56] MASL World, "Quantum Computing in 2025: The Dawn of a New Tech Era," LinkedIn Pulse, March 2025, https://www.linkedin.com/pulse/quantum-computing-2025-dawn-new-tech-era-maslworld-ltumc/.

[57] Brian Blum, "Israel Building Leading-Edge Quantum Computing Center," Israel21c, July 25, 2022, https://www.israel21c.org/israel-building-leading-edge-quantum-computing-center/.

[58] Massachusetts Institute of Technology, "iQuHACK 2024," accessed July 14, 2025, https://www.iquise.mit.edu/iQuHACK/2024-02-02.

[59] Rachel Yang, "Hackathon Unlocking Quantum Future," MIT News, March 18, 2024, https://news.mit.edu/2024/hackathon-unlocking-quantum-future-0318.

[60] Quantum Zeitgeist, "MIT's iQuHACK 2024: Quantum Computing Hackathon Goes Hybrid with In-Person and Virtual Participation," January 24, 2024, https://quantumzeitgeist.com/mits-iquhack-2024-quantum-computing-hackathon-goes-hybrid-with-in-person-and-virtual-participation/.

Investing in quantum computing is undeniably a long-term proposition. The technology is still maturing, and widespread "quantum advantage"—where quantum systems consistently outperform classical ones on real-world problems—is still on the horizon. However, the foundational work is being done now. Venture capital is flowing to companies with clear road maps and credible technology, from hardware developers to software and component providers. Those who invest strategically in this nascent stage are not just backing a new technology; they are securing a foothold in a future where computational limits are redrawn, and industries are reshaped from the ground up.[61,62]

Space Technology and Aerospace

For the next generation of data-driven startups, the defining frontier may not be found on any terrestrial map but in the vast, untapped territory orbiting our world. The space technology and aerospace industry is undergoing a profound transformation, fueled by the integration of artificial intelligence, which turns the cosmos into a fertile ground for investment. The deluge of data from orbit has historically been a challenge to process efficiently. Now, AI-powered satellite data analytics is unlocking its value. AI is becoming essential for processing Earth observation data, enabling applications from tracking natural disasters and monitoring climate change to remotely managing shipping fleets. Space agencies like NASA and ESA are pioneering these efforts, using AI to sift through satellite imagery to identify burn scars from wildfires, monitor greenhouse gas concentrations, and even spot illegal fishing operations. This extends to space-based communications, where AI-driven cognitive radios can autonomously find and use open frequencies, making networks more efficient, resilient, and reliable without constant

---

[61] Ryan Morrison, "Emerging Quantum Computing Start-Ups," Tech Monitor, August 15, 2022, https://www.techmonitor.ai/technology/emerging-technology/quantum-computing-start-ups.

[62] Seedtable, "30 Best Active Quantum Computing Investors in 2025," July 8, 2025, https://www.seedtable.com/investors-quantum-computing.

human oversight. For startups, this means an explosion of opportunities to build services on top of this newly accessible and analyzable space data.[63,64,65,66,67,68,69]

Beyond data analysis, AI is reshaping the physical logistics of space operations, heralding an era of unprecedented efficiency and autonomy. Launch vehicle technology is moving from simple automation to true artificial intelligence, with systems capable of performing their own check-ups and making real-time trajectory adjustments during flight. This reduces reliance on human ground control, minimizes the risk of human error, and promises to make space access more affordable and reliable. Looking further ahead, the most transformative shift may be in space manufacturing and in-orbit services. The concept of an in-orbit factory, once the domain of science fiction, is now being actively developed. These facilities will use AI-guided robotics for the assembly, integration, and testing of satellites directly in space. This approach promises to drastically reduce launch costs and risks, enabling on-demand deployment and

[63] European Space Agency, "Working Towards AI and Earth Observation," accessed July 13, 2025, https://www.esa.int/Applications/Observing_the_Earth/Working_towards_AI_and_Earth_observation.

[64] Zoe Hobbs, "As Global Satellite Data Expands, Space Companies Rely on AI," AIThority, November 23, 2023, https://aithority.com/machine-learning/as-global-satellite-data-expands-space-companies-rely-on-ai/.

[65] Curt Hall, "AI Takes Orbit: Transforming Satellite Data into Environmental Action," Cutter, May 21, 2025, https://www.cutter.com/article/ai-takes-orbit-transforming-satellite-data-environmental-action.

[66] Justin Goodwill, Christopher Wilson, James MacKinnon, "Current AI Technology in Space" (NASA briefing document), July 2023, https://ntrs.nasa.gov/api/citations/20240001139/downloads/Current%20Technology%20in%20Space%20v4%20Briefing.pdf.

[67] Marcin Frąckiewicz, "Artificial Intelligence in Satellite and Space Systems," TS2, June 12, 2025, https://ts2.tech/en/artificial-intelligence-in-satellite-and-space-systems/.

[68] Financial Express, "NASA looks at Artificial Intelligence to communicate with space," accessed July 13, 2025, https://www.financialexpress.com/life/science-nasa-looks-at-artificial-intelligence-to-communicate-with-space-967483/.

[69] Airport Technology, "NASA Explores AI for Space-Based Communications," accessed July 13, 2025, https://www.airport-technology.com/news/nasa-explores-ai-space-based-communications/.

servicing of space assets. For investors, this represents a ground-floor opportunity in building the infrastructure for a permanent, self-sustaining economy in orbit.[70,71]

Key Takeaways

In this new era, the old rules no longer apply. The AI revolution has triggered a fundamental economic shift, one where an individual's drive, curiosity, and will to act are becoming more valuable than traditional credentials. This is a world where **agency surpasses qualifications**. AI acts as a "force multiplier," empowering individuals and small teams to achieve what once required entire departments. The idea of a one-person, billion-dollar company is no longer a fantasy but an emerging reality.

As the old startup playbook of growth and distribution becomes obsolete, a new competitive advantage has emerged: **the data moat**. Enduring value now flows from "full-stack AI" ventures that build proprietary data flywheels. This creates a self-improving loop where unique data continuously enhances the product, making it difficult for competitors to catch up. This evolution is mirrored in the technology itself, as the modern data stack is rebuilt with AI at its core. The future is one of composable, AI-native architectures, with the **AI-optimized data lakehouse** as the foundational platform. The most significant opportunities for this new paradigm lie in applying specialized AI to solve challenges in the physical world. Industries like agriculture, defense, and biotechnology are ripe for disruption by startups that can generate unique data from real-world interactions.

Recommendations for Startup Founders

Founders should embrace AI not as a tool but as a co-founder. From day one, leverage AI to validate ideas, conduct market research, and formulate business strategies with a speed and at a cost that was previously unimaginable. This allows for extreme agility and

---

[70] Adam Hadhazy, "Artificially Intelligent Rockets Could Slash Launch Costs," Space.com, March 21, 2011, https://www.space.com/11181-rocket-launches-artificial-intelligence-japan.html.

[71] Florian Leutert, David Bohlig, Florian Kempf, Klaus Schilling, Maximilian Mühlbauer, Bengisu Ayan, Thomas Hulin, Freek Stulp, Alin Albu-Schäffer, Vladimir Kutscher, Christian Plesker, Thomas Dasbach, Stephan Damm, Reiner Anderl, Benjamin Schleich, AI-enabled Cyber–Physical In-Orbit Factory - AI approaches based on digital twin technology for robotic small satellite production, Acta Astronautica, Volume 217, 2024, Pages 1-17, ISSN 0094-5765, https://doi.org/10.1016/j.actaastro.2024.01.019.

the ability to operate with a lean, effective team. However, it is crucial to choose your path strategically, understanding the distinct trade-offs of the new AI landscape.

Before diving into specific strategies, it's important to recognize that the AI startup landscape can be broadly categorized into three main types of software companies. Each path presents a different set of challenges and opportunities, and your choice will depend on the level of technical risk you are willing to undertake and the capital you are prepared to invest. Whether you aim for rapid market entry or deep technological defensibility, one thing is certain: to succeed in this new era, you will leverage AI to a significant extent.

One path is to build **AI Front Ends**. These are often "wrappers" around existing models and represent the fastest route to market with lower capital needs. However, this path comes with the inherent risk of being easily replicated. Success here depends on dominating distribution in hard-to-access niche markets where a unique brand or community can be built.

A more demanding path is **AI Infrastructure**. This requires deep technical expertise and significant capital. It is a long-term play focused on building the foundational tools for the AI ecosystem. While the barriers to entry are high, the reward is a powerful, defensible moat built on platform lock-in.

The most ambitious and potentially rewarding path is creating **AI Full-Stack Products**. This approach focuses on solving a problem and using proprietary data that no one else possesses. This data flywheel becomes your most defensible asset. These opportunities are often found in physical-world sectors like robotics, biotech, or agriculture, where data is scarce and immensely valuable. In an era where AI can replicate features with ease, your competitive advantage can be found in deep technical risk and unique data. A product that becomes smarter with every user interaction creates a barrier that cannot be overcome by marketing budgets or larger teams. Therefore, do not treat your data infrastructure as an afterthought. Architect it from the start on a composable, AI-native foundation, prioritizing an AI-optimized lakehouse to manage both structured and unstructured data, control costs, and build the trust necessary for your AI systems to succeed.

Recommendations for Professionals Joining Startups

For professionals, the new landscape demands a shift in mindset. Your most valuable asset is no longer your ability to simply write code or analyze data—AI can perform those tasks. Instead, your value lies in your **agency**: your drive to identify problems, orchestrate technologies to solve them, and take ownership of the outcomes. A bias for action has become the new currency. The role of a data professional is moving up the value chain, from analyst to strategist. The focus must be on strategic functions like AI system design, optimizing data infrastructure, and ensuring data accuracy. The most critical human role is to "socially construct truth" by validating AI-generated insights and aligning them with business reality.

As AI handles increasing levels of automation, the need for human oversight in high-stakes industries will only grow. Specialize in becoming an expert in the **"human-in-the-loop,"** ensuring accountability, security, and reliability in domains where imperfect AI is not an option. This is a critical and defensible human skill. To support this, you must master the next-generation data stack. Develop expertise in composable architectures, open lakehouse formats like Apache Iceberg and Delta Lake, and tools for data observability. As consumption-based pricing becomes standard, your ability to manage and prove the ROI of the data stack will be indispensable.

Recommendations for Anyone Navigating the New Economy

Success in this rapidly changing world requires a commitment to **lifelong, AI-powered learning**. The skills required for success are evolving at an unprecedented pace. Use AI as a personal tutor to acquire new knowledge quickly. Your ability to adapt and learn is now more important than what you already know. Cultivate a **founder's mindset**, regardless of your role. Identify inefficiencies and opportunities around you and consider how emerging AI tools could be orchestrated to create value. The power to build is now accessible to everyone.

Look beyond the digital hype and understand that AI's most profound impact will be in the physical world. Pay attention to how it is transforming foundational industries like agriculture, manufacturing, healthcare, and logistics, as these shifts will create new careers and investment opportunities. In this new era, the bias is toward doing. The accessibility of powerful tools means you can build and test ideas faster than ever before. Do not wait for the perfect plan; **prioritize action over perfection**. Start building, learn from the results, and iterate.

Summary

This chapter has charted the course of a new startup era, one forged by the powerful convergence of data and artificial intelligence. We began by exploring how AI acts as a "force multiplier," empowering determined individuals and lean teams with unprecedented capabilities and placing a premium on personal agency over traditional credentials. This shift marks the end of the old growth-and-distribution playbook, replacing it with a new imperative for technical defensibility.

The cornerstone of this new landscape is the **"full-stack AI"** product. We examined how the most enduring companies will be those that build proprietary data flywheels—self-improving systems where unique data continuously enhances the product, creating a powerful and defensible competitive moat. This strategy is especially potent when applied to physical-world industries where data is scarce and valuable.

Fueling this revolution is the **evolution of the modern data stack**, which is being fundamentally rebuilt to be more intelligent, composable, and economically efficient. At its heart, there will be solutions like the AI-optimized lakehouse, an architecture designed to handle the diverse data demands of the AI era, from structured analytics to unstructured data processing. This intelligent foundation enables startups to manage costs effectively, ensure data governance, and unlock insights that were previously out of reach.

Finally, we surveyed the most promising frontiers for this new generation of data-driven startups. The greatest opportunities lie beyond pure software and in the **application of specialized AI to the complex challenges** of the physical world. From revolutionizing agriculture and discovering new medicines to advancing autonomous vehicles and building an economy in space, data and AI are unlocking transformative innovations across sectors. This chapter has provided a comprehensive road map for founders, professionals, and investors to navigate this dynamic landscape and build the next generation of industry-defining companies.

Index

A

A/B testing
- AI-driven platforms, 405
- analyzing impact, 410
- back-end algorithms, 411
- Bayesian approach, 404
- brainstorming, 410
- definition, 399
- disciplined approach, 412–414
- email campaigns, 411
- experiment funnel, 410
- fake door test, 409
- foundational projects, 407
- frequentist approach, 403
- high-impact hypotheses, 412
- key benefits, 400
- key scenarios, 406–408
- multivariate testing, 405–406
- peeking, 403
- preference tests, 409
- primary challenge, 412
- prioritizing, 410
- product features and design, 411
- product-market fit, 407
- psychological trigger, 414
- qualitative methods, 408
- radical redesigns, 408
- randomization, 401
- run A/A test, 423–424
- sample size, 402–403
- sequential testing, 401
- statistical methods, 401
- stratified randomization, 402
- structured approach, 409–411
- subscription plan, 399–400
- sufficient sample size, 401
- test card, 425
  - competitor pages, 426
  - conversion rate, 428
  - core components, 425
  - hypothesis, 426, 427
  - metric/criteria, 427
  - results, 428
  - subscription app, 425–426
- test duration, 402
- usability tests, 408
- user funnel, 413
- website, 411

Acquisition, Activation, Retention, Referral, Revenue (AARRR), 130
- acquisition, 146–148
- activation rate, 148–150
- conversion funnel, 147
- conversion rate, 146
- framework, 160
- frameworks comparison, 161–162
- free-to-paid conversion rate, 150
- functions, 145
- fundamentals, 163
- pirate metrics, 145
- RACE framework, 161
- referrals, 156–157
- retention, 150–156, 160
- revenue, 157–160

543

B

<u>GPSR Compliance</u>

*The European Union's (EU) General Product Safety Regulation (GPSR)
is a set of rules that requires consumer products to be safe and our
obligations to ensure this.*

*If you have any concerns about our products, you can contact us on
ProductSafety@springernature.com*

In case Publisher is established outside the EU, the EU authorized
representative is:

Springer Nature Customer Service Center GmbH
Europaplatz 3
69115 Heidelberg, Germany

Batch number: 10145973

Printed by Printforce, the Netherlands